Wei Yang · W.B. Lee

Mesoplasticity and its Applications

With 197 Figures

Springer-Verlag Berlin Heidelberg GmbH

Wei Yang

Department of Engineering Mechanics
Tsinghua University
Beijing 100084
China

W.B. Lee

Department of Manufacturing Engineering
Hong Kong Polytechnic, Hung Hom
Kowloon, Hong Kong

Series Editors

Prof. Bernhard Ilschner

Polytechnique Fédérale de Lausanne
Laboratoire de Métallurgie Mécanique
MX-D Ecublens Ecole
CH-1015 Lausanne/Switzerland

Prof. Kenneth C. Russel

Department of Materials Science and Engineering and
Department of Nuclear Engineering
Room 8-411
Massachusetts Institute of Technology
Cambridge, MA 02139/USA

ISBN 978-3-642-50042-8   ISBN 978-3-642-50040-4 (eBook)
DOI 10.1007/978-3-642-50040-4

Library of Congress Cataloging-in-Publication Data
Yang, Wei, 1954-
Mesoplasticity and its applications / Wei Yang, W.B. Lee.
p. cm.
Includes indexes.

New York Berlin Heidelberg : acid-free paper) : $ 169.00
1. Plasticity. I. Lee, W.B. II. Title.

Data Conversion by Danny L. Lewis, Berlin;
Printing: Color-Druck Dorfi GmbH, Berlin; Binding: Lüderitz & Bauer, Berlin

61/3020-5 4 3 2 1 0 – Printed on acid-free paper

Editor's Preface

This monograph written by two Chinese scientists of the younger generation opens a window into the world of thoughts on Mechanical Metallurgy in this fascinating area of our world, characterized by age old cultural heritage as well as by its dynamic evolution into the future. Based on notions and names all so familiar to the western scientist, and regarding the subject from the point of view of the theoretical mechanical engineer (Yang) as well as that of the materials and manufacturing engineer (Lee), the authors present a synthesis of both approaches and endeavour to guide the reader from basic theory to engineering applications. Between structural defects in the micrometer scale and the meter-measures of engineering components, the term of mesoplasticity is meant to place the reader right in the center: This is certainly a challenging enterprise, and the editor expresses his sincere wishes as to enrichment and stimulation which will emanate from this interesting book and its creative perspective.

March, 1993.   Prof.B. Ilschner

Preface

In the past two decades, enormous advances in materials and manufacturing technology have been achieved, which upgrade the material design, processing and precision manufacturing as quantitative and concise scientific disciplines. Rapid improvements on mechanics understanding have been instrumental in the above-mentioned development. A topic of great interest and importance in plasticity research has been the design and processing of materials themselves on the mesoscale to achieve the desired macroscopic properties. In recent years, the studies on the plastic behaviour of various materials and their constitutive representations have been the focus point in the field of plasticity. Several excellent survey articles focused on the subsets of these knowledge data appeared. However, near the end of 1989, we still can not find a book which overviews the whole subject from a mesoplasticity viewpoint.

As a mechanist and a material/manufacturing scientist, the two authors of the present book happened to meet together in Beijing and discuss their common interests in the interdisciplinary field where their expertness overlaps. After several rounds of communication back and forth, an outline of the book was set. Through a Hong Kong Polytechnic Special Fund for Academic Exchange with China, WY was able to visit Hong Kong in early 1990 for two months, and in that period of time a substantial part of the book came out as the product of our collaboration. Later on, our writing project was encouraged by Prof. B.Ilschner of Ecole Polytechnic Federale de Lausanne, the Editor-in-Chief of Material Science and Engineering Series, and the Engineering Editor of the Springer-Verlag Press. We cannot thank them more for their sincere supports to a book on a relatively new field by two relatively young authors. To meet their expectations, WBL travelled back to Beijing in the autumn of 1991 to finalize the manuscript. We tried our best to write out whatever we know about mesoplasticity, and needless to mention that we ourselves learned a lot during the preparation of the manuscript.

The intention of this book is to treat the subject of plasticity as well as its engineering applications through a somewhat unconventional angle, in which both mathematical rigor and engineering appeal could be maintained. In the book, the scope of macroplasticity is firstly reviewed and a systematic framework of mesoplasticity is laid down from a combined approach of solid mechanics and materials science, encompassing various physical deformation mechanisms. Besides a systematic discussion on the constitutive formulation derived from crystalline and geological materials, other interesting aspects of mesoplasticity , like the hardening mechanism and the meso-damage theory are also discussed in details. The applications of mesoplasticity in materials science and manufacturing engineering are examined from the author's own work as well as from other researchers' findings. This book will be suitable for research scientists and engineers, and postgraduate and senior undergraduate students working in the field of materials science, solid mechanics and physical theory of plasticity. We would be overjoyed if the readers could find this book interesting and beneficial to their own studies and researches.

The authors are deeply indebted to many organizations and individuals who made this book possible. We would like to acknowledge the supports from the Research Sub-Committee of the Hong Kong Polytechnic, Fok Ying-Tung Education Foundation, and the State Education Commission of China for the encouragement in mesoplasticity research in the recent years. We are also very grateful to Professor K.C. Hwang of Tsinghua University for the stimulating discussions and sincere advices during the preparation of the manuscript. Special thanks are due to Dr. B.J.Duggan of the University of Hong Kong for introducing texture work to one of the authors, and Dr. K.C.Chan for his help in computer programming and the preparation of some of the drawings used in the book.

Our wives – Cheng Li and Louisa Lee – are the greatest support behind the work and without their patience and endurance, this book would not have been materialized.

January, 1992.　　　　　Yang Wei　　　　　W.B.Lee
　　　　　　　　　　　　Tsinghua University,　　Hong Kong Polytechnic
　　　　　　　　　　　　Beijing, China.　　　　Hong Kong.

Contents

1 Introduction

1.1 What Is Mesoplasticity?

Metallurgists, mechanical engineers and civil engineers are all fascinated by the great variety of plastic behavior of materials. The plastic deformation of solids can be studied from different viewpoints and their classification is by no means universal. Loosely speaking, two general categories emerge from the vast content of plasticity. One is the phenomenological approach within the framework of continuum mechanics, with particular emphasis on the mathematical formulation, boundary value problem solutions and practical applications. The other is rather physically motivated as to deal with the physical background of materials subject to plastic deformation, especially devoted to the actual plastic deformation mechanisms and their interrelationship with the evolution of microstructure. Conventionally, the former approach is termed mathematical theory of plasticity, or simply *macroplasticity*; whereas the latter is termed physical theory of plasticity, or simply *microplasticity,* in which the fields of microplasticity and *mesoplasticity* overlap.

From the old era in the development of plasticity, the macroplasticity approach was largely enjoyed by mechanical and civil engineers, and established mostly by peoples engaged in solid mechanics. On the other hand, the microplasticity was usually favored and developed by metallurgists and physicists. This situation has been changed gradually during the past two decades by the need of interdisciplinary researches and the powerful intervention of computers. Nowadays, the essential knowledge of microplasticity is in great demand not only from scientists but also from engineers. The new area created by mesoplasticity stimulates a close workmanship among metallurgists, machanists and engineers, and promotes combined approach of solid mechanics and material science. As a central link between micro- and macro-scales, mesoplasticity aims at introducing the essential microplasticity concepts to various intermediate (or meso) scales where the quantitative theory of continuum mechanics is still applicable in describing the evolution of material microstructures during the plastic deformation processes. Mesoplasticity represents an important connection between the continuum-based macroplasticity and the atomistic physical theory of microplasticity, a connection in both length scales and the method of investigations. Furthermore, mesoplasticity would keep the mechanists and engineers informed about the state-of-art development of microplasticity and provide endless opportunities to the material scientists at the same time. Beside the above-mentioned role as the linkage between continuum and microstructures, mesoplasticity has a merit of its own. It has already branched out from conventional plasticity and mechanical metallurgy as an independent and concise discipline. In fact, mesoplasticity is acknowledged as one of the most fast growing inter-disciplinary fields. The *meso-level continuum based microstructure formulation*, a term which will be specified later, provides unique scientific attraction to the persons endeavored in both solid mechanics and material science.

1.2 Scales for Plasticity Investigation

Plastic deformation of solids can be explored from different *length scales*. The research subjects and related academic disciplines embodied in various length scales are summarized in Table 1.1. A comprehensive discussion on the plasticity mechanisms and representations in different length scales was given by Drucker[1,2]. The discrete microscopic models under the knowledge of solid state physics can be roughly classified into two length scales, namely the angstrom scale (10^{-10} m) and the atomic particle scale (10^{-9} to 10^{-8} m). Investigations under the former scale rely on quantum mechanics formulations, and they provide, through calculations on electron cloud interaction, the bonding energy expressions among atoms aggregated in an atomic array. Investigations under the latter scale, however, simply regard those atoms (of certain nuclear physics structures) as particles in a periodic crystalline lattice undergoing randomly-like thermal fluctuation and orderly-like mechanical deformation when subject to a prescribed thermal-mechanical environment. The approach of statistical mechanics, both classical and quantum, has been used to study the constitutive responses of solids from the very fundamental root. Considerable progresses have been reported, e.g. in the book of Weiner[3], for the microstructural insensitive responses of solids. For the intrinsically inhomogeneous behavior of material deformation and failure induced by non-uniformly distributed defects, however, the statistical treatment is still in the stage of early development. Examples belong to that category include phenomena of plastic deformation, ductile fracture, material damage and fatigue, all of them are of significant engineering values. Except sparse success in the qualitative physical explanations of limited plastic phenomena, the statistical approach has not yet achieved theoretical completeness and computational feasibility for the description of plastic constitutive responses.

As an interesting compensation to the above-mentioned inadequacy of the statistical approach, the studies in a microstructural or meso-scale has spearheaded the research frontier of physical theory of plasticity. These studies form the bulk of mesoplasticity research. Mesoplasticity encompasses four different length scales listed in Table 1.1, namely the scales of dislocation, continuous slip, meso-structure and grain. With the help of rapid upgrading of computing facilities, the constitutive simulation of a realistic single crystal or even a polycrystal aggregate under a selected mesoplasticity model has become as frequent practices to scientists and en-

Table 1.1. Plastic deformation approached by various scales

Scales	Research subjects	Academic disciplines
Å	Atomic bonding, interaction between electron clouds	Quantum mechanics
Atomic	Thermal fluctuation, diffusion, rate processes	Statistical mechanics
Dislocation	Dislocation motion and interaction, plastic flow and various strengthening mechanisms	Dislocation theory, micromechanics
Slip	Slip, texture, geometric softening	Crystal plasticity
Meso-structure	Precipitates, meso-damage (voids & microcracks), phase transformation	Physical metallurgy, meso-damage mechanics
Grain	Grain boundaries, lattice orientations, twinning	Crystallography
Continuum	Ductility, flow localization (necking and macro-shear bands), macroscopic fracture	Continuum mechanics
Structure	Structure geometries, environmental effects, integrity assessment	Computation mechanics

gineers. The generation of workable constitutive relations for polycrystal aggregates by mesoplasticity signifies a breakthrough on human's understanding and mastery of the nature, not to mention the reverse process on the design of mesoplastic behavior to achieve better material performance and component manufacturing.

As reported in the past decade, computer simulations with the incorporation of mesoplasticity models have achieved great successes in reproducing complicated deformation patterns in the finest detail as those observed experimentally under the identical testing conditions and material prescriptions. Cases for those successful predictions include extremely large deformation, flow localization, diffusive and concentrating damage, and ductile fracture, as would be elaborated in detail on the text to follow. Since the birth of dislocation theory more than half century ago, the development of the physical theory of plasticity has undergone two distinct stages. The first stage is concentrated on various aspects and physical consequences of dislocations. It was highlighted in the fifties by the direct experimental observation on dislocations. The second stage was started roughly at the late sixties from a modern reformulation of the continuous slip theory of Taylor, who laid down the general methodology of mesoplasticity. Major advances on the actual simulation of plastic deformation during the past twenty years have been rather focused on the approach from a microstructural or meso-level as defined above. Accordingly, our attention will be largely confined in this aspect to catch the excitement of the recent development.

1.3 Methods for Mesoplasticity Investigation

Mesoplasticity investigation requires a cooperated approach of micro-, meso- and macro-mechanics. The methodology of continuum mechanics is employed in different length scales for the characteristic material configurations which reflect the essential microstructural features exhibited in the length scale considered. To illustrate this point, we list in Table 1.2 the continuum assumptions and the corresponding structural characteristics for the four length scales crucial to mesoplasticity analysis, namely dislocation, slip, meso-structure and grain scales. In the dislocation scale, the crystal lattice outside the dislocation cores is regarded as an elastic continuum while the dislocations are treated as line defects situated in the elastic body. The self-induced stress-strain fields dominate the static and dynamic responses of the dislocations, as well as their interactions to each other or to the applied loads. In the continuous slip scale, the dislocations are no longer handled discretely. Instead, the plastic deformation accumulated by massive dislocation motion along specific slip system is concerned and the lattice outside the slip bands is still modeled by continuum elasticity. The computation in this scale will provide a constitutive law incorporating the essential features of single crystal plasticity. It can be further employed as the ground rule for meso-structures of larger length scales, or for polycrystal aggregates when the misorientations between the neighboring grains are considered. The following procedure for mesoplasticity modelling is put forward as a summary of the above descriptions.

(1) Selecting an appropriate length scale at which the phenomenon of interest occurs or can be best described;

(2) highlighting the essential microstructures for consideration, while the background material is treated as continuum;

(3) employing the constitutive relations obtained in the previous finer length scale for the background continuum.

Table 1.2. Basic assumptions in each length scale

Scales	Microstructures	Continuum
Dislocation	Dislocations modeled as line defects with induced stress fields	Elastic lattice outside dislocation cores
Slip	Continuous slip in discrete slip planes along fixed directions	Elastic lattice outside intensive shear bands
Meso-structure	Precipitates or meso-damage embedded as second phases with distinct constitution	Grain interior with meso-structures excluded is modeled by crystal plasticity
Grain	Grains of distinct orientations are separated by grain boundaries	Grain interior is modeled by single crystal plasticity

The deformation mechanisms in various length scales not only depend on the material background, but also relate to the thermal and mechanical loading environments. A spectrum of phenomena can occur in each length scale by the different excitation of stress-temperature combinations, plus the effect of chemical environment in processes such as stress corrosion, fatigue and diffusion. Influence on plastic deformation by stress environment can be served as an good example to explain this point. In the dislocation level, the stress exerts its influence in the form of Peach-Koehler force, as will be described in subsection 4.3.5. The stress influence to continuous slip, interesting enough, is dominated by Schmid resolved shear stress which can be regarded as an integral mean of the above dislocation gliding force. In the meso-structure length scale, however, the driving force for the meso-structures (such as voids and inclusions) exerted by the stress environment is different from the two mentioned for finer length scales, and assumes different forms for different meso-structures. Those driving forces, although different in forms, can be acconnodated in the general framework of energetic forces and can be evaluated by conservation integrals[4].

Another important aspect on method of mesoplasticity investigation is how to make a transition from one length scale to another length scale. The transition from a finer length scale to a coarser one is obtained by averaging over a volume with the presence of sufficient number of repeating microstructures. This process is termed *homogenization* in the literature. Various methods and algorithms have been developed for homogenization. Some of them, e.g., the self consistent method and the statistic assembling on grain orientations over pole figures via orientation distribution function, will be described in the present book. The reverse transition from a coarser length scale to a finer one, termed *heterogenization*, has also attracted the recent interest of research. The links among various length scales greatly multiply the interchanges of research data obtained under different length scales and strengthen the integrity of mesoplasticity as a unified academic discipline.

1.4 Layout of the Book

As a multi-step link from microstructures to engineering applications, the present book is designed to carry the readers from macroplasticity to mesoplasticity and finally to the applications of mesoplasticity, as explicitly implied by the arrangement of two basic parts in the book, namely I: Fundamentals of Mesoplasticity, II: Engineering Applications. The theoretical framework of mesoplasticity is laid down in the first part which travels from macroplasticity to the micro- and meso-aspects of plasticity. In the next chapter, the essential features of the macroplasticity is reviewed to the extent appropriate for a comparison to its mesoplasticity counterpart. The tensor notation, as well as the concept of finite deformation, are also briefly introduced. At the end of Chapter 2, the limitations of macroplasticity are addressed, so that a mesoplasticity approach, which will be treated in detail in the forthcoming 6 chapters, is beneficial and inevitable. As a starting point of mesoplasticity, a route along the material description rather than mechanics analysis is taken in Chapter 3, where both qualitative and quantitative features on various material structures are attempted. The latter includes the latest developments such as the more precise mathematic representation of the morphological features of microstructures, grain boundary structures and crystallographic textures. The mechanics route and the material route merge together from Chapter 4 to Chapter 8 where the essential contents of mesoplasticity theory gradually unfold. We begin from the essential knowledge of dislocations and then cast the dislocation fields in elegant mathematical formulation. The description on dislocation mechanism is later blended with other interesting plastic deformation mechanisms such as twinning, shear banding, grain boundary sliding and phase transformation in Chapter 5. The knowledge on plastic deformation mechanisms serves as the background to the discussion on various strengthening mechanisms in Chapter 6, and provides guidelines to the constitutive descriptions for crystalline and geological materials in Chapter 7. The content in the latter chapter is still in the verge of rapid development, so attention is paid to a careful selection of the coverage. The emphases there are imposed on the essential structures of constitutive framework developed from the average form of mesoplasticity and the physical assumptions underline those features. A similar treatment is also applicable for the case of meso-damage crafted by micro-cavities or micro-cracks, as proceeded in Chapter 8. The discussion in that chapter is centered around the general methodology, the void model of Gurson type and self consistent theory for microcracks. These topics have received more and more attention during the past years. In both Chapter 7 and Chapter 8, we frequently travel from mesoplasticity to macroplasticity by the homogenization procedure to reveal their intrinsic similarity. Upon the established principles of mesoplasticity, the second part of the book collects the highlights of its engineering applications. A wide class of application issues, ranging from metal forming to ultra-precision machining, could be viewed by means of the mesoplasticity methodology. We tentatively group those applications into three aspects, namely modelling of sheet metal textures, prediction of formability, grain boundary engineering and related topics. In particular, Chapter 9 is devoted to the texture modelling in sheet metals. The issues facilitated by mesoplasticity approach include the prediction of lattice rotation and the generation of new orientations in recrystallization. Another important aspect in metal forming, namely the formability prediction, will be treated in length in Chapter 10. The mesoplasticity formulation will be proved useful in the calculation of plastic strain ratio, in earing control, for analysis of forming limit, as well as to the revelation on the role of microstructural design. The last chapter is focused on grain boundary engineering, the most dynamically advancing applications of mesoplasticity. Interesting problems such as the design of single crystal turbine blades, micrograin superplasticity and ultra-precision machining will be illustrated there to show our readers that mesoplasticity is essential in the understanding of modern precision manufacturing. The last three chapters as a whole would provide engineers the accomplishments and the application schemes of mesoplasticity. Some of those applications are chosen from the best examples in the literatures to demonstrate the usage of mesoplasticity, and others come from the authors' own recent research activities. It is our believe that engineers and scientists can both enjoy this subject and make their own contributions in the near future.

1.5 References

1. Drucker, D.C., Material response and continuum relation; or from microscales to macroscales, ASME Annual Winter Meeting, New Orleans, (1984).

2. Drucker, D.C., Conventional and unconventional plastic response and representations, Appl. Mech. Rev., 41(1988), p.151.
3. Weiner, J.H., Statistical Mechanics of Elasticity, Wiley-Interscience, (1983).
4. Rice, J.R., in Fundamentals of Deformation and Fracture, Bilby, B.A., Miller, K.J. and Willis, J.R. eds., Pergamon Press, (1985).

General Reading Materials

1. Asaro, R.J., Micromechanics of crystals and polycrytals, in Adv. in Appl. Mech., 23(1983), p.1.
2. Haritos, G.K., Hager, J.W., Amos, A.K. and Salkind, M.J., Mesomechanics : the microstructure-mechanics connection, Int. J. Solids Structures, 24(1988), p.1081.
3. Mura, T., Micromechanics of Defects in Solids, 2nd Ed., Martinus Nijhoff Publishers, (1987).
4. Nemet-Nasser, S., J. Appl. Mech., 50(1983), p.1114.
5. Hirth, J.P. and Lothe, J., Theory of Dislocations, 2nd Ed., John & Wiley Pub., (1982).
6. Mayers, M.A. and Chawla, K.W., Mechanical Metallurgy, Prentice-Hall Inc., (1984).
7. Dieter, G.E., Mechanical Metallurgy, 3rd Ed., McGraw-Hill Book Company, (1986).
8. Cocks, A.C.F. and Leckie, F.A., Creep constitutive equations for damaged materials, in Adv. in Appl. Mech., vol.25, (1987), p.239.
9. Hill, R., Mathematical Theory of Plasticity, Clarendon Press, Oxford, (1950).

Part I
Fundamentals of Mesoplasticity

2 Scope and Limitations of Macroplasticity

2.1 Macroplastic Observations

2.1.1 Introduction

The studies on the plastic behavior of various materials and their constitutive representations have been the focus point of mechanics community in the recent years. The following sentences

"Perhaps the most active and controversial area in modern solid mechanics is that of modeling finite strain plastic and viscoplastic response. There are presently a number of competing theories and a number of important questions are being debated .... . The resolution of such questions has characterized research in solid mechanics at foundational level, and must continue to do so."

were quoted from a conclusion remark made by Carroll[1] at the ASME Applied Mechanics Division Report entitled <Trends and Opportunities of Solid Mechanics Research>. The reasons for the current boom on plasticity study, from both macroscopic and microscopic approaches, can be listed as follows:

(1) Standing at the crossroad of solid state physics, solid mechanics and material science, the plasticity theory has constantly been driven by impetus from the above-mentioned three major disciplines. The mathematical and physical complexity inhabited in plasticity theory, as well as the multi-disciplinary demand, put it as a spectacular research field endeavored by the scientists.

(2) During the past years, a general modern framework of continuum mechanics has almost been accomplished by the mechanists, except in the area of continuum plasticity. Stimulated by the rapid progress of computer technology and instrumented material testing facilities, the theory and applications of plasticity assume a unique challenge (as well as the single most important enterprise) in the field of solid mechanics.

Metal plasticity has played a central and dominant role in the historic development of plasticity. Various early observations of plastic phenomena were recorded in metallic materials. Plastic deformation can be identified in almost all FCC and BCC crystalline metals as it constitutes a substantial portion of total deformation before the failure of metals. On the other hand, the microplasticity theory was also deeply rooted in metallic solids, through the anticipation of dislocation mechanism as its basic foundation. The present knowledge on metal plasticity, however, is still far from satisfactory. The mystery of metal plasticity still puzzles scientists from understandings of the most fundamental level to practical applications. It seems to be appropriate for us to concentrate on metal plasticity, at least in a majority of the content. Deviations to non-metallic materials are also planned to notify their similarity and distinction to the plastic behavior of metals.

Metal plasticity is characterized by the metallic bonds in the atomic level and its microplastic deformation is typically portrayed by dislocation glide within crystalline lattice. The dislocation gliding movement proceeds successively in crystalline lattice without causing decohesion of the lattice structure. This deformation mechanism imposes the following characteristics to metal plasticity , as readily observed from marcoscopic measurements :

(1) The mathematical characteristics of plasticity is its *nonlinearity*, which gives tremendous challenge as well as difficulties to mathematicians and mechanists.

(2) The mechanics characteristics of plasticity is its *energy dissipation* and the creation of *residual or permanent deformation*. These features, rather than non-linearity mentioned above, separate plastic behavior from elastic behavior.

(3) The thermodynamic characteristics of plasticity lies in the *irreversibility* of its processes. Consequently, its behavior should be presented mathematically by history dependent functionals rather than by functions of state variables unless under certain postulates, such as that discussed in section 2.8.

A phenomenological description on the above-mentioned features (with essential mathematical representations) is attempted in this chapter which starts from the experimental observations obtained from uniaxial stress-strain curves.

2.1.2 Uniaxial Stress-strain Curve

We begin with the familiar uniaxial stress-strain curve of mild steel, as plotted in Fig. 2.1(a). The abscissa in the graph is the uniaxial strain ϵ and the ordinate is the uniaxial stress σ. For simplicity, the assumption of infinitesimal deformation is adopted and we do not distinguish different stress and strain measures at this stage. The initial response on mild steel below a critical stress level σ_s is featured by an apparent *elastic range* where identical linear response can be recorded in both loading and unloading events. The identification of initial *plastic yielding* is rather controversial, due to the possible local adjustment of dislocation lines at very low stress level. If one instead adopts the concept of *global yielding* (which refers to the occurrence of massive dislocation glide in the sense of mesoplasticity), then a pronounced upper yield point and a lower yield plateau are usually observed from the uniaxial stress-strain curve of mild steel. The cause of the yield point drop is related to dislocation break-away from Cottrell atmospheres, and will be discussed in subsection 6.2.5. The formation of macroscopic slip bands, or *Lüders bands*, is detected during the yield plateau, and is responsible for the slight fluctuation of stress.

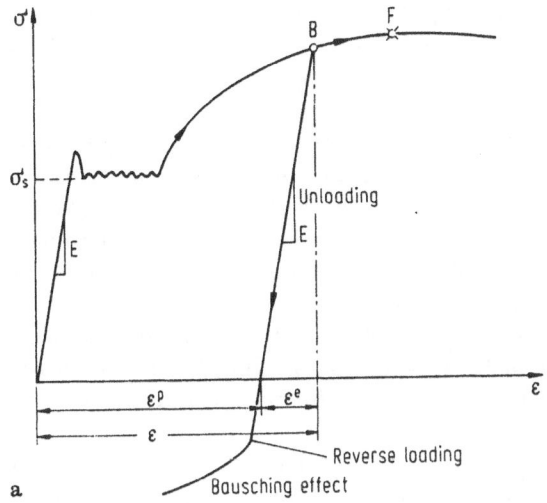

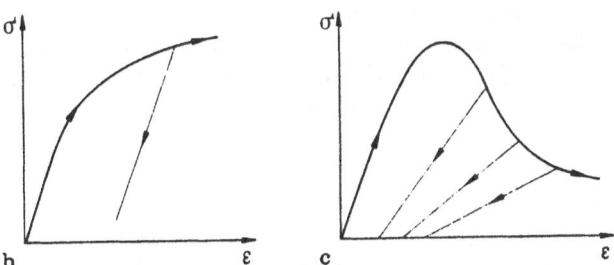

Fig. 2.1. Uniaxial stress-strain curves for mild steel, aluminium alloy and concrete. a Uniaxial tensile curve of mild steel, featured by yield plateau and Bauschinger effect; b Uniaxial tensile curve of aluminium alloy, featured by continuous strain hardening; c Uniaxial compressional curve of concrete, with softening and material damage

As the extent of plastic deformation advances passing through the yield plateau, *strain hardening* comes into effect which causes a continuous elevation of the stress level to accomplish further plastic deformation. This strain hardening phenomenon is due to the mutual interaction among dislocations and will be discussed in detail in section 5.1.

Wherever in the plastic regime of the uniaxial stress-strain curve, two distinct paths of *continued loading* or *elastic unloading* can be imposed to the material. This branching phenomenon, as exhibited in Fig. 2.1, is of essential significance so far as energy dissipation, permanent deformation and irreversibility of the deformation process are concerned. Furthermore, it is also served as an apparatus to decompose the total strain measurement into an elastic part and a plastic part as shown schematically in the figure. This decomposition will be cast into precise mathematical formulation in the sequel. During the process of elastic unloading, almost all dislocations cease to move and materials deform only by lattice distortion with the macroscopic response described by the same Young's modulus as recorded in the initial elastic loading. The area enclosed by the stress-strain curve and a straight elastic unloading path to the stress-free state represents the energy dissipation for the undergone plastic deformation. If one traces the unloading path further, a point of *reverse loading* is encountered at which the dislocations inside the specimen begin to glide in the opposite direction. If the reverse loading occurs at the stress level equal in magnitude but different in sign to the previous unloading stress, the material response is termed *isotropic hardening*. In a majority of cases, reverse loading takes place at the stress level inferior to the isotropic hardening prediction, sometimes even under a positive stress. This deviation from isotropic hardening is called *Bauschinger effect*. An *anisotropic hardening* law should be employed for its description.

If one varies the stress between certain positive and negative values, material will display a *cyclic plastic response*. Important features in cyclic plasticity include the evolution and stabilization of *hysteresis loops*.

Now we come back to the initial branching point, as marked by B in Fig. 2.1(a). If one chooses an increase of stress rather than unloading, continued monotonic plastic deformation is anticipated with further strain hardening. Starting from the peak point marked F in Fig. 2.1(a), a *softening* response appears for the uniaxial stress versus total strain curve. This softening behavior could come from the three sources listed below:

(1) *Material softening* due to the evolution of voids. This interesting subject will be examined carefully in section 8.2.

(2) *Macroscopic geometric softening* due to the reduction on the effective load carrying cross-section, especially accompanying neck formation. A macroscopic flow localization is about to form near point F in the stress-strain curve. Before flow localization, this pseudo-softening effect can be eliminated under a finite deformation framework with the engagement of true (Cauchy) stress measure, as described in subsection 2.5.3.

(3) *Microscopic geometric softening* due to the rotation of the slip system, as discussed in subsection 7.1.7. This phenomenon, however, is more likely to happen in single crystals, and does not substantiate in the overall deformation response of commercial mild steel.

Somewhat different stress-strain curves are recorded for other metallic materials such as aluminum alloys. As shown in Fig. 2.1(b), the yield plateau is absent in the stress response of aluminum alloys, signifying continuous strain hardening from the very beginning of plastic yielding. Another different stress-strain response is exhibited in Fig. 2.1(c) for concrete under compression, where the uniaxial curve demonstrates a pronounced stage of softening due to the formation of micro-cracks and localized shear bands inside the concrete. This stage is usually termed *cracking damage* process in the literature. Another feature of concretes is the gradual declination of the unloading modulus during cracking damage stage, caused by the weakening of micro-crack array developed in the damage process. A quantitative representation of this behavior will be given in section 8.3 for meso-damage by progressive cracking.

2.2 One-Dimensional Elastic-Plastic Formulation

Attention is now turned to a mathematical formulation of one dimensional elastic-plastic response. We start from the case of infinitesimal deformation and then generalize to the finite deformation case. As shown in Fig. 2.1, the only observable quantities through uniaxial test are the total elongation of the tensile bar (related to the engineering strain ϵ) and the sustained load (related to the nominal stress σ). Under infinitesimal strain assumption, all the differences between various stress and strain measures are negligible. The deformation starts from an elastic response, where stress σ is proportional to total strain ϵ, with the proportionality constant E signifying Young's modulus of uniaxial elastic deformation. When the specimen comes to the elastic-plastic regime, the stress would be balanced by the material flow stress Y and the strain would be a mixture of elastic and plastic contributions. An imaginary (or actual in some cases) elastic unloading process to the stress free state can help us to resolve the plastic strain, ϵ^p, as delineated in Fig. 2.1(a). Consequently, the following *additive decomposition law of strain* is established under infinitesimal deformation

$$\epsilon = \epsilon^e + \epsilon^p \tag{2.1}$$

where the elastic strain ϵ^e relates to the stress by a uniaxial Hooke's law

$$\epsilon^e = \sigma/E \tag{2.2}$$

as easily seen from the elastic unloading process in Fig. 2.1(a). A specific expression for the plastic strain, however, is difficult to elaborate because it is generally influenced by the entire stress history, $\sigma(t)$. The parameter t here is a monotonically increasing time-like parameter. The rate of any quantity is defined as its time derivative at a fixed material point and is denoted by a superimposed dot,

$$(\dot{\ }) = \frac{\partial}{\partial t}(\ ) \text{ at fixed material point} \tag{2.3}$$

Generally speaking, the *plastic strain rate*, $\dot{\epsilon}^p$, is regarded as a complicated function of both current stress and current stress rate. The latter represents the *rate sensitivity* of constitutive responses. If our attention is temporarily confined in *rate independent processes*, then its constitutive response should be invariant with respect to any linear transformation of t. Consequently, a rate independent hypothesis is justified if and only if the plastic strain rate is a homogeneous function of the stress rate of degree one. Here we would like to emphasize that a homogeneous function of degree one does not necessarily imply a linear homogeneous function. The latter would rather be viewed as a further constitutive assumption to the rate independent solids. More complicated case than linear homogeneous dependence of stress rate will be discussed in section 2.7. Under the linearity assumption between the plastic strain rate and the stress rate, we can write down the first equality in the formula below

$$\dot{\epsilon}^p = \dot{\sigma}/h(\sigma) = \beta(1/E_t - 1/E)\dot{\sigma} \tag{2.4}$$

where $h(\sigma)$ denotes a plastic hardening function of stress. Two explanations are given for the second expression in (2.4):

(1) The term $(1/E_t - 1/E)$ in (2.4), as a substitute for $1/h$, is derived from the rate forms of (2.1) and (2.2), with $E_t = d\sigma/d\epsilon$ representing the uniaxial *tangential modulus* at the current stress level σ. The value of E_t is measurable from the stress versus total strain curves plotted in Fig. 2.1, and varies with the stress level.

(2) The factor β in (2.4) corresponds to a loading parameter. It assumes a value of unity for plastic loading and switches to zero whenever elastic unloading takes place, because no further plastic deformation is accumulated within the elastic range.

The criterion for loading or unloading is an important issue in macroplasticity. For the present uniaxial case, its determination appears to be simple. For a deformation process controlled by total strain, loading or unloading is distinguished by whether the total strain rate is strictly positive or not, regardless the material behavior. For a stress controlled deformation, loading or unloading is decided by whether the stress rate is non-negative or not for a non-softening material, or in the non-softening regime of material response. For the softening material, the strain criterion of loading would be much more convenient to use than the stress criterion.

2.3 Three-Dimensional Formulation for Infinitesimal Deformation

In this section, we intend to generalize the previous uniaxial results to a three dimensional formulation of elastic-plastic deformation under the framework of infinitesimal deformation. For convenience and compactness of the description, tensor notations will be used hereafter throughout the text. A very brief explanation on tensor notations is given below.

2.3.1 Tensor Notations

The tensor quantities, such as stress, strain, displacement and constitutive moduli, are denoted collectively by the corresponding bold letters such as σ, ϵ, $\mathbf{u}$ and $\mathbf{L}$. The inner product, double inner product and vectorial product between two tensors are denoted by a single dot, $\cdot$, a double dot, :, and a product symbol, $\times$, respectively. A blank symbol will be used for a tensorial product between two tensors.

Alternatively, the index notation of tensors will be used whenever convenient for the illustration. Cartesian coordinates are employed in most part of the present book unless otherwise notified, which renders the simplified index notation by subscripts. The number of indices (subscripts) of the associated tensor quantity signifies the *rank of tensor*. For example, displacement u_i is a tensor of rank one (or simply a vector), stress σ_{ij} and strain ϵ_{ij} are tensors of rank two, and a constitutive modulus L_{ijkl} is a tensor of rank four, etc.. For distinction, the Latin indices have a range from 1 to 3, and the range of tensors with Greek indices would be specified for each individual cases. In a tensorial expression, some indices are termed *free indices* which appear only once in the either sides of the equation. The repeated indices are termed *dummy indices* which signify the summation over the corresponding ranges of indices, such as

$$\sigma_{ij}\epsilon_{ij} = \sigma_{11}\epsilon_{11} + \sigma_{12}\epsilon_{12} + \sigma_{13}\epsilon_{13} + \sigma_{21}\epsilon_{21} + \sigma_{22}\epsilon_{22} +$$
$$\sigma_{23}\epsilon_{23} + \sigma_{31}\epsilon_{31} + \sigma_{32}\epsilon_{32} + \sigma_{33}\epsilon_{33} = \boldsymbol{\sigma} : \boldsymbol{\epsilon}$$

$$\sigma_{ij}u_j = \sigma_{i1}u_1 + \sigma_{i2}u_2 + \sigma_{i3}u_3 = \boldsymbol{\sigma} \cdot \mathbf{u}. \tag{2.5}$$

The readers should also keep in mind that each dummy index cannot repeat more than once. The number of the component equations in a tensorial expression is given by the range of the free indices raised to the power of the number of distinct free indices.

After the above preparation, we are at a stage to formulate three dimensional constitutive relations.

2.3.2 Additive Strain Decomposition

The basic question asked in a three dimensional elastic-plastic constitutive formulation can be stated as follows :

Under a prescribed stress history $\sigma(t)$, what will be the strain response $\epsilon(t)$, especially the plastic strain response $\epsilon^p(t)$ for the material under consideration?

The time-like, monotonically increasing parameter t in the above description represents the elapse of history. It can be taken as the actual time (Newton time), as the arc length of the loading path (see Pipkin and Rivlin[2]), or as the endochronic time (see Valanis[3]). The last one is still related to the material property for the memory of plastic deformation.

The basic conservation laws in continuum mechanics, including the conservations of mass, linear momentum and angular momentum, are unconditionally obeyed for continuum plasticity, as well as the basic kinematic relation between total strain tensor and displacement vector. The conservation of angular momentum and the kinematic relation jointly require that both stress and strain are symmetric tensors of rank two, with six independent components for each of them. Alternatively, they can be represented by a "hyper-vector" in a stress or strain space of 6-dimension. Conventionally, the 6-dimension stress space formulation is used in macroplasticity with axes relating to the six independent stress components. Any stress state can be represented by a point in the stress space. A reformulation in a strain space will be pursued later in subsection 2.4.4.

An *elastic range*, designated by É, usually appears in the stress space. It consists of all stress states at which the plastic strain rates vanish. For the materials capable of purely elastic deformation, É is non-empty and connected. The bounding surface of the elastic range in the stress space, in a sense parallel to flow stress in uniaxial loading, is called *yield surface* and abbreviated as Y.S. in the sequel. The following remarks can be made to the yield surface :

(1) Yield surface exists as long as the elastic range is non-empty. Cases of plastic deformation response without elastic range have also been addressed in the literature. The possible mesoplastic explanation for such a treatment will trace back to dislocation dynamics curve as discussed in subsection 4.4.2. On the other hand, the plasticity theories with two or more yield surfaces have received some attention in the contemporary research of plasticity since the pioneer work of Mróz[4]. In most of the multiple yield surface theories, the inner yield surface serves to confine the stress points and to provide information for the plastic strain rate whereas the other yield surface(s) outside is(are) rather artificially devised to control the evolution of inner surface.

(2) The yield surface should be a closed surface except possibly unboundedness along the hydrostatic stress direction.

(3) Under the continuity requirement of stress history, $\sigma(t)$, with respect to t, the yield surface must be simply connected.

Both the elastic range É and the yield surface Y.S. vary with the history variable t. The current stress points can be either in É or on Y.S. at any time. The *stress states outside Y.S. are inaccessible* under a rate independent formulation. When the stress point is strictly inside É, an instantaneous elastic response between stress and strain increment tensors is expected. When the stress point is on Y.S., the total strain increment could be a mixture of elastic and plastic parts. By a unloading process to the previous stress state which is inside or on the current yield surface, the total strain rate can be resolved as

$$\dot{\epsilon} = \dot{\epsilon}^e + \dot{\epsilon}^p \tag{2.6}$$

and the elastic strain rate links with the stress rate by the well-known generalized Hooke's law

$$\dot{\epsilon}^e = \mathbf{L}^{-1} : \dot{\sigma} \tag{2.7}$$

where $\mathbf{L}$ is the fourth rank elasticity tensor with the following Voigt symmetry

$$L_{ijkl} = L_{jikl} = L_{ijlk} = L_{klij}. \tag{2.8}$$

Thus, the elasticity tensor $\mathbf{L}$ at most has 21 independent components for general elastically anisotropic solids. For infinitesimal deformation, the additive decomposition of (2.6), as a generalization of equation (2.1) for the uniaxial case, holds true not only for an infinitesimal stress increment but also for any finite increment of stress. The proof for the latter assessment is again facilitated by the application of an elastic unloading path to a stress free state. This unloading is attainable only if the origin of stress space is covered by the current elastic range. Otherwise, this unloading process is merely a hypothetical mean for strain decomposition. The plastic strain rate in (2.6) relates to the yield surface by a postulate on maximum plastic work as stated in the next subsection.

2.3.3 Principle of Maximum Plastic Work

In macroplasticity, further illustration on plastic response would require ad hoc postulate with heuristic thermodynamic or physical background. For metal plasticity, such postulates include the Drucker's postulate of non-negative stress work, the Ilyushin's postulate of non-negative work of an infinitesimal strain cycle, and Bishop-Hill postulate of maximum plastic work (the macroscopic form)[5]. The last one is often regarded as quasi-thermodynamic in nature for metal plasticity induced by slip, and can be easily generalized to the case of finite deformation, as shown in the next section and section 2.8. Accordingly, it will be quoted as a fundamental postulate in the sequel. The *Principle of Maximum Plastic Work* (PMPW) can be stated mathematically as follows

$$(\sigma - \sigma^*) : \dot{\epsilon}^p \geq 0 \quad \text{for any } \sigma \text{ on Y.S. and any } \sigma^* \text{ in É,} \tag{2.9}$$

which signifies the fact that among all statically admissible stress states σ^* in the elastic range (with the points on Y.S. included), the actual stress σ (at which the plastic strain rate $\dot{\epsilon}^p$ is created) would produce the maximum plastic work. Here we would like to point out that PMPW is not always true for arbitrary materials. Counter examples include rock masses with internal friction, and metals of non-Schmid type. Those exceptions would be addressed in section 2.9 and section 7.4. As a basic postulate, an appropriate proof of PMPW cannot be constructed under the framework of macroplastcity. A formal proof of PMPW under the continuous slip model and a Schmid-type shear criterion, however, will be furnished in subsection 7.3.2.

Following consequences are derivable from the principle of maximum plastic work :

(1) Normality of the plastic strain rate tensor to the smooth yield surface, as described mathematically by

$$\dot{\epsilon}^p_{ij} = \lambda(\partial F / \partial \sigma_{ij}) \tag{2.10}$$

where F denotes the yield function with stress tensor and a collection of material hardening parameters $Y_i, i = 1, \ldots, n$, as its arguments.

(2) Convexity of smooth yield surface.

(3) Corner convexity of yield surface. Whenever a corner is formed by several smooth pieces in a yield surface, it must be convex.

(4) Constraining cone for plastic strain rate on a yield surface corner. The plastic strain rate must be confined in a cone formed by the normals of the smooth yield surface pieces adjacent to the corner.

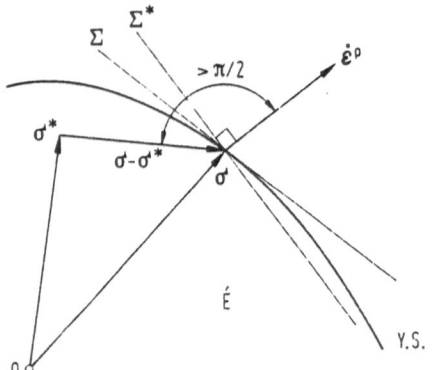

Fig. 2.2. Demonstration on normality of plastic strain rate to the smooth portion of a yield surface and the convexity of the yield surface

Proof. Here we would like to prove the first two consequences and leave proofs for the others to our readers. Referred to Fig. 2.2, at any stress point σ on Y.S., a tangential hyperplane Σ in stress space can be constructed and so is another plane Σ^* normal to the plastic strain rate $\dot{\varepsilon}^p$. If the two planes Σ and Σ^* do not coincide, then there must exist a stress point σ^* in the elastic range such that the stress differential vector $\sigma - \sigma^*$ forms an acute angle with $\dot{\varepsilon}^p$, resulting in the violation of PMPW. By contradiction, the plane Σ must coincide with plane Σ^* and consequently leads to desired normality.

The proof of convexity is straightforward after the establishment of normality. An arbitrary stress point σ under consideration is fixed on Y.S. with an associate plastic strain rate tensor $\dot{\varepsilon}^p$ directed along outward normal of Y.S. (as well as the hyperplane Σ). If the stress point σ^* is further chosen at another generic point on Y.S., then the definition of PMPW as described mathematically in (2.9) becomes the necessary and sufficient conditions for the convexity (global and local) of a yield surface. Q.E.D.

Beside the formal form of (2.10), further specification for the plastic strain rate can be obtained from PMPW and a *consistency condition*. The latter states that the yield surface, as described mathematically by the yield function F, should be identically zero during plastic loading process. That is

$$F(\sigma, Y_i) = 0 \quad \text{for any } t. \tag{2.11}$$

Combining the rate form of (2.11) and the normality rule (2.10), an expression for the pre-gradient factor (termed *flow factor* in macroplasticity) can be obtained as

$$\lambda = -\frac{\dot{\varepsilon}^p : \dot{\sigma}}{\sum_{i=1}^{n} \frac{\partial F}{\partial Y_i} \dot{Y}_i}. \tag{2.12}$$

Another expression of the flow factor can be obtained through taking self inner-product of the both sides of (2.10). That expression, when combines with (2.12), leads to the following explicit form of the plastic strain rate

$$\dot{\epsilon}^p = (3/2h)(\mathbf{P} : \dot{\sigma})\mathbf{P} \equiv \sqrt{3/2}\dot{\bar{\epsilon}}^p \mathbf{P}. \tag{2.13}$$

In the canonical expression (2.13) for the plastic strain rate,

$$\mathbf{P} = \frac{\partial F}{\partial \sigma} \bigg/ \sqrt{\frac{\partial F}{\partial \sigma} : \frac{\partial F}{\partial \sigma}} \quad \text{with} \quad \mathbf{P} : \mathbf{P} = 1 \tag{2.14}$$

is the normalized plastic strain, and

$$h = -\sqrt{\frac{3}{2}} \sum_{i=1}^{n} \frac{\partial F}{\partial Y_i} \frac{dY_i}{d\bar{\epsilon}^p} \bigg/ \sqrt{\frac{\partial F}{\partial \sigma} : \frac{\partial F}{\partial \sigma}} \tag{2.15}$$

is the plastic hardening modulus as in the uniaxial case of (2.4). The partial derivatives $\partial F/\partial \sigma$ and $\partial F/\partial Y_i$ can be obtained from the specification of the yield function F, while evolution of the yield surface parameters $dY_i/d\bar{\epsilon}^p$ is inferred from the experimental measurement. Additionally, the $\dot{\bar{\epsilon}}^p$ in (2.13)

$$\dot{\bar{\epsilon}}^p = \sqrt{2/3}\,\dot{\epsilon}^p : \mathbf{P} = \sqrt{3/2}(1/h)\,\mathbf{P} : \dot{\sigma} \tag{2.16}$$

represents the effective plastic strain rate. $\dot{\bar{\epsilon}}^p$ acts as the three dimensional counterpart of the plastic strain rate in uniaxial case. It can be viewed as the length of $\dot{\epsilon}^p$ when projected along its own direction, and then scaled by a factor $\sqrt{2/3}$ introduced to agree with the plastic strain rate in uniaxial loading. The physical significance of $\dot{\bar{\epsilon}}^p$ can also be observed from the expression of plastic work dissipation

$$\dot{W}^p \equiv \sigma : \dot{\epsilon}^p = \sqrt{3/2}(\sigma : \mathbf{P})\,\dot{\bar{\epsilon}}^p = \bar{\sigma}\dot{\bar{\epsilon}}^p \tag{2.17}$$

where $\bar{\sigma}$, usually termed as effective stress in macroplasticity, is defined as the projection of the stress tensor σ in the direction of plastic strain rate $\mathbf{P}$ and then scaled by an inverse factor $\sqrt{3/2}$ to agree with the tensile stress in uniaxial loading

$$\bar{\sigma} \equiv \sqrt{3/2}\,\sigma : \mathbf{P}. \tag{2.18}$$

As shown in (2.17), $\bar{\sigma}$ forms the work conjugate of the effective strain rate $\dot{\bar{\epsilon}}^p$.

To conclude this subsection, we remark that the above expressions for the plastic strain rate is only appropriate for plastic loading. The plastic strain rate will vanish during elastic unloading.

2.3.4 Metal Plasticity

Plastic incompressibility is usually assumed to metallic materials after the historic hydrostatic testing done by Bridgeman. The volume or mass conservation during purely plastic deformation can also be justified from the continuous slip model (see Sect. 7.1) and from dislocation glide motion (see Sect. 4.2). Violations to this assumption are occasionally observed from the pressure sensitivity and shear dilatancy phenomena observed in geological materials (see Sect. 2.9 and 7.4), and even for metals in the regime of substantial void growth as described in section 8.2. Nevertheless, plastic incompressibility is assumed in most content of the present book and is characterized mathematically by

$$\epsilon_{ii}^{p} = 0. \tag{2.19}$$

Consequently, both plastic strain and plastic strain rate tensors are deviatoric. From the normality rule (2.10), condition (2.19) requires that the yield function F cannot rely on the hydrostatic stress. Thus, it is sensible to decompose the stress tensor σ into a hydrostatic spherical part and a deviatoric part, with the latter defined by

$$\mathbf{S} = \sigma - (tr\sigma/3)\,\mathbf{1} \quad \text{with} \quad tr\,\mathbf{S} = 0 \tag{2.20}$$

where $\mathbf{S}$ denotes for stress deviator, $\mathbf{1}$ represents a second rank identity tensor and the symbol "tr" in front of a tensor represents a trace operation. Therefore, in the case of metal plasticity, the yield function is reduced to the first relation of the following equations

$$F(\sigma) = F(\mathbf{S}) = F(J_2, J_3) = F(J_2, J_3^2) \tag{2.21}$$

where the second and the third equalities correspond to further assumptions on isotropic material response and indifference between tension and compression, respectively. For brevity, the hardening parameters Y_i are suppressed in (2.21). The symbols J_2 and J_3 in (2.21) denote the second and third invariants of the stress deviator, defined as

$$J_2 = S_{ij}S_{ij}/2, J_3 = \det \mathbf{S} \tag{2.22}$$

where $\det \mathbf{S}$ implies the determinant of tensor $\mathbf{S}$. The simplifications made in various steps in (2.21) can be stated as follows. First, the assumption of isotropic response leaves the yield function $F(\mathbf{S})$ as a function of the non-trivial invariants of the stress deviator $\mathbf{S}$, namely J_2 and J_3. The indifference between tension and compression naturally indicates that F must be an even function of J_3, so comes the last specification in (2.21). Two special yield conditions are frequently quoted in the literature, they are

(1) *Mises yield criterion*, in which the yield function F only relies on J_2 by

$$F \equiv J_2 - Y^2/3 = 0 \tag{2.23}$$

where Y is the flow stress for uniaxial loading. The above yield criterion is also referred in the literature as J_2 type yield criterion.

(2) *Tresca yield criterion*, in which the yield function relies on the maximum shear stress inside the solids

$$F \equiv \text{Max}\,\{|S_1 - S_2|, |S_2 - S_3|, |S_3 - S_1|\} = 2k = Y \tag{2.24}$$

where S_1, S_2 and S_3 are principal deviatoric stresses, and k is the shear flow stress under simple shear deformation. The last equality in (2.24) is arrived by requiring an identical flow stress prediction from both Mises and Tresca yield criteria for the special case of uniaxial tension.

It is shown in any textbook of macroplasticity that in the case of plane stress, the Mises yield criterion can be represented graphically by an ellipse in the plane of in-plane principal stresses. While the Tresca yield criterion, described by a hexagon, inscribes the Mises ellipse. The two yield surfaces touch each other at the corners of the Tresca hexagon, and their largest deviation is 15.5%. For the other extreme two dimensional case of plane strain incompressible deformation, however, the Mises yield criterion becomes indistinguishable to the Tresca one. Both criteria can be described analytically as

$$(\sigma_{11} - \sigma_{22})^2 + 4\sigma_{12}^2 = (4/3)Y^2. \tag{2.25}$$

As an exercise, the readers are encouraged to derive the above equation from (2.23) and (2.24) under plane strain incompressible constraint. Identical representation of Mises or Tresca yield criterion for this particular case is attributed to the shear dominated deformation in a volume preserving solid.

2.4 Three-Dimensional Formulation for Finite Deformation

2.4.1 Elastic-Plastic Decomposition

A kinematic illustration on the decomposition of elastic and plastic deformations is given in Fig. 2.3(a) and (b) for both uniaxial and three dimensional cases. Let us start from the simpler uniaxial case demonstrated in Fig. 2.3(a), where a one dimensional bar of a unit length in the stress free *reference configuration* is elongated to a length λ in the *current configuration* by the application of load. When this deformation is carried out, the only measurable quantities are the current total elongation and the current load at various stages of the deformation process. To resolve the elastic and plastic portions entangled within the total elongation λ, a hypothetical (or sometimes actual) elastic unloading like that described in section 2.2.1 is engaged to the specimen which has already undergone a pre-elongation λ. This elastic unloading process can be regarded as an inverse (in the sense of unloading) length transformation λ^{e-1} on the stretched length λ. After the elastic unloading to a stress free state, the residual (or permanent) elongation, λ^p, is given by a combination of two operations as $\lambda^{e-1}\lambda$. Alternatively, the total elongation λ is decomposed into an elastic part λ^e and a plastic part λ^p by a multiplication rule as

$$\lambda = \lambda^e \cdot \lambda^p. \tag{2.26}$$

An additive strain decomposition can be obtained by taking the logarithm of the above equation and by using the *logarithmic strain* measure

$$\epsilon \equiv \ln \lambda = \ln \lambda^e + \ln \lambda^p = \epsilon^e + \epsilon^p. \tag{2.27}$$

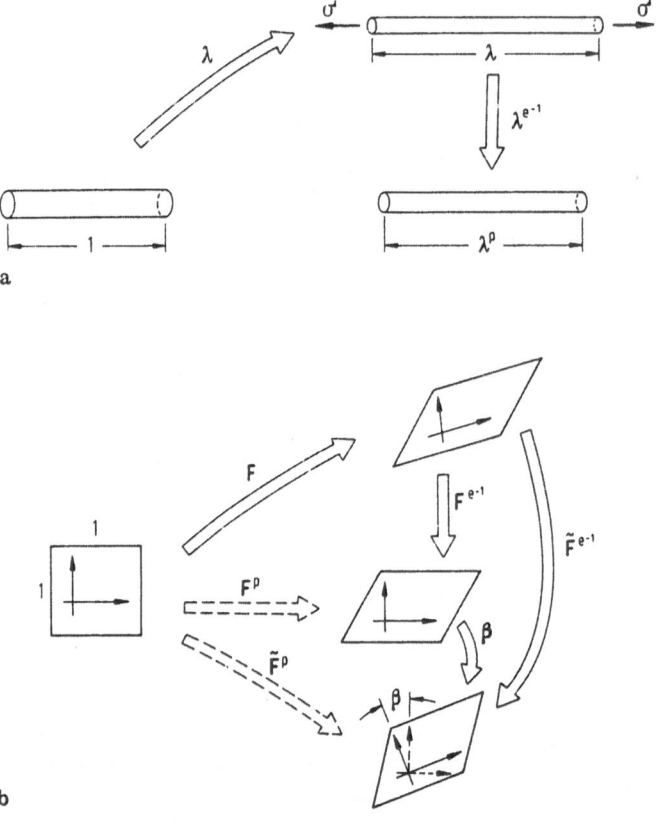

Fig. 2.3. Illustration of elastic-plastic decomposition of deformation at finite strain. a Uniaxial deformation; b General three dimensional deformation

For the general case of three dimensional deformation, the total deformation gradient $\mathbf{F}$ defined as

$$\mathbf{F} = \partial\mathbf{x}/\partial\mathbf{X} \qquad (2.28)$$

is usually used as a measure for the extent of deformation. Capital vector $\mathbf{X}$ represents the position vector in the reference configuration and lowercase vector $\mathbf{x}$ denotes the position vector in the current configuration. The difference between $\mathbf{x}$ and $\mathbf{X}$ signifies the displacement vector $\mathbf{u}$. Following a similar process of unloading to a stress-free *intermediate configuration* as delineated in Fig. 2.3(b), the total deformation gradient $\mathbf{F}$ can be decomposed into two parts, one for the elastic deformation as an inverse transformation to elastic unloading from the current configuration to the intermediate configuration and the other for the permanent deformation left behind in the intermediate configuration after complete elastic unloading. This process motivated E.H. Lee[6] to propose in 1969 a multiplicative decomposition of the total deformation gradient as follows

$$\mathbf{F} = \mathbf{F}^e \cdot \mathbf{F}^p. \tag{2.29}$$

The decomposition formula, as presented in (2.29), is nevertheless non-unique. An arbitrary rigid body rotation β can be imposed on the intermediate configuration, as shown in Fig. 2.3(b). Hence, a proper definition should be carefully proposed to preserve the so-called β-invariance condition[7] from the continuum plasticity point of view. A more enlightening explanation for the decomposition formula (2.29) is offered by crystalline plasticity as will be presented in section 7.1, where the choice of intermediate configuration is facilitated by the inherent orientation of crystalline lattices.

The multiplicative decomposition for the deformation gradient will lead to an additive decomposition of the velocity gradient tensor $\mathbf{l}$ defined by

$$\mathbf{l} = \partial \mathbf{v}/\partial \mathbf{x} = \dot{\mathbf{F}} \cdot \mathbf{F}^{-1} \tag{2.30}$$

where $\mathbf{v}$ is the particle velocity vector. Substituting (2.29) into (2.30), one obtains the following decomposition formula after straightforward algebras

$$\mathbf{l} = \mathbf{l}^e + \mathbf{l}^p \tag{2.31}$$

where

$$\mathbf{l}e = \dot{\mathbf{F}}^e \cdot \mathbf{F}^{e-1}, \quad \mathbf{l}^p = \mathbf{F}^e \cdot (\dot{\mathbf{F}}^p \cdot \mathbf{F}^{p-1}) \cdot \mathbf{F}^{e-1}. \tag{2.32}$$

It is relatively easier to comprehend that $\mathbf{l}^e$ in (2.32) is referred to the elastic deformation. But the visualization for the second term $\mathbf{l}^p$ in (2.32) as purely plastic deformation appears to be rather difficult from continuum plasticity viewpoint. As will be shown in section 7.1 from a mesoplasticity argument that the $\mathbf{l}^p$ term in (2.32), under appropriate definition of the intermediate configuration, does correspond to slip induced plastic deformation along the *current slip system*.

The general second rank tensor $\mathbf{l}$ can always be written as a sum of a symmetric tensor $\mathbf{D}$ and a skew-symmetric tensor $\mathbf{W}$ such that

$$\mathbf{D} = (\mathbf{l} + \mathbf{l}^T)/2, \quad \mathbf{W} = (\mathbf{l} - \mathbf{l}^T)/2 \tag{2.33}$$

where the superscript T indicates the transpose operation on the attached tensorial matrix. In (2.33), $\mathbf{D}$ is termed the *deformation rate tensor* which will be used as a primitive quantity in the formulation of constitutive relations at finite deformation, and $\mathbf{W}$ is termed *material spin* signifying the local spinning rate of material fibres. The additive elastic-plastic decomposition as illustrated in equation (2.31) is also applicable to tensors $\mathbf{D}$ and $\mathbf{W}$.

2.4.2 Selection of Stress and Strain Measures

Various stress and strain measures have been proposed under finite deformation theory. The readers not acquainted with them are advised to read an introductory book on continuum mechanics. Nevertheless, a self-inclusive treatment is attempted here with necessary background information.

Let us begin from the stress measures. The *Cauchy stress*, σ, is the true stress defined per unit area in the current configuration. It is symmetric and has the physical significance in relation to current traction. Despite its popularity in continuum mechanics, the Cauchy stress suffers from the disadvantage that it would lead to a non-symmetric formulation, as discussed by Hutchinson[8]. This undesirable outcome would jeopardize everything from the variational principle to an efficient numerical algorithm for finite deformation plasticity. As a remedy to this problem, the *Kirchhoff stress* τ defined by

$$\tau = J\sigma, \quad \text{where} \quad J = \det \mathbf{F} \tag{2.34}$$

is employed instead in the formulation of elastic-plastic constitutive equation at finite deformation[8]. The Kirchhoff stress τ can be regarded as the stress per unit mass and is obviously symmetric. The difference between Cauchy stress and Kirchhoff stress is usually negligible even in finite elastic-plastic deformation because the elastic compressibility is comparatively small and plastic incompressibility is assumed for metal plasticity. The other two popular stress measures, namely the *Piola-Kirchhoff stress* of the first kind (or nominal stress), designated by $\mathbf{N}$, and the Piola-Kirchoff stress of the second kind, designated by $\mathbf{T}$, are defined through the Kirchhoff stress as

$$\mathbf{N} = \mathbf{F}^{-1} \cdot \tau, \quad \mathbf{T} = \mathbf{F}^{-1} \cdot \tau \cdot \mathbf{F}^{T-1} \tag{2.35}$$

These stress measures are defined in the reference configuration and often used in the elasticity theory. As will be shown later for the simple case of uniaxial tension that a pseudo-softening behavior could develop by the employment of Piola-Kirchhoff stress tensors. This artificial softening behavior would hinder the implementation of routine incremental type numerical scheme even if the basic material behavior is deformation hardening.

The plastic work per unit volume in the current configuration, as designated by $\dot{W}_v$, is formed by the inner product of the deformation rate tensor $\mathbf{D}$ with its work conjugate, Cauchy stress σ. More conveniently, people use the plastic work per unit volume in the reference configuration designated simply by $\dot{W}$. From (2.34), one easily writes down

$$\dot{W} = \tau : \mathbf{D} \tag{2.36}$$

which can be alternatively regarded as the work per unit mass through dividing both sides in (2.36) by a reference mass density. The formula (2.36) and the definitions of Piola-Kirchhoff stresses of the first and the second kind give rise to the following work conjugate strain measures of them

$$\dot{W} = \tau : \mathbf{D} = \mathbf{N} : \dot{\mathbf{F}} = \mathbf{T} : \dot{\mathbf{E}}. \tag{2.37}$$

That is, the deformation gradient $\mathbf{F}$ is the work conjugate of the first Piola-Kirchhoff stress $\mathbf{N}$, and the work conjugate of the second Piola-Kirchhoff stress $\mathbf{T}$ is the Green strain tensor defined by

$$\mathbf{E} = (\mathbf{F}^T \cdot \mathbf{F} - \mathbf{1})/2. \tag{2.38}$$

However, the additive elastic-plastic decompositions of $\mathbf{F}$ and $\mathbf{E}$ usually assume very complicated forms involving pull-back manifold operations. This fact supports the choice of Kirchhoff stress τ and deformation rate $\mathbf{D}$ as the basic work conjugate pair in finite deformation elastic-plastic constitutive formulation.

By the additive decomposition of $\mathbf{D}$, an expression for *plastic work* can by written as

$$\dot{W}^p = \tau : \mathbf{D}^p = \bar{\sigma} d^p. \tag{2.39}$$

This is exactly the finite deformation version of plastic work recorded in (2.17), except the specifications on Kirchhoff stress τ in the place of stress and on $\mathbf{D}^p$ in the place of plastic strain rate. Accordingly, the principle of maximum plastic work (as previously stated in (2.9) for infinitesimal deformation) can be recast into a finite deformation form

$$(\tau - \tau^*) : \mathbf{D}^p \geq 0 \quad \text{for any } \tau \text{ on Y.S. and any } \tau^* \text{ in } \acute{E}. \tag{2.40}$$

Up to now, the previous results from (2.13) to (2.18) on plastic deformation rate can be generalized into a finite deformation version

$$\mathbf{D}^p = \sqrt{3/2}\, d^p \mathbf{P} \tag{2.41}$$

where $\mathbf{P}$ is given by (2.14) except that the stress tensor there would be specified as the Kirchhoff stress τ. The effective plastic strain rate, d^p, in (2.41) is defined as

$$d^p = \sqrt{2/3}\, \mathbf{D}^p : \mathbf{P} = \sqrt{3/2}\,(1/h)\, \mathbf{P} : \breve{\tau}. \tag{2.42}$$

The meaning of the corotational stress rate, $\breve{\tau}$, in (2.42) will be explained in the next subsection. The plastic hardening function h can be obtained from (2.15), with the equivalent plastic strain $\bar{\epsilon}^p$ there defined by

$$\bar{\epsilon}^p = \int_{t_o}^{t} d^p\, dt \tag{2.43}$$

where t_o represents the instant of initial plastic yielding. The physical significance of d^p is described in the second equality of (2.39), with its work conjugate stress $\bar{\sigma}$ given by (2.18) except the employment of Kirchhoff stress.

2.4.3 Objective Corotational Rate

Attention is now focused on the formulation of elastic deformation rate, $\mathbf{D}^e$, in terms of appropriate stress rate. As is well-known from continuum mechanics, a straightforward application of ordinary material derivative would lead to the loss of objectivity. An objective rate can be constructed as a rate in corotation to a skew-symmetric spin tensor Ω. As established by Lee, Mallet and Wertheimer[9], if the spin tensor Ω satisfies the frame transformation law during a frame rotation $\mathbf{Q}$ carrying frame F to frame F', i.e.

$$\Omega' = \mathbf{Q} \cdot \Omega \cdot \mathbf{Q}^T + \dot{\mathbf{Q}} \cdot \mathbf{Q}^T \tag{2.44}$$

while frame rotation $\mathbf{Q}$ in (2.44) is obviously orthogonal, then any stress rates corotational with Ω will be objective

$$\check{\tau} = \dot{\tau} - \Omega \cdot \tau + \tau \cdot \Omega \tag{2.45}$$

and termed *objective corotational rates*. The Kirchhoff stress in (2.45) can be replaced by any objective symmetric tensor of second rank for its respective corotational rate. It is straightforward to prove that

$$\check{\tau} : \mathbf{B} = \dot{\tau} : \mathbf{B} \tag{2.46}$$

for any symmetric, second rank tensor $\mathbf{B}$ commutable to τ, with $\mathbf{B}$ equal to τ as a special case. Various objective corotational rates have been employed in constitutive formulation, such as Jaumann's rate (while Ω equals to material spin $\mathbf{W}$), relative spin rate by Dienes[10], and spin rate of crystalline lattice as will be described in section 7.1. The corotational spin Ω is by no means unique. Its general form can be determined by an induction approach, as in the work of Fardshisheh and Onat[11], and recently by Agah-Teharni et al.[12]. Alternatively, Hwang and Cheng[13] showed that Ω can be described via a construction approach in terms of the frame spin executed by the principal axes of arbitrary, second rank symmetric objective tensor $\mathbf{B}$

$$\Omega^B = \sum_{\substack{i \neq j}}^{3} \left(\lambda_j^B - \lambda_i^B \right)^{-1} (\dot{\mathbf{B}})_{(ij)} \, \mathbf{n}_i^B \mathbf{n}_j^B \tag{2.47}$$

following the methodology of principal axes first invented by Hill[14]. In (2.47), λ_i^B (i = 1, 2, 3) are principal values of $\mathbf{B}$ with associated eigen-vectors (along the principal axes) $\mathbf{n}_i^B$.

Under a suitable definition of objective corotational rate, the elastic constitutive equation in a rate form can be written as

$$\check{\tau} = \mathbf{L} : \mathbf{D}^e = \mathbf{L} : (\mathbf{D} - \mathbf{D}^p) \tag{2.48}$$

where $\mathbf{L}$ is the fourth rank elasticity tensor with Voigt symmetry as shown in equation (2.8). The above elastic constitutive law characterizes the finite deformation version of (2.7). Distinctions, however, should be pointed out between (2.7) and (2.48). First, the work conjugate variables for the finite deformation are specified to Kirchhoff stress and deformation rate tensor. Second, the corotational rate as defined in (2.45) is used instead of simple material derivative used in (2.7), introducing the interplay of corotational spin in the constitutive formulation. The last point lies in the rate form expression of equation (2.48) with the modulus $\mathbf{L}$ possibly relevant to deformation (as specified in subsection 7.1.2) in contrast to the integrable form of (2.7). The form as presented in (2.48) is termed *hypoelastic* behavior which does not necessarily guarantee the existence of a path independent integral form.

2.4.4 Elastic-Plastic Constitutive Equations

The elastic-plastic constitutive equations can be inferred by combining the plastic constitutive relation (2.41) with the hypoelasticity law (2.48). After straightforward manipulations, one arrives

$$\overset{\triangledown}{\tau} = \mathbf{L}^t : \mathbf{D} \tag{2.49}$$

where the elastic-plastic tangential stiffness moduli tensor $\mathbf{L}^t$ can be expressed as

$$\mathbf{L}^t = \mathbf{L} - (3\beta/2H)\,\mathbf{R}\,\mathbf{R} \tag{2.50}$$

featuring the elastic moduli reduced by a fourth rank correction tensor formed by the self tensorial product of a symmetric second rank tensor

$$\mathbf{R} = \mathbf{L} : \mathbf{P}. \tag{2.51}$$

The form of (2.50) clearly suggests the Voigt symmetry of tangential moduli $\mathbf{L}^t$. The tensor $\mathbf{R}$ given in (2.51) has a dimension of stress and a magnitude comparable to Young's modulus. The scalar quantities H and β in (2.50) also have their respective physical significances. The value of H is given by

$$H = h + \frac{3}{2}\,\mathbf{P} : \mathbf{L} : \mathbf{P} \tag{2.52}$$

portraying a newly defined uniaxial "over-unloading hardness" function. This function is dictated by the second term in (2.52) and always has a positive value except the impossible case of Fig. 2.4(b) in which the strain softening is so severe that its tangential modulus is even less than the elastic unloading slope! The likely positiveness of H greatly facilitates the calculation for conventional strain softening case as shown in Fig. 2.4(a). For a stress controlled loading, the plastic hardening function h becomes negative whenever strain softening occurs. But if the formulation from (2.49) to (2.52) is followed, the adjusted modulus H will remain positive unless situation like Fig. 2.4(b) occurs. The latter case can actually be ruled out under thermodynamics ground. The formulation developed above is featured by evaluating the stress increment from a given incremental response in the strain space. This approach is often termed *strain space formulation* in the literature.

The symbol β in (2.50) is the loading type parameter in the sense of (2.4). The value of β can switch from one, pertinent to plastic loading, to zero for the case of elastic unloading. The loading or unloading criterion is given by

$$\begin{aligned} \beta = 1 \quad &\text{if} \quad \mathbf{R} : \mathbf{D} > 0 \\ \beta = 0 \quad &\text{if} \quad \mathbf{R} : \mathbf{D} \leq 0 \end{aligned} \tag{2.53}$$

when a strain space formulation is adopted.

The above results establish a general framework for three dimensional elastic-plastic constitutive relations with the capacity to deal with arbitrarily large deformation. Specifications for various simple forms will be given in the sections to follow.

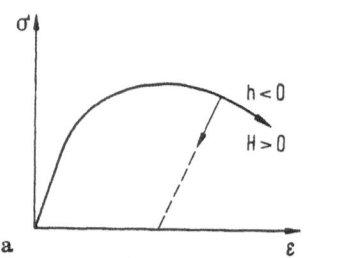 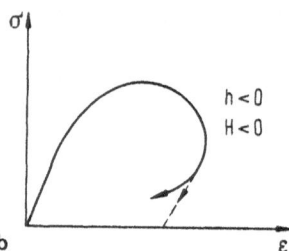

Fig. 2.4. A uniaxial demonstration for treatment of strain softening by strain space formulation. a Typical case of strain softening in which h is negative but H is positive; b the case of negative H in which the loading slope is even less than the elastic unloading slope

2.5 J_2 Flow Theory

2.5.1 Isotropic Hardening

As the first example of elastic-plastic constitutive formulation, let us consider the case of J_2 flow theory under isotropic hardening. The discussion on the case of infinitesimal deformation will be conducted first, then move on to its generalization to finite deformation.

A von-Mises yield criterion, as quoted from (2.22) and (2.23) is incorporated in the J_2 type theory

$$F(\mathbf{S}) \equiv S_{ij}S_{ij}/2 - Y^2/3 = 0 \qquad (2.54)$$

where flow stress is determined via a uniaxial test as shown in Fig. 2.5

$$\bar{\sigma} = Y(\bar{\epsilon}^p). \qquad (2.55)$$

The essence of J_2 theory is to extend the uniaxial testing data to the three dimensional deformation with an equivalent J_2 quantity. The identification is made in (2.55) by taking the effective stress $\bar{\sigma}$ and the effective plastic strain $\bar{\epsilon}^p$ as the uniaxial stress σ and uniaxial plastic strain ϵ^p, respectively. Equation (2.55) is sometimes called *single curve postulate* in macroplasticity literature, implying the anticipation of three dimensional plastic response from a single uniaxial curve.

The direction tensor $\mathbf{P}$ for plastic strain rate can be specified under J_2 flow theory by substituting (2.54) into the general expression (2.14)

$$\mathbf{P} = \mathbf{S}/\sqrt{2J_2} \qquad (2.56)$$

which indicates the *coaxiality* between plastic strain rate and the stress deviator under J_2 flow theory with isotropic hardening. Substituting the just obtained expression of $\mathbf{P}$ back to (2.18), one discovers that effective stress links with J_2 simply

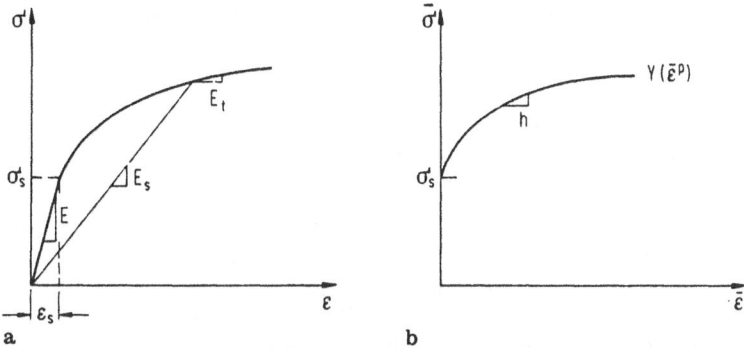

Fig. 2.5. Single curve postulate under J_2 type theory. **a** Uniaxial stress versus strain curve, featured by elastic modulus E, tangential modulus E_t and secant modulus E_s. **b** Effective stress versus effective plastic strain curve, viewed as identical to the uniaxial stress versus plastic strain curve from the "single curve postulate"

by

$$\bar{\sigma} = \sqrt{3J_2}. \tag{2.57}$$

Consequently, the plastic strain rate under infinitesimal deformation as expressed by (2.13) and (2.16) can be specified as

$$\dot{\epsilon}^p = [(3\dot{\bar{\sigma}})/(2h\bar{\sigma})]\,\mathbf{S}. \tag{2.58}$$

The plastic hardening function h in (2.58) may be determined from (2.55) as $dY/d\bar{\epsilon}^p$. Alternatively, the factor $1/h$ in (2.58) can be determined from the monotonic uniaxial loading curve of Fig. 2.5(a) as $(1/E_t - 1/E)$, yielding

$$\dot{\epsilon}^p = (3/2)(1/E_t - 1/E)(\dot{\bar{\sigma}}/\bar{\sigma})\,\mathbf{S}. \tag{2.59}$$

Because the above expression of plastic strain rate is deviatoric, it is convenient to break the elastic strain rate into a sum of a deviatoric, designated by $\dot{\mathbf{e}}^e$, and a volumetric part

$$\dot{\epsilon}^e = \dot{\mathbf{e}}^e + (\dot{\epsilon}_v/3)\,\mathbf{1}, \quad \dot{\epsilon}_v = \mathrm{tr}\,\dot{\epsilon}. \tag{2.60}$$

If an isotropic elastic response is further imposed, then the elastic constitutive equation can be cast in a form of generalized Hooke's law

$$\dot{\mathbf{e}}^e = [(1+\nu)/E]\,\dot{\mathbf{S}}, \quad \dot{\epsilon}_v = [(1-2\nu)/E]\,\mathrm{tr}\,\dot{\sigma}. \tag{2.61}$$

For the volumetric dilation ϵ_v is purely elastic, attention here is rather focused on deviatoric deformation. Combining (2.61) and (2.59), one obtains the following result for total deviatoric strain rate

$$\dot{\mathbf{e}} = \dot{\mathbf{e}}^e + \dot{\mathbf{e}}^p = [(1+\nu)/E]\,\dot{\mathbf{S}} + (3/2)(1/E_t - 1/E)(\dot{\bar{\sigma}}/\bar{\sigma})\mathbf{S}. \tag{2.62}$$

The above result can be inverted into a strain space formalism similar to equation (2.49)

$$\dot{S} = L' : \dot{e} \tag{2.63}$$

where the tangential stiffness tensor can be written in an explicit form

$$
\begin{aligned}
L^t_{ijkl} &= [E/(1+\nu)]\{\delta_{ik}\delta_{jl} - \\
&\quad (3\beta/2)(S_{ij}S_{kl}/\bar{\sigma}^2)/[1 + 2(1+\nu)E_t/3(E - E_t)]\} \\
&= (2E/3)\delta_{ik}\delta_{jl} - \beta(E - E_t)(S_{ij}S_{kl}/\bar{\sigma}^2)
\end{aligned}
\tag{2.64}
$$

where the first equality is referred to an arbitrary Poisson's ratio (elastic compressible) and the second equality only to the extreme case of Poisson's ratio being one half (elastic and total incompressible). The equations from (2.62) to (2.64) recapitulate the famous *Prandtl-Reuss equation*, which played an important role in the human's exploration of macroplasticity.

The above Prandtl-Reuss equation will be reformulated under the framework of finite deformation. A comparison between (2.63) and (2.49) leads to the pertinence of equation (2.49) as a general form of Prandtl-Reuss equation in finite deformation. Same conclusion may be reached for equation (2.50) except the specification on R and H under a J_2 flow, isotropic hardening framework. Particularly, R is now expressed as

$$\mathbf{R} = \sqrt{3/2}\,\mathbf{L} : (\mathbf{S}/\bar{\sigma}) = \sqrt{3/2}\,[E/(1+\nu)]\,(\mathbf{S}/\bar{\sigma}) \tag{2.65}$$

where the last step is only valid for the case of elastic isotropy. It is interesting to note that R becomes coaxial to the stress deviator S for an isotopic solid. Moreover, under the recognition that plastic hardening function h now represents the slope of a uniaxial Kirchhoff stress versus logarithmic strain curve, general expression (2.52) for H can be specified as

$$
\begin{aligned}
H &= EE_t/(E - E_t) + (9/4\bar{\sigma}^2)\mathbf{S} : \mathbf{L} : \mathbf{S} \\
&= E\{E_t/(E - E_t) + 3/[2(1+\nu)]\} \\
&= E^2/(E - E_t)
\end{aligned}
\tag{2.66}
$$

where the second and the last equalities correspond to further assumptions on elastic isotropy and elastic incompressibility, respectively. It is observed from (2.66) that H will be positive except for the impossible case of E_t is even less than $-E$. The switching value of β in (2.50) for the isotropic hardening J_2 flow theory is still dictated by (2.53) with the substitution of R given in (2.65). For the particular case of isotropic elastic response, (2.53) is simplified to

$$
\begin{aligned}
\beta &= 1 \quad \text{if} \quad \mathbf{S} : \mathbf{D} > 0 \\
\beta &= 0 \quad \text{if} \quad \mathbf{S} : \mathbf{D} \leq 0
\end{aligned}
\tag{2.67}
$$

namely the deviatoric stress should form an acute angle with D to sustain plastic loading.

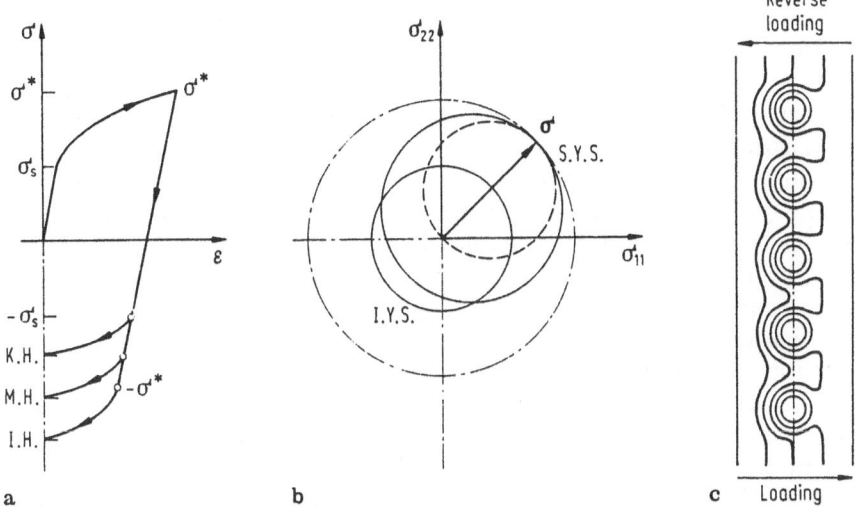

Fig. 2.6. Mixed hardening. a Bauschinger effect for an earlier reverse loading; b evolution of subsequent yield surfaces from an isotropic initial yield surface according to isotropic hardening (dot-dashed contour), kinematic hardening (dashed contour) and mixed hardening (solid contour) laws; c physical origin for mixed hardening by dislocation loops trapped around the hard second phase particles

2.5.2 Mixed Hardening

It was mentioned in subsection 2.1.2 that the uniaxial prediction based on an isotropic hardening law frequently over-estimates the elastic range and causes delayed reverse loading response. In reality, *Bauschinger effect* appears to recognize the difference between stress magnitudes at unloading and reverse loading incidents, as depicted in Fig. 2.6(a). Consequently, *anisotropic hardening laws* should be developed to count for this biased situation. When generalizing the above idea to three dimensional formulation, a simplified treatment attempted by scientists consists of incorporating Bauschinger effect by a *mixed hardening model*. This model visualizes the evolution of yield surfaces as the combination of a self (isotropic) expansion and a translation carried by the movement of its center, as shown in Fig. 2.6(b). Movement of the yield surface center is described mechanistically by the evolution of *back stress* **b**, whose physical origin is attributed to accumulated dislocation loops around the hard second phase particles dispersed in the alloys. These dislocation loops are trapped by the second phase particles when dislocation lines swept by, as delineated in Fig. 2.6(c) and explained in section 6.3. Those loops exert a global resistant force against the oncoming dislocation movement to further the plastic deformation, and exert a global attractive force to the dislocation regression excited by reverse plastic deformation. This mechanism explains both the strain hardening behavior, as detailed in Sect. 5.3, and the early occurrence of reverse loading.

Beside Bauschinger effect, mixed hardening theory is also applicable to pre-dictions of cyclic stress-strain relationship, as employed by Drucker and Palgen[15] for the general constitutive framework and by Yang, Brown and Miller[16] for the description of biaxial stabilized hysteresis loops. The interesting fact for cyclic plasticity is that stabilized hysteresis loops, as frequently observed for metals dur-ing uniaxial or multiaxial cyclic testings, cannot be predicted from an isotropic hardening theory or from a mixed hardening theory with non-saturated isotropic hardening part. A case of *elastic-shakedown* will be predicted instead with gradu-ally diminished hysteresis loop. Stabilized hysteresis loops are only predicted under a *kinematic hardening law*, in which the size of yield surface will not change, or under a mixed hardening theory with a saturated isotropic part.

Another important aspect of mixed hardening theory relates to the effect of yield surface curvatures. As easily observed from Fig. 2.6(b), the radii of curva-ture for the subsequent yield surfaces (S.Y.S.) developed from a common isotropic initial yield surface (I.Y.S.) take a variety of values from the maximum prediction of isotropic hardening to the minimum estimate of kinematic hardening. Recent researches confirmed the importance of yield surface curvature to the simulated constitutive response for bifurcation problems involving sudden change on the loading path. Examples belong to this category range from plastic buckling, neck formation and shear bands to other flow localization phenomena. Preliminary stud-ies also indicated that the use of mixed hardening type formulation[17,18] could provide better description for those phenomena.

The theoretical framework of mixed hardening constitutive law will be eluci-dated here for finite deformation situation. To begin with the illustration, we notice that the elastic constitutive law as presented in (2.48) is still valid with appropriate selection of the corotational rate, but the plastic deformation is no longer governed by an isotropically expanding yield surface. In the conventional mixed hardening theory, the evolution of yield surfaces is characterized by the combined motion of an isotropic expansion of the yield surface radius and a kinematic translation of yield surface featured by the movement of its center, as indicated in Fig. 2.6(b). A *normalized active stress* α defined as

$$\alpha = (\mathbf{S} - \mathbf{b})/\sigma_F \qquad (2.68)$$

is used to describe this evolution of yield surface

$$F \equiv \alpha : \alpha - 2/3 = 0. \qquad (2.69)$$

In (2.68), σ_F represents the radius of yield surface and $\mathbf{S}$ is the deviator of Kirchhoff stress. The normalized active stress α and the back stress $\mathbf{b}$ are all deviatoric stresses. It should be mentioned that the yield surface motion governed by (2.68) and (2.69) does not correspond to a general anisotropic hardening law. The latter additionally requires the ability to change the shape of yield surfaces. In contrast to the spherical shape of J_2 type yield surface, an elliptical yield surface capable of describing certain anisotropic hardening behavior during the texture formation in metal forming plasticity was given by Hill[19] to achieve an empirical correlation with the experimental data. Criticisms on the mixed hardening law are also raised on its non-tensorial appearance[20] stemmed from the prescribed translation of yield surfaces. This stringent limitation precludes it to fit into a general anisotropic hardening framework with close tensorial alliance.

Despite the above mentioned drawbacks, mixed hardening theory is still regarded as the simplest effective mean to deal with anisotropic hardening. The single curve postulate (2.55) is still used to get a J_2 generalization between the Kirchhoff stress and the deformation rate tensors. The value of Y in (2.55) now represents a collection of the yield surface radius, σ_F, and the kinematic shifting of the yield surface center. In the special case proposed by Kadaschevich and Novozhilov[21], σ_F is assumed to evolve in proportion to the evolution of flow stress Y in uniaxial tension

$$\sigma_F = cY_o + (1 - c)Y. \qquad (2.70)$$

Parameter c varies between zero and one. Extremes $c = 0$ and $c = 1$ correspond to special cases of *isotropic hardening* and *kinematic hardening*, respectively, where $Y_o = Y(0)$ is the radius of initial yield surface.

The normality structure as anticipated in (2.41) is preserved under mixed hardening law through the application of principle of maximum plastic work. The direction of plastic deformation rate is given by

$$\mathbf{P} = \sqrt{3/2}\,\alpha. \qquad (2.71)$$

Although the coaxiality to the stress deviator $\mathbf{S}$ is lost, $\mathbf{P}$ is coaxial to the active stress tensor. If one substitutes the above expression back to equation (2.41), he obtains the following expression of plastic deformation rate

$$\mathbf{D}^p = (9/4h)(\check{\mathbf{S}} : \alpha)\,\alpha \qquad (2.72)$$

where h is again given by the slope of $Y(\epsilon^p)$ curve.

Evolution of back stress $\mathbf{b}$ is governed by the Prager-Ziegler law, see Ziegler[22]

$$\check{\mathbf{b}} = (2h_b/3)\,\mathbf{D}^p \qquad (2.73)$$

where the choice of appropriate corotational rate remains a controversial issue of macroplasticity and will be elaborated later. The function h_b in (2.73) represents the plastic tangential modulus of $(3/2)b_{11}$ versus ϵ^p in a uniaxial (but not necessarily monotonic) loading. Further specification of back stress evolution is benefited from a consistency condition of $f = 0$ as observed from (2.11)

$$\check{\mathbf{b}} = [h/(h - h_b)]\dot{\sigma}_F\alpha. \qquad (2.74)$$

Equations (2.73), (2.72), (2.69) and (2.46) are utilized in the above derivation. More explicit formula for back stress evolution can be obtained for the particular case of the "proportional isotropic-kinematic hardening" law (2.70). Its engagement yields

$$\check{\mathbf{b}} = c\dot{Y}\alpha, \quad h_b = ch. \qquad (2.75)$$

We next consider the specification of the general constitutive equation (2.49) and (2.50) for mixed hardening case. The **R** tensor in (2.51) is reduced to

$$\mathbf{R} = \sqrt{3/2}\,\mathbf{L} : \alpha = \sqrt{3/2}\,[E/(1+\nu)]\,\alpha \tag{2.76}$$

where the last step is only valid for elastically isotropic materials, in which **R** is coaxial to **P** as well as to active stress α. Consequently, the tangential stiffness matrix $\mathbf{L}'$, as given in (2.50), can be written down explicitly as

$$\begin{aligned}
\mathbf{L}' &= \mathbf{L} - \beta\{[3E/2(1+\nu)]^2/[h + (9/4)\alpha : \mathbf{L} : \alpha]\}\,\alpha\alpha \\
&= \mathbf{L} - \beta\{[3E/2(1+\nu)]^2/[h + (3E)/2(1+\nu)]\}\,\alpha\alpha \\
&= \mathbf{L} - \beta[E^2/(h+E)]\,\alpha\alpha
\end{aligned} \tag{2.77}$$

where the second and the third equalities are only valid under further assumptions of elastic isotropy and elastic incompressibility, respectively, with relevant expressions of **L** for the first term on the right hand sides.

2.5.3 Examples

Several simple examples will be worked out in this subsection to demonstrate some essential features of J_2 flow theory.

Proportional loading. As the first demonstrative example, let us consider a proportional (radial) loading path starting from the origin of stress space. For this case, the deviatoric stress and its rate are both coaxial to the normalized active stress, with the latter becoming a constant tensor irrelevant to loading time

$$\mathbf{S}(t) = Y(t)\,\alpha, \quad \check{\mathbf{S}}(t) = \dot{Y}(t)\,\alpha. \tag{2.78}$$

Furthermore, the corotational rate coincides with the conventional material derivative for proportional loading. As easily concluded from (2.72), the plastic deformation rate can be simply written as

$$\mathbf{D}^p = (3/2h)\,\dot{\mathbf{S}} \tag{2.79}$$

regardless the form of hardening law. This result comes as no surprise to us. Because a proportional loading can be considered as a directional monotonic "uniaxial" loading in which the Bauschinger effect connected to reverse loading would never take into effect.

Uniaxial loading. We next consider the case of uniaxial tension in X_1 direction to show the pertinence of choosing Kirchhoff stress tensor, instead of Piola-Kirchhoff stresses, as the basic stress measure. As shown in Fig. 2.7(a), a bar of unit initial length is pulled along X_1 direction by a Cauchy stress σ to reach a current length λ. Various deformation and stress tensors for this uniaxial case can be simply worked out as

$$F = \begin{bmatrix} \lambda & & \\ & \lambda^{-\frac{1}{2}} & \\ & & \lambda^{-\frac{1}{2}} \end{bmatrix} \quad D = \begin{bmatrix} \dot{\lambda}/\lambda & & \\ & -\dot{\lambda}/2\lambda & \\ & & -\dot{\lambda}/2\lambda \end{bmatrix}$$

$$\tau = \sigma = \begin{bmatrix} \sigma & & \\ & 0 & \\ & & 0 \end{bmatrix} \tag{2.80}$$

where the elastic volumetric change is neglected for simplicity. The total constitutive response along the tensile direction, in terms of Cauchy stress σ and logarithmic strain $\epsilon = \ln \lambda$, is

$$\dot{\epsilon} = \dot{\sigma}/E_t \tag{2.81}$$

where the tangential modulus E_t is assumed as always positive.

The corresponding "11" components of the Piola-Kirchhoff stresses N and T are

$$N_{11} \equiv N = \sigma/\lambda, \quad T_{11} \equiv T = \sigma/\lambda^2 \tag{2.82}$$

with the corresponding work conjugates, as delivered by (2.37), being

$$F_{11} = \lambda, \qquad E_{11} = (\lambda^2 - 1)/2. \tag{2.83}$$

Combining expressions from (2.81) to (2.83), we arrive the following uniaxial stress-strain relations for Piola-Kirchhoff stresses

$$\dot{N} = (\dot{\lambda}/\lambda^2)(E_t - \sigma)$$

$$T = [\dot{E}/(1 + 2E)^2](E_t - 2\sigma). \tag{2.84}$$

The above expressions clearly demonstrate that *pseudo-softening* behavior occurs for N versus λ curve at $\sigma = E_t$, and for T versus E curve at $\sigma = E_t/2$, respectively. This undesirable feature associated with Piola-Kirchhoff stresses would cause a false breakdown, or an occurrence of pseudo-instability, for the solution algorithm based on a tangential modulus approach, rendering the Piola-Kirchhoff stresses undesirable for very large elastic-plastic deformation. Those pseudo-instability points are indicated in Fig. 2.7(c) and Fig. 2.7(d). For a concrete demonstration to clarify the above remarks, let us take a simple power-law hardening curve between Cauchy stress σ and logarithmic strain ϵ

$$\begin{aligned} \sigma/\sigma_s &= \epsilon/\epsilon_s & \text{when} \quad \sigma < \sigma_s \\ \sigma/\sigma_s &= (\epsilon/\epsilon_s)^n & \text{when} \quad \sigma \geq \sigma_s. \end{aligned} \tag{2.85}$$

Then the pseudo-instability points for Piola-Kirchhoff stresses occur when

$$\begin{aligned} \sigma &= \sigma_s(n/\epsilon_s)^n & \text{for} \quad N \\ \sigma &= \sigma_s(n/2\epsilon_s)^n & \text{for} \quad T \end{aligned} \tag{2.86}$$

justifying the claim we made before on the unsuitability of Piola-Kirchhoff stress measures for large deformation elastic-plastic formulation.

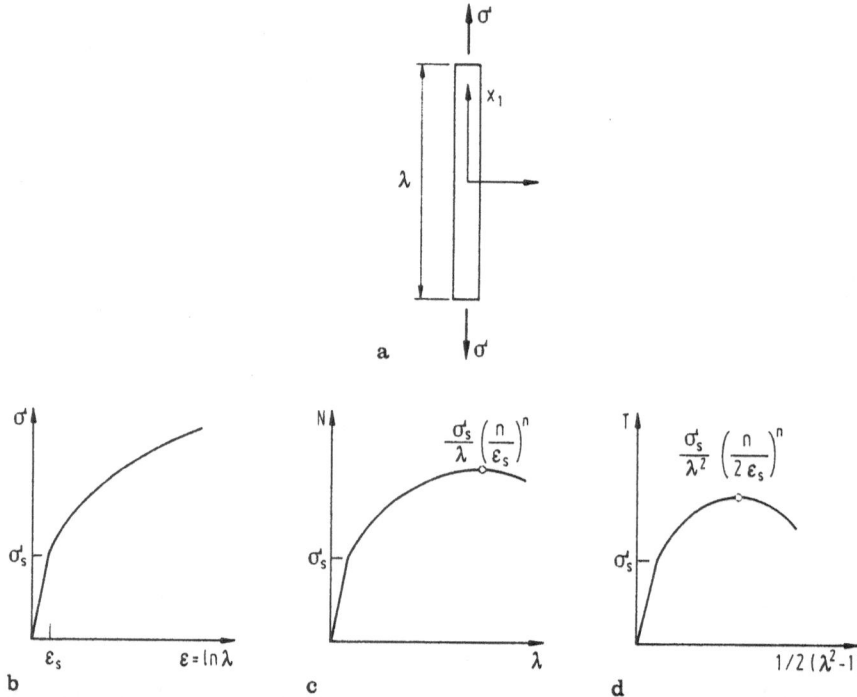

Fig. 2.7. Uniaxial tension curves in terms of different work conjugate pairs. **a** Uniaxial tension along X_1 direction; **b** non-softening Cauchy stress versus logarithmic strain curve; **c** same curve described by Piola-Kirchhoff stress of the first kind; **d** again described by Piola-Kirchhoff stress of the second kind. Pseudo-softening behavior is exhibited for the last two cases

Simple shear problem and stress oscillation. Our last example consists of simple shear deformation. The kinematics of simple shear motion is described in Fig. 2.8(a). The only non-trivial components of the deformation rate and the corotational spin tensors are

$$D_{12} = D_{21} = \dot{\gamma}/2$$
$$\Omega_{12} = -\Omega_{21} = \dot{\phi}/2 = \phi'(\gamma)\dot{\gamma}/2 \tag{2.87}$$

where ϕ is the corotational angle induced by shear amount γ. A complete analysis for this problem under a general elastic-plastic mixed hardening constitutive law has been accomplished recently by Yang, Cheng and Hwang[23]. They are able to show, through an asymptotic analysis, that the rigid plastic behavior dominates the shear response when the current shear strain γ far exceeds the shear strain at initial yield. Furthermore, their rigid plastic analysis predicts the following result on the evolution of back stress **b**, recording the trajectory of yield surface center as shear proceeds

$$b_{11} + ib_{12} = i\left(cY_o/\sqrt{3}\right) \int\limits_0^{\phi(\gamma)} R(\phi')\exp\{-i(\phi - \phi')\}d\phi' \qquad (2.88)$$

where $i = \sqrt{-1}$ is the imaginary number and function R in the integrand of (2.88) denotes the derivative of material hardening response $y = Y/Y_o$ with respect to corotational angle ϕ

$$R(\phi) = dy/d\phi. \qquad (2.89)$$

A variety of shear oscillatory (or non-oscillatory) behavior is displayed in terms of the back stress trajectories governed by (2.88) for monotonically increasing shear amount γ, albeit the common initial response under infinitesimal shear strain. If the basic material hardening behavior is prescribed by a power-law hardening rule with an exponent n

$$Y = Y_o(1 + \alpha\gamma^n) \qquad (2.90)$$

then entirely different back stress responses will take place under different corotational rates characterized by function $\phi(\gamma)$, as shown in Fig. 2.8(b).

For the corotational rates of power-law type

$$\phi(\gamma) = \Phi_o\gamma^m \quad 0 < m \le 1 \qquad (2.91)$$

including the particular case of Jaumann rate of $\Phi_o = m = 1$, the oscillatory behavior of back stress only relies on the exponent ratio n/m. When $n < m$, back stress oscillates to approach a limit circle in the complex back stress plane. This limit circle, as drawn by the dashed curve in Fig. 2.8(b), is centered at the origin and has a radius of magnitude $(c\alpha n/m)\Phi_o^{-n/m}\Gamma(n/m)$, where Γ denotes the Gamma function. When $n > m$, the back stress (or trajectory of yield surface center) wriggles away towards b_{11} direction as γ increases. For the coincidental case of $n = m$, interesting enough, a periodical solution tangent to the origin emerges, which means the solution has a trifurcation with respect to parameter m/n in the neighborhood of unity. None of the three cases discussed above can be considered as satisfactory for simple shear response, as observed from Fig. 2.8(b).

Another type of corotational rates, characterized by a trigonometric expression of $\phi(\gamma)$ given below

$$\phi(\gamma) = \phi_o \arctan(\gamma/k) \qquad (2.92)$$

with a bounded maximum value of $\phi_o\pi/2$, was proposed by several scholars like Dienes[10] ($\phi_o = k = 2$) and Lee, et al.[9] ($\phi_o = 2, k = 1$) as we derived from their works. It can be shown that Euler spin also corresponds to (2.92) with $\phi_o = 1$ and $k = 2$. The substitution of (2.92) into (2.88) would lead to a non-oscillatory response as plotted also in Fig. 2.8(b), and predict correct asymptotic responses at extremely large shear strain. In the intermediately large shear strain range, however, the back stress response by the incorporation of (2.92) (with ϕ_o

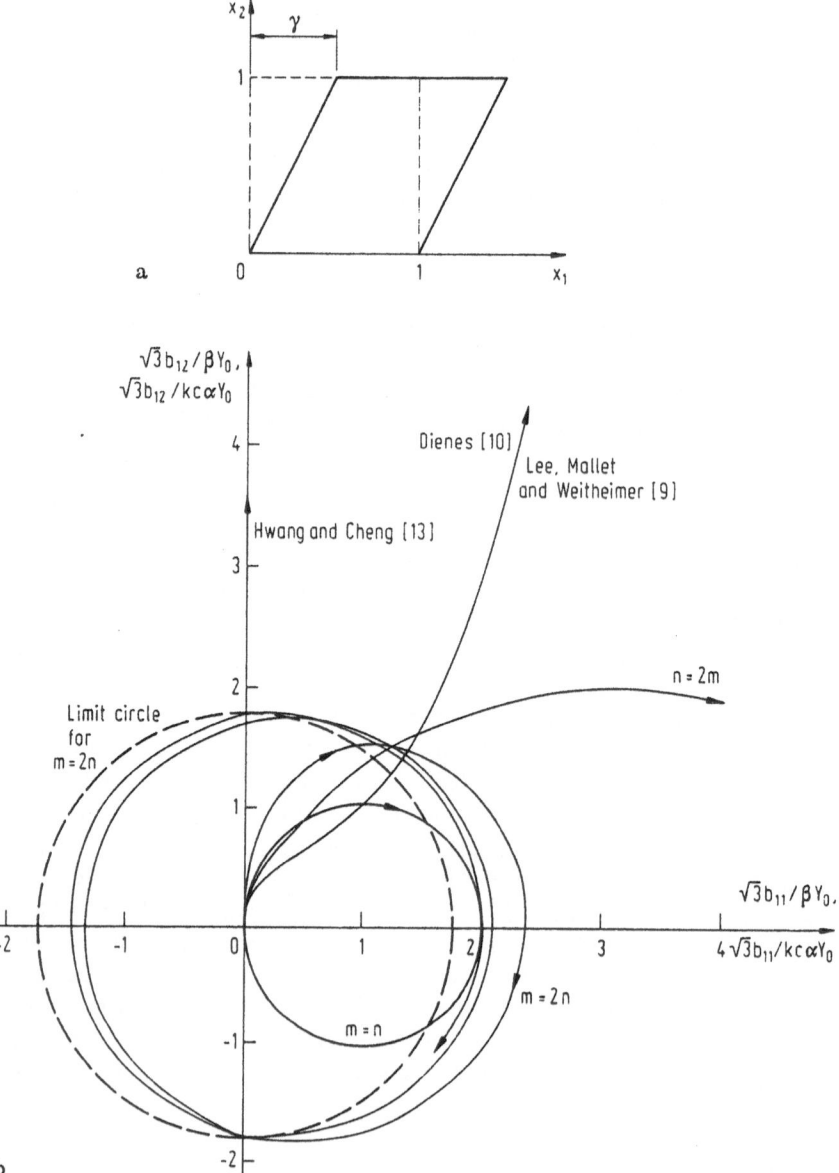

Fig. 2.8. Simple shear deformation. **a** Kinematics of simple shear deformation; **b** trajectories of yield surface center under different corotational rates. $\beta = (c\alpha n/m)\Phi_0^{-n/m}$

taken as 2) still substantially deviates from the shear dominated stress response when material hardening is relatively weak, as shown from Fig. 2.8(b) and the numerical calculation in reference[23]. Hwang and Cheng[13] adopted a stress rate corotational to the frame of deformation rate tensor **D**. As shown in Fig. 2.8(b), their result predicts a monotonic increase of back stress along b_{12} axis.

2.6 J_2 Deformation Theory

2.6.1 Formulation of J_2 Deformation Theory

J_2 deformation theory has a close resemblance to that of nonlinear elasticity. The attention here is only focused on the infinitesimal strain version of the theory. Various variants exist for its finite deformation counterpart, e.g. see Neale[24], and an appropriate choice between them is still elusive. Besides, all the analytical and numerical advantages of J_2 deformation theory do not preserve in any of its finite deformation versions, so the macroplasticity research toward this direction seems to be less motivated in the past years. The infinitesimal version of J_2 deformation theory is based from the assumption that the *total plastic strain* is coaxial to the stress deviator

$$\epsilon^p = \Phi(J_2)\mathbf{S} \tag{2.93}$$

where the Nadai factor Φ is a function of J_2, as suggested by the name of J_2 theory, or a function of effective stress $\bar{\sigma}$ defined in (2.57). Comparing the above assumption with the integration form of equation (2.79), featuring the J_2 flow theory prediction under proportional loading condition, we find these two predictions coincide with each other. Therefore, J_2 deformation theory is equivalent to J_2 flow theory under the restriction of proportional loading. For the general non-proportional loading cases, J_2 deformation theory is different from J_2 flow theory. The former is characterized by the proportionality between stress deviator and total plastic strain, so it is also named as *total theory*, whereas the latter is featured by the proportionality between stress deviator and the incremental plastic strain, as shown in (2.59), so named as *incremental theory*.

The Nadai factor Φ in (2.93) can be determined from the uniaxial stress strain curve exhibited in Fig. 2.5(a) by the "single curve postulate". For uniaxial tension in X_1 direction, we have

$$\epsilon^e_{11} = \sigma/E, \quad \epsilon^p_{11} = (2\sigma/3)\Phi(J_2), \quad \epsilon_{11} = \sigma/E_s \tag{2.94}$$

where E_s represents the secant modulus of uniaxial stress strain curve measurable from Fig. 2.5(a). Combining the above three expressions, one finds the Nadai factor can be evaluated from the secant modulus as

$$\Phi = (3/2)(1/E_s - 1/E). \tag{2.95}$$

Therefore, the plastic strain expression (2.93) can be specified into

$$\epsilon^p = (3/2)(1/E_s - 1/E)\,\mathbf{S}. \tag{2.96}$$

Under the further assumption of isotropic elastic response characterized by

$$\mathbf{e}^e = \mathbf{S}/2G \tag{2.97}$$

where $G = E/2(1 + \nu)$ is the elastic shear modulus, one is able to cast the deformation theory into the following elegant "pseudo elastic" form

$$\mathbf{e} = \mathbf{S}/2G_s, \quad \mathrm{tr}\,\epsilon = [(1 - 2\nu_s)/E_s]\,\mathrm{tr}\,\sigma \tag{2.98}$$

where the secant Poisson's ratio ν_s and the secant shear modulus G_s are defined by

$$(1 - 2\nu_s)/E_s = (1 - 2\nu)/E$$

$$G_s = E_s/[2(1 + \nu_s)] \tag{2.99}$$

in the sense that the elasticity equation for isotropic material can be obtained simply by changing the secant moduli in (2.98) to the elastic moduli. The derivation of equation (2.98) is left to our readers.

2.6.2 Comparison to J_2 Flow Theory

We now compare the J_2 deformation theory just obtained to the previous J_2 flow theory. This comparison is easily made by taking the rate form of (2.96) as follows

$$\dot{\epsilon}^p = (3/2)\left[(1/E_s - 1/E)\dot{\mathbf{S}} + (1/E_t - 1/E_s)(\dot{\bar{\sigma}}/\bar{\sigma})\,\mathbf{S}\right] \tag{2.100}$$

where the first term is due to the variation of stress deviator and the second term is caused by the change of secant modulus E_s. From (2.100), several remarks can be made to the deformation theory of plasticity

(1) Two terms containing secant modulus E_s in (2.100) cancel each other for the case of proportional loading, reducing the rate form of J_2 deformation theory to a form identical to J_2 flow theory as presented in (2.59). Therefore, the J_2 deformation theory is justified by the principle of maximum plastic work in the special case of proportional loading.

(2) For non-proportional loading, the first term in (2.100) for $\dot{\epsilon}^p$ is non-coaxial to stress deviator $\mathbf{S}$. In fact, it is not perpendicular to the yield surface and consequently the principle of maximum plastic work is violated. Some mechanists believe that deformation theory cannot be extended to cases beyond proportional loading due to the lack of physical foundation.

(3) Let us examine the responses predicted under J_2 flow and J_2 deformation theories for the extreme case consisting of a stress increment $d\mathbf{S}$ parallel to the current yield surface. Under the flow theory, the response would be purely elastic because the resultant increment of effective stress, $d\bar{\sigma}$, in (2.59) is identically zero. Hence, the flow theory anticipates a rather stiff constraint to sudden variation of loading path away from the previous proportional loading course, unfavorable to phenomena such as plastic buckling, deformation bifurcation and flow localization. On the other hand, the deformation theory offers a

much softer response for this deliberately selected stress change, due to the inclusion of the first term in (2.100) signifying the plastic response to dS parallel to the yield surface. The total deviatoric strain increment for a stress increment dS is in fact approximately equal to $(3/2E_s)\,dS$, so this response is roughly characterized by the secant modulus in contrast to the elastic modulus for the case of flow theory. In practices, the secant modulus is far less than the elastic modulus for metal plasticity especially under substantial pre-bifurcation plastic deformation, rendering the relative ease to the occurrence of plastic buckling, deformation bifurcation and flow localization under a J_2 deformation theory.

2.6.3 Reassessment of J_2 Deformation Theory

A reassessment of J_2 deformation theory for loading paths not far from the proportional loading is offered by B. Budiansky[25] by considering a non-smooth yield surface. It was demonstrated in the previous subsection that the PMPW is violated by J_2 deformation theory in non-proportional loading, as put more precisely by Budiansky in the following sentences:

Suppose the yield surface were smooth at σ, then, in accordance with PMPW, the unique normal to the surface would have the direction of the plastic strain rate associated with proportional loading. But the directions of $\dot{\epsilon}^p$ for non-proportional loadings are different, and hence violate PMPW, which requires a unique direction of plastic strain rate for all $\dot{\sigma}$.

Budiansky further attempted to solve this dilemma (at least partially) by considering the possibility of corners on the yield surface, he reasoned:

For yield surface with corner, a $\dot{\epsilon}^p$ can be taken from a set of normals lying within a cone, which leads to a class of loading path (non-radial) which can be applied without violating PMPW.

The extent of deviation from the proportional loading path for deformation theory without violating PMPW is connected to the corner structure of the yield surface. The geometry of a yield surface in deviatoric stress space with a corner of cone angle 2β is portrayed in Fig. 2.9(a). Here we mention that the experimental evidences in macroplasticity can neither establish nor eliminate the existence of yield surface corners, so the corner in Fig. 2.9(a) is assumed *a priori*. The loading is first carried out along a radial path from stress origin up to the point marked by S in the figure with the appearance of a corner of cone angle 2β in the current yield surface. Attention is then focused on the succession of a non-proportional loading increment dS upon S. As argued by Budiansky, two restrictions on the acceptable non-proportional loading paths must be imposed

(1) The stress increment dS must be confined in the extension cone (with cone angle 2β) from the yield surface corner.
(2) The plastic strain rate $\dot{\epsilon}^p$ should be confined in the normal cone (with cone angle $\pi - 2\beta$) formed by yield surface normals adjacent to the corner.

The second requirement listed above is obviously due to the principle of maximum plastic work. Some explanations, however, have to be made for the first restriction. Referred to Fig. 2.9(b), when stress changes from **S** to **S** + d**S**, the corresponding movement of the yield surface should be able to cover the previous stress state **S** to maintain plastic loading. If this requirement is observed, as indicated by the left graph of Fig. 2.9(b), then a *total loading* response (a terminology which will be further elaborated in the next subsection) is anticipated. The violation of this requirement, as portrayed from the right graph of Fig. 2.9(b), would result in a non-acceptable case demonstrated on the right graph which leaves the previous stress states inaccessible and outside the current yield surface.

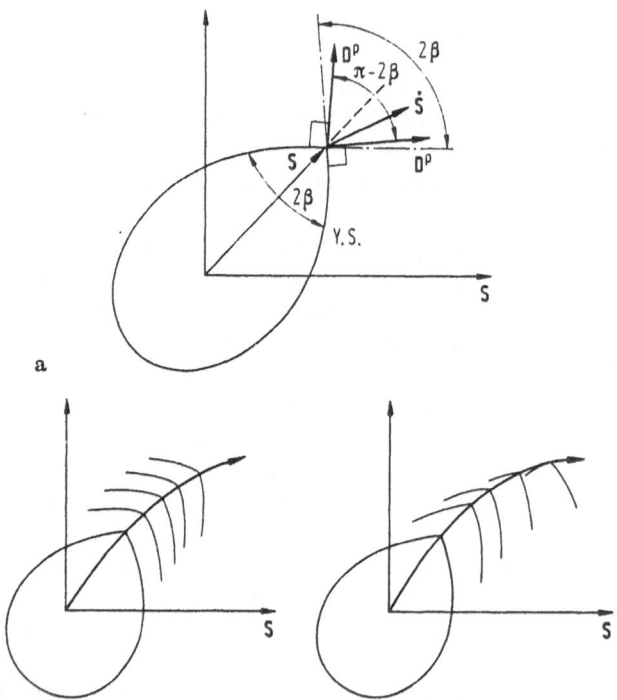

Fig. 2.9. Reassessment of deformation plasticity theory by introducing a corner on the yield surface. **a** Geometry of yield surface corner; **b** acceptable and non-acceptable loading paths

Explicit mathematical deductions for the above two restrictions are arrived from the rate form of deformation theory stated in (2.100). If the direction cosines of the stress rate and the plastic strain rate with respect to deviatoric stress are denoted as Θ and δ, respectively, i.e.

$$\cos \Theta = \mathbf{S} : \dot{\mathbf{S}} / \left[\sqrt{\mathbf{S} : \mathbf{S}} \sqrt{\dot{\mathbf{S}} : \dot{\mathbf{S}}} \right]$$

$$\cos \delta = \mathbf{S} : \dot{\boldsymbol{\varepsilon}}^p / \left[\sqrt{\mathbf{S} : \mathbf{S}} \sqrt{\dot{\boldsymbol{\varepsilon}}^p : \dot{\boldsymbol{\varepsilon}}^p} \right] \qquad (2.101)$$

where Θ serves as a controllable measure of non-proportional loading and δ describes the deviation from a normality response as measured by the angle between plastic strain rate and cone axis. Substituting the plastic strain rate (2.100) into the second expression of (2.101) and performing lengthy algebras, one can get a relationship between $\cos\Theta$ and $\cos\delta$

$$\cos\delta = M\cos\Theta / \left[1 + (M^2 - 1)\cos^2\Theta\right]^{\frac{1}{2}} \qquad (2.102)$$

where the symbol M is related to the material uniaxial behavior, and is defined by

$$M \equiv (1/E_t - 1/E)/(1/E_s - 1/E) \approx E_s/E_t > 1. \qquad (2.103)$$

Accordingly, the two restrictions listed above for total loading path can be interpreted in terms of the stress increment angle Θ as

$$\Theta \leq \beta \quad \text{and} \quad \Theta \leq \arctan(M/\tan\beta). \qquad (2.104)$$

As long as these conditions are satisfied by the step loading increment, the justification of deformation theory can be made on the ground of PMPW for a non-smooth yield surface.

2.7 J_2 Corner Theory

2.7.1 Essential Features of Corner Theory

The unsuitability of classical flow theory of plasticity for the bifurcation phenomena, such as buckling and necking, promotes active research in yield surface with corners, as pioneered by the works of Batdorf and Budiansky[26] for the bifurcation evaluation by simple corner theory of flow type, and that of Budiansky[25] on the reassessment of deformation theory just described in the previous subsection. A verdict has not yet been reached by the experimentalists on the existence of yield surface vertices, and conflicted data have been documented on this controversial issue. Nevertheless, almost all specialists in this field do agree on the existence of high curvature zone on the yield surface, albeit the similarity between a high curvature zone and a sharp corner has not yet been fully appreciated. On the other hand, the existence of sharp corners in a yield surface is justified under a rate independent single crystal slip model, as will be elucidated in section 7.1. The content in the present section is rather focused on the phenomenological J_2 corner theory of Christoffersen and Hutchinson[27], which highlighted the essential features of the modern corner theory of plasticity.

Although corner-plasticity theory was first established under the framework of infinitesimal deformation, it is not difficult, as pointed out in reference [27], to generalize it to the case of finite deformation by utilizing Kirchhoff stress and corotational rate. The loading paths applicable to a corner theory may encompass abrupt direction change of stress rate or even unloading in the stress space. However, the

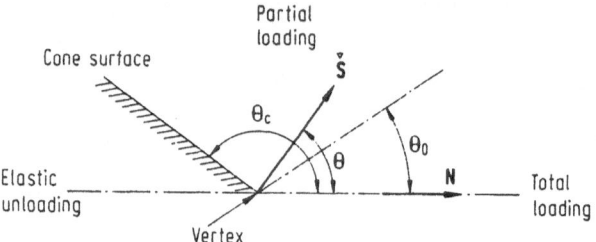

Fig. 2.10. Geometry of yield surface corner centered along axis N, the lower half of the corner can be furnished by symmetry

possibility of reverse loading is presently excluded because the law governing the re-development of yield surface corner is still elusive. The essential features of corner theory are

(1) Corner in the yield surface is inevitably developed during almost proportional loading, like the evolution of a membrane (resembling the yield surface) under the push of a stick (implied to the loading stress vector). The geometry of yield surface corner in stress space is illustrated in Fig. 2.10. The cone angle Θ_c in Fig. 2.10 is usually required to be no more than $135°$.
(2) The plastic deformation rate associated with further stress increment (in various possible directions) is governed by a plastic rate potential function Ω^p whose existence can be deduced from an incremental form of the plastic work $\dot{W}^p$ introduced before. For corner plasticity, the proportionality between the plastic strain rate and the stress rate is no longer valid for a rate independent solid. The function Ω^p is still homogeneous but nevertheless highly nonlinear.
(3) Different loading zones with respect to corner axis (designated by N) exist and distinct deformation responses appear within each individual zones.

2.7.2 Nonlinear Rate Form

From the additive decomposition of deformation rate tensor, the total stress rate potential Ω can be expressed as the sum of an elastic rate potential and a plastic rate potential

$$\Omega = \Omega^e + \Omega^p \tag{2.105}$$

where the elastic part is given by a quadratic form of corotational stress rate

$$\Omega^e = \frac{1}{2}\overset{\triangledown}{\tau} : \mathbf{C} : \overset{\triangledown}{\tau} \tag{2.106}$$

by means of the elastic compliance modulus $\mathbf{C} = \mathbf{L}^{-1}$, while the plastic part is given by a highly nonlinear rate form

$$\Omega^p = \frac{1}{2}F(\Theta)\,\check{\mathbf{S}} : \mathbf{C}^d : \check{\mathbf{S}} \tag{2.107}$$

proposed by Christoffersen and Hutchinson[27] as will be closely examined in the next subsection. In (2.107), the notation $\mathbf{C}^d$ represents the tangential compliance modulus under a deformation plasticity formulation, as previously assumed for the case of total plastic loading. From (2.100), one easily writes it down as

$$\mathbf{C}^d = (3/2) \left[(1/E_s - 1/E)\mathbf{I}_4 + (3/2\bar{\sigma}^2)(1/E_t - 1/E_s)\mathbf{S}\mathbf{S} \right] \qquad (2.108)$$

where $\mathbf{I}_4$ is the fourth rank identity tensor such that $\mathbf{I}_4 : \mathbf{A} = \mathbf{A}$ for any second rank tensor $\mathbf{A}$. The Θ in (2.107) relates to the inclination angle between the stress rate and the corner axis $\mathbf{N}$ as delineated in Fig. 2.10. The definition of this inclination angle, as advanced by Christoffersen and Hutchinson, however, is weighted by the material compliance $\mathbf{C}^d$ for total plastic loading

$$\cos\Theta \equiv \check{\mathbf{S}} : \mathbf{C}^d : \mathbf{N} / \left[\check{\mathbf{S}} : \mathbf{C}^d : \check{\mathbf{S}} \right]^{\frac{1}{2}}$$

$$\mathbf{N} \equiv \mathbf{S} / \left[\mathbf{S} : \mathbf{C}^d : \mathbf{S} \right]^{\frac{1}{2}}. \qquad (2.109)$$

Therefore, $F(\Theta)$ represents the *cone transition function* which introduces complicated nonlinear dependence on the stress rate beside the conventional quadratic rate form.

The plastic strain rate $\mathbf{D}^p$ can be written in terms of Ω^p as

$$\mathbf{D}^p = \partial\Omega^p / \partial\check{\mathbf{S}} \qquad (2.110)$$

via an incremental form of PMPW, where $\mathbf{S}$ again is the deviatoric Kirchhoff stress.

The rate independent requirement would leave $\mathbf{D}^p$ as a homogeneous function of stress rate $\check{\mathbf{S}}$ of degree one, and consequently Ω^p represents a homogeneous function of stress rate $\check{\mathbf{S}}$ of degree two, or expressed mathematically

$$\Omega^p(k\check{\mathbf{S}}) = k^2\Omega^p(\check{\mathbf{S}}). \qquad (2.111)$$

We should point out here that $\mathbf{D}^p$ could be a highly nonlinear function of the stress rate regardless its homogeneity of degree one with respect to $\check{\mathbf{S}}$, due to the complication of $F(\Theta)$. This highly nonlinear feature of $\mathbf{D}^p$, however, does not influence the following relations

$$\Omega^p = \frac{1}{2}\check{\mathbf{S}} : \mathbf{D}^p$$

$$\mathbf{D}^p = \mathbf{C}^p : \check{\mathbf{S}} \qquad \mathbf{C}^p = \partial^2\Omega^p(\check{\mathbf{S}})/\partial\check{\mathbf{S}}^2 \qquad (2.112)$$

as can be proved by utilizing the homogeneity condition (2.111).

2.7.3 Cone Transition Function

As argued by Christoffersen and Hutchinson[27] that any loading increment $d\mathbf{S}$ could fall into three different cones with distinct material descriptions. They are classified as

(1) *Total loading cone* for the angle Θ ranging from 0 to Θ_o. In which the plastic response of deformation theory, as described simply by (2.108), is expected.

(2) *Elastic unloading cone* for the angle Θ larger than Θ_c, where a purely elastic response as described by (2.106) appears to be the case.

(3) *Partial loading cone* for Θ ranging between Θ_o and Θ_c in which the complicated loading response of (2.107) is confronted.

According to above classification, the cone transition function $F(\Theta)$ assumes the following values in the total loading and elastic unloading cones

$$
\begin{aligned}
F(\Theta) &= 1 \qquad \Theta \leq \Theta_o \\
F(\Theta) &= 0 \qquad \Theta \geq \Theta_c.
\end{aligned}
\tag{2.113}
$$

In the nonlinear transition region of $F(\Theta)$, there appears to be no unique way for its determination from macroplasticity argument. $F(\Theta)$ should be non-negative, and vary continuously and monotonically from one to zero as Θ increases from Θ_o to Θ_c. Further restrictions to $F(\Theta)$ were pointed out in reference [27] to guarantee the convexity of the plastic rate potential Ω^p

$$
1 - K(\Theta)\cot\Theta \geq 0, \quad 1 + K(\Theta)^2 - K'(\Theta) \geq 0
\tag{2.114}
$$

where $K(\Theta)$ is equal to $-F'(\Theta)/2F(\Theta)$. Some empirical choices of $F(\Theta)$ were listed in reference[27].

2.7.4 Constitutive Laws

The general constitutive framework for J_2 corner theory was described by equations from (2.105) to (2.108). Complete set of constitutive relations for the cases of total loading and elastic unloading were obtained by further incorporation of equation (2.113). The constitutive behavior for the partial loading zone appears to be more complicated and we will approach its resolution from equation (2.107) and the last expression of (2.112). In the differentiation operations of (2.112), the variation of stress rate angle Θ with respect to $\check{S}$ should not be overlooked, and its calculation is facilitated by equation (2.109). After lengthy algebras, the plastic compliance in the partial loading zone is arrived

$$
\mathbf{C}^p = F(\Theta)\mathbf{C}^d + \frac{1}{2}F'(\Theta)\{\cot\Theta(\mathbf{C}^d - \mathbf{PP} - \mathbf{QQ}) + \mathbf{PQ} + \mathbf{QP}\}
$$
$$
+ \frac{1}{2}F''(\Theta)\mathbf{QQ}
\tag{2.115}
$$

where in this section, symbols $\mathbf{P}$ and $\mathbf{Q}$ are reserved for

$$
\mathbf{P} = \mathbf{C}^d : \check{S} / \left[\check{S} : \mathbf{C}^d : \check{S}\right]^{\frac{1}{2}}
$$

$$
\mathbf{Q} = (\cos\Theta\mathbf{P} - \mathbf{C}^d : \mathbf{N}) / \sin\Theta.
\tag{2.116}
$$

It is easy to observe from the above two expressions that the plastic compliance modulus $\mathbf{C}^p$ possesses the desired Voigt symmetry. It would be also positive definite if conditions listed in (2.114) are satisfied.

For the abrupt change of loading path associated with bifurcation problems, the responses predicted by slip theory or J_2 corner theory can correlate the experimental data better than the J_2 flow theory based on a smooth yield surface. This fact is illustrated in Fig. 2.11, as quoted from the research work of Pan and Rice[28].

Some problems still remain unsolved for the J_2 corner theory. Noticeably the choice of the $F(\Theta)$ function and the evolution law of the yield surface. Should the yield surface with corner evolve isotropically, kinematically or by keeping a fixed cone angle? The resolution for these problems are beyond the realm of macroplasticity.

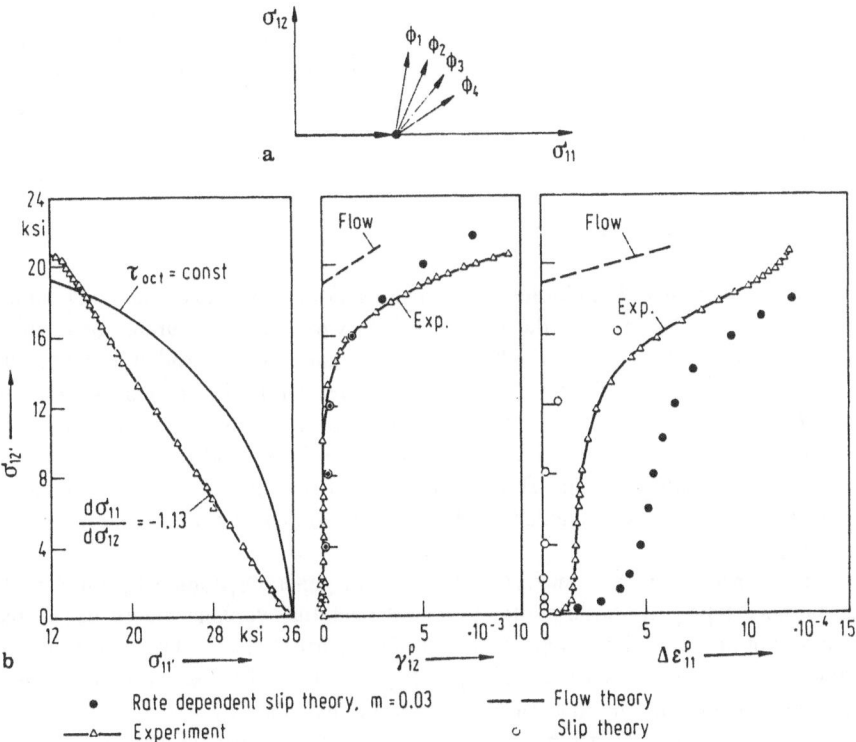

Fig. 2.11. Slip theory vs. J_2 flow theory predictions for biaxial responses with abrupt change on stress history. **a** Biaxial loading path; **b** predictions for J_2 flow and slip-type theories. After Pan and Rice[28]

2.8 Theory of Internal State Variables

The theory of internal state variables was first advanced by Biot and Meixner for visco-elasticity problems, where a set of *internal state variables* (abbreviated as internal variables in the sequel) was selected to characterize approximately the effect of functionals of memory type. This idea was followed up and generalized to the finite deformation regime by Coleman and Gurtin[29] in the late sixties.

The Coleman-Gurtin theory was featured by a rigorous mathematical deduction from the basic approximation of internal variables and basic laws in continuum thermodynamics, following the route of *Coleman-Noll formalism*. Motivated by the development of general framework on internal variable theory, its applications in *macroplasticity* were pursued by various distinguished scholars in plasticity, including Mandel, Hill, Rice and Nemat-Nasser. The readers can consult the classical papers, from reference [29] to [33], for more detailed account beyond the present outline.

The popularity of internal variable theory has boosted in the recent years by the emerge of *continuum damage mechanics* which can be cast and thermodynamically reasoned under the general framework of internal variable theory. According to its growing importance in nowadays macroplasticity theory, a brief introduction of internal variable theory is attempted here to describe its essential framework and major results.

2.8.1 Internal Variables and Free Energy

External and internal variables. In mechanics, all field variables can be classified into two categories, namely the field variables which are measurable or controllable and the field variables which cannot be measured or controlled directly. The first type of variables is termed *external variables* while the second the *internal variables*. The external variables include stress σ, total strain ϵ, temperature T, etc. The stress and strain used here are required to be second rank tensors and to form a work conjugate pair. Their choice should be otherwise arbitrary under a correct formulation as demanded by the *principle of measure invariance* proposed by Hill[14].

The plastic state of a material sample is in general specified by functionals of memory type. The basic assumption in the internal variable theory lies in the proposal of an approximate representation of the current plastic state by the current values of a set of internal variables $\eta = \{\eta_1, \ldots, \eta_n\}$, where n can be an infinite or finite integer. The internal variables could be described by

(1) *Structural internal variables*, representing the plastic structural patterns of various individual micro-elements inside the material. They are variables of extensive type and their number n is proportional to the volume occupied by the material. Examples of structural internal variables include slip amount γ^α in various slip systems and grain boundary sliding on different grain boundaries.

(2) *Averaging internal variables*, representing the macroscopic averaging property of the material plasticity state. They are variables of intensive type and their number is irrelevant to the volume occupied by the material. Examples of averaging internal variables are plastic strain ϵ^p, plastic deformation rate $\mathbf{D}^p$ and void volume fraction f.

The choice of employing either structural or averaging internal state variables as the primitive field variables partially reflects the philosophy to approach the plasticity formulation. The formal approach, as adopted by Hill and Rice[30,31,32], embeds more mesoplastic insights to the plasticity formulation and also enables us to deal with the possible non-differentiable behavior of internal variables. On the other hand, the averaging internal variable approach, as followed by many scholars engaged in continuum plasticity and continuum damage mechanics, will lead to more compact information. Moreover, the differentiability of internal variables with respect to plastic deformation history is required by the latter approach.

Free energy. With the inclusion of a set of internal state variables designated collectively by η, the *Helmholtz free energy* of the material in an elastic-plastic state can be written approximately as a state function of ϵ, T and η

$$\Phi(\epsilon, T, \eta) = U - TS \tag{2.117}$$

where U the internal energy possessed by the material, T the absolute temperature and S the associated entropy function. The total variation of Φ is composed of an elastic variation, $d^e\Phi$, caused by the changes of external variables ϵ and T, and a plastic variation, $d^p\Phi$, caused by the changes in internal variables η

$$d\Phi = d^e\Phi + d^p\Phi \tag{2.118}$$

where

$$d^e\Phi = \Phi(\epsilon + d\epsilon, T + dT, \eta) - \Phi(\epsilon, T, \eta) \equiv \sigma : d\epsilon - S\, dT$$

$$d^p\Phi = \Phi(\epsilon, T, \eta + d\eta) - \Phi(\epsilon, T, \eta) \equiv -\sum_{\alpha=1}^{n} f_\alpha d\eta_\alpha. \tag{2.119}$$

In the last steps of (2.119), the definitions of σ, $-S$ and $-f_\alpha$ are introduced as the work conjugates of ϵ, T and η_α. The symbol f_α represents the generalized *energetic force* conjugate to the internal variable η_α. The minus signs in (2.119) taken in front of S and f_α signify their *energy dissipation* effects. If the Helmholtz free energy function Φ in (2.118) is viewed as a state function of the state variables ϵ, T and η, and $d\Phi$ as a total differential, then their conjugate forces can be derived from the partial derivatives of Φ as observed from (2.119)

$$\sigma = \partial\Phi/\partial\epsilon, \quad S = -\partial\Phi/\partial T, \quad f_\alpha = -\partial\Phi/\partial\eta_\alpha \tag{2.120}$$

Alternatively, one can formulate the problem in terms of the complementary energy potential, ψ, by a Lagendre transformation

$$\psi(\sigma, T, \eta) = \sigma : \epsilon - \Phi \tag{2.121}$$

which uses stress σ instead of strain ϵ among the list of primitive variables. The work conjugates of σ, T and η can be expressed through ψ by equations similar to (2.120)

$$\epsilon = \partial\psi/\partial\sigma = \epsilon(\sigma, T, \eta), \quad S = \partial\psi/\partial T = S(\sigma, T, \eta_\alpha)$$

$$f_\alpha = \partial\psi/\partial\eta_\alpha = f_\alpha(\sigma, T, \eta). \tag{2.122}$$

The plastic strain can be obtained from the first expression of (2.122) through a hypothetical unloading process to relieve the stress σ

$$\epsilon^p = \epsilon(0, T, \eta). \tag{2.123}$$

Furthermore, equations in (2.122) also provide us several Maxwell relations, one of them is

$$\partial\epsilon/\partial\eta_\alpha = \partial f_\alpha/\partial\sigma \quad \alpha = 1,\dots,n. \tag{2.124}$$

The strain rate, $\dot\epsilon$, can be obtained by taking the total differential of the first expression in (2.122), which leads to the following additive decomposition

$$\dot\epsilon = \dot\epsilon^e + \dot\epsilon^p \tag{2.125}$$

where

$$\dot\epsilon^e = \mathbf{C} : \dot\sigma + (\partial\epsilon/\partial T)\dot{T} \quad \mathbf{C} = \partial^2\psi/\partial\sigma\partial\sigma$$

$$\dot\epsilon^p = \sum_{\alpha=1}^{n}(\partial\epsilon/\partial\eta_\alpha)\dot\eta_\alpha = \sum_{\alpha=1}^{n}(\partial f_\alpha/\partial\sigma)\dot\eta_\alpha. \tag{2.126}$$

The Maxwell relation (2.124) is engaged in the last step of derivation for the plastic strain rate.

2.8.2 Thermodynamic Restrictions

Attention is then focused on the restrictions to elastic-plastic constitutive relations imposed by the first and the second laws of thermodynamics. If the heat supply into the system under consideration is denoted by Q, the first law of thermodynamics can be written as

$$\sigma : \dot\epsilon + Q = \dot{U} \tag{2.127}$$

where U again represents the internal energy possessed by the system. The second law of thermodynamics, in the form of entropy production inequality, can be written as follows

$$\dot{S} \geq Q/T. \tag{2.128}$$

Combining the above two laws with previous equations from (2.117) to (2.119), one obtains

$$d^p\Phi \leq 0 \quad \text{or} \quad d^p\psi = \sum_{\alpha=1}^{n}f_\alpha d\eta_\alpha \geq 0. \tag{2.129}$$

That is, the plastic variation of the Helmholtz free energy should be non-positive, or equivalently the plastic dissipation, expressed by the changes of internal state variables multiplied by the work conjugate force applied to them, must be non-negative. An expression for the complementary energy function was given by Rice[32]

$$\psi = -\Phi^o(\eta) + \sigma : \epsilon^p(\eta) + \frac{1}{2}\sigma : \mathbf{C}(\eta) : \sigma \tag{2.130}$$

where Φ^o represents the *lock-in free energy* at completely unloading sress (zero stress) state. Substituting the above expression of ψ into (2.129) and neglecting the second order terms, one can interpret the non-negative dissipation requirement from the viewpoint of macroscopic (or averaging) internal variables as

$$\sigma : \dot{\epsilon}^p \geq \dot{\Phi}^o \qquad (2.131)$$

for an isothermal process.

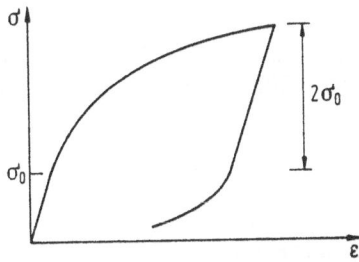

Fig. 2.12. Uniaxial demonstration for negative rate of lock-in free energy, with the possible occurrence of a negative macroscopic plastic dissipation

It is noticed the above restriction does not necessarily conform to the requirement of a non-negative macroscopic plastic work, as the changing rate of lock-in free energy could be negative in accordance to the possible loss of lock-in free energy during the plastic deformation processes. A simple uniaxial demonstration to the negative rate of $\dot{\Phi}^o$ is given in Fig. 2.12 for the case of reverse plastic loading under a still positive stress level. Typical examples for this situation include :

(1) extremely pronounced Bauschinger effect, possibly occurs in the case of kinematic hardening with extensive elevation of stress level since initial yield;

(2) time dependent strain recovery after partial unloading, which possibly occurs under positive sustained stress level;

(3) pseudo-elastic behavior as exhibited in certain structural ceramics and shape memory alloys.

2.8.3 Flow Potential and Normality

As we mentioned before, the variations of internal variables cannot be directly monitored or controlled from the mechanical testing, and they are governed by the *evolution equations* instead. If the evolution of η is determined by its conjugate energetic force, namely

$$\dot{\eta}_\alpha = H_\alpha(f_1, \ldots, f_n) \equiv H_\alpha(f) \qquad \alpha = 1, \ldots, n \qquad (2.133)$$

then a question can be raised concerning under what circumstances an integrating factor $\lambda(f)$ and a flow potential $\Omega(f)$ would exist such that the above general evolution low could be expressed as

$$\dot{\eta}_\alpha = \lambda(\partial\Omega/\partial f_\alpha) \qquad \alpha = 1,\ldots,n. \tag{2.134}$$

The existence of an integration factor for the Pfaffian form (2.133) can be assessed from Carathèodary theorem, as done in classical thermodynamics, see Buchdahl[34]. Here we would not step into the mathematical details of this assessment, but rather point out three useful particular cases

(1) *Non-interacted mechanisms.* Non-interacted assumption on various deformation mechanisms could be traced to W. Thomson in 1882 who first stated the hypothesis on separation of individual processes. The application of this hypothesis to metal plasticity was due to Rice[30]. Under this assumption, equation (2.133) is replaced by

$$\dot{\eta}_\alpha = H_\alpha(f_\alpha, T, \eta) \qquad \alpha = 1,\ldots,n \tag{2.135}$$

namely, the evolution of the internal variable η_α will depend on stress only through its work conjugate f_α. This assumption is pertinent for metal plasticity because the plastic slip in each slip system is dominated by the resolved shear stress (the work conjugate of slip rate) on the same slip system, as will be elucidated in section 7.1. Under the assumption (2.135), it is straightforward to show that

$$\dot{\eta}_\alpha = \partial\Omega/\partial f_\alpha \text{ where } \Omega = \sum_{\alpha=1}^{n} \int_0^{f_\alpha(\sigma,T,\eta)} H_\alpha(f_\alpha, T, \eta)\, df_\alpha. \tag{2.136}$$

For the rate independent case, the above assumption could further lead to the existence of a yield surface $F(\sigma, T, \eta) = 0$, such that the plastic strain rate can be derived through a normality flow rule[33].

(2) *Two internal variables,* for the special case of $n = 2$, it is well-known that an integration factor λ would exist. So the form of (2.134) is always attainable.

(3) *Linear evolution law,* in some cases, the evolution of internal variables is governed by the linear combination of the work conjugate forces as

$$\dot{\eta}_\alpha = \sum_{\beta=1}^{n} L_{\alpha\beta}(T, \eta) f_\beta \tag{2.137}$$

where $L_{\alpha\beta}$ is symmetric by Onsager reciprocal theorem on the evolution laws. Through a variable transformation, the linear matrix $L_{\alpha\beta}$ can always be diagonalized to become the form of (2.135), then it is also representable by the potential form of (2.134).

Hereafter, the existence of flow potential as presented in (2.134) is assumed a priori, then a general expression for the flow potential is arrived as

$$\Omega(\sigma, T, \eta) = \sum_{\alpha=1}^{n} \int_0^{f_\alpha(\sigma,T,\eta)} [H_\alpha(x, T, \eta)/\lambda(x, T, \eta)]\, dx. \tag{2.138}$$

Based on this flow potential, we are able to obtain an expression for the macroscopic plastic strain rate tensor

$$\lambda[\partial\Omega/\partial\sigma] = \sum_{\alpha=1}^{n} H_\alpha[\partial f_\alpha/\partial\sigma] = \sum_{\alpha=1}^{n} [\partial f_\alpha/\partial\sigma]\dot{\eta}_\alpha = \dot{\epsilon}^p. \qquad (2.139)$$

Established in (2.139) also the normality structure of the macroscopic plastic strain rate tensor.

As proved by Hill and Rice[31], the following two generalizations can be imposed on the normality structure exhibited in (2.139)

(1) If the above normality structure applies for any one choice of conjugate stress and strain measures and choice of reference state, then it necessarily applies for every choice of conjugate variables and reference state. Consequently, the principle of measure invariance of Hill[14] is observed under above-mentioned formulation.

(2) If a composite material is assembled by subelements that can be modelled as continua in which the above normality structure applies to the local stress-strain relations, then the same normality structure is necessarily transmitted to the overall stress-strain relations of the composite, when these are phrased in the work conjugate variables. The examples signifying this composite operation include the model for polycrystal aggregates (see Sect. 7.2) and Gurson model for void damage behavior (see Sect. 8.2).

As the conclusion of this section, we examine the basic macroplasticity postulates which are claimed to lead to the above-mentioned normality structure. Two separate quasi-thermodynamic postulates (Drucker, 1951; Il'yushin, 1961) were proposed in the rate independent plasticity literature, and lead to normality structure of (2.139) for infinitesimal deformation. The postulate by Il'yushin[35] stated that the net stress work done in enforcing an infinitesimal isothermal strain cycle that begins and ends at the same arbitrary strain state should be non-negative. Phrased in the above way, the Il'yushin postulate seems to be a separate assumption rather than a natural consequence of the second law of thermodynamics. Because the restoration of ϵ and T, as required by Il'yushin postulate, does not necessarily fully restore the plastic state. In operational aspect, Il'yushin postulate is suitable for rate independent plastic deformation and has the advantage of measure invariance with respect to any work conjugate pairs of stress and strain for the case of finite deformation. The normality structure can be obtained pointwise in the constrained infinitesimal deformation cycles, as well as the corner convexity of the yield surface. The general convexity of the yield surface, however, cannot be established by Il'yushin postulate due to the limitation of required infinitesimal loading cycle.

Drucker's postulate[36] deals with a material element under an arbitrary prestress and postulates that the net work done by an external agency in a cycle of imposing and removing same additional set of loadings is non-negative. It was shown by Hill[37] in a thorough and comprehensive manner that the Drucker's postulate does not possess the measure invariance when generalized to finite deformation, which precludes its unambiguous interpretation in general. However, if limited to cycles involving only an infinitesimal accumulation of plastic straining, it becomes invariant to stress measures and leads to the normality structure dictated in (2.139). Furthermore, Drucker postulate can be extended to finite deformation formulation provided the special choice of Kirchhoff stress and logarithmic strain is made as the work conjugate pair, see Hill[37].

2.9 Non-Associated Flow Rules

Non-associated flow rules are frequently observed in geological and geotechnical materials, such as rock masses, concretes, soil, etc. A brief outline is provided in this section to contrast the constitutive structures discussed earlier for metal plasticity.

2.9.1 Pressure Sensitive Yielding

We first examine the case of plastic slip with the participation of friction. As shown in Fig. 2.13, the yield stress for plastic slip not only depends on the tangential stress τ but also relies on the normal stress σ by the notion of friction. Two cases demonstrated in Fig. 2.13 correspond

(1) *Pressure sensitivity*, referred to the case that plastic yielding is dependent on hydrostatic pressure as shown in Fig. 2.13(a). This behavior is observed for geological materials with internal friction effect, or even for metals under the occurrence of voids, as discussed in detail in section 8.2.

(2) *Columb friction*, referred to the case that plastic slip is influenced by the normal stress acting on the slip plane in the form resembling Columb friction law. This behavior occurs for geological and geotechnical materials in which plastic slip proceeds along rough slip surfaces.

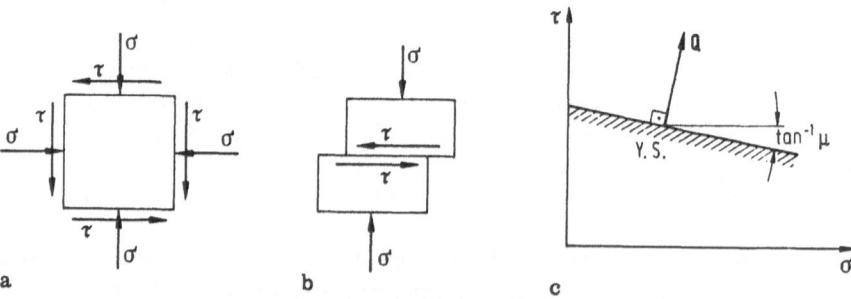

Fig. 2.13. Schematic illustration on the causes of non-associated flow rules. **a** Pressure sensitive yielding behavior; **b** Columb friction mechanism; **c** yield surface and its normal vector **Q**

As a macroplasticity interpretation of the above phenomena, the constitutive modelling responsible for plastic slip (in particular the yield surface or yield function) should not only be described by tangential stresses, such as the J_2 type theory and Tresca yield function, but also be effected by normal stress. Accordingly, for the simple case depicted in Fig. 2.13(a) and (b), a friction coefficient (presumably constant) μ can be introduced in the following *Columb law* for plastic loading criterion

$$\dot{\tau} + \mu\dot{\sigma} > 0 \qquad \text{plastic loading}$$

$$\dot{\tau} + \mu\dot{\sigma} < 0 \qquad \text{elastic unloading} \tag{2.140}$$

where tension stress of σ is taken as positive. The above criterion can be easily generalized to the three dimensional case by replacing τ by its J_2 generalization $\bar{\sigma}/\sqrt{3}$ and replacing σ by the mean Kirchhoff stress $\tau_{kk}/3$, yielding

$$\tau_e \equiv \dot{\bar{\sigma}}/\sqrt{3} + (\mu/3)\dot{\tau}_{kk} > 0 \qquad \text{plastic loading}$$

$$\tau_e \equiv \dot{\bar{\sigma}}/\sqrt{3} + (\mu/3)\dot{\tau}_{kk} < 0 \qquad \text{elastic unloading.} \tag{2.141}$$

Then the equation of yield surface becomes

$$F(\tau_e) = 0 \tag{2.142}$$

as draw schematically in Fig. 2.13(c). Here we would like to point out that the generalization from one dimensional loading criterion (2.140) to three dimensional loading criterion (2.141) are only suitable for

(1) the case of internal friction as demonstrated in Fig. 2.13(a);
(2) an isotropic hardening law; and
(3) a J_2 type generalization.

The normal of this yield surface in stress space is denoted by $\mathbf{Q}$,

$$\mathbf{Q} = (\sqrt{3}/2\bar{\sigma})\,\mathbf{S} + (\mu/3)\,\mathbf{1} \tag{2.143}$$

which is only slightly different from unit length, and defined such that

$$\partial F/\partial \tau = F'(\tau_e)\,\mathbf{Q} \tag{2.144}$$

then it is straightforward to show that the loading criterion (2.141) can be reinterpreted in terms of $\mathbf{Q}$ as

$$\mathbf{Q} : \dot{\tau} > 0 \qquad \text{plastic loading}$$

$$\mathbf{Q} : \dot{\tau} < 0 \qquad \text{elastic unloading.} \tag{2.145}$$

2.9.2 Plastic Dilatancy

Beside the deviatoric plasticity dominated in the conventional metal plasticity theory, dilatational plasticity recently received a great deal of attention acknowledged partially to the advance of constitutive modelling for non-metallic materials and damage mechanics. Plastic dilatancy can occur in a variety of situations. Two classes of phenomena among them are shown in Fig. 2.14(a) and (b). They are

(1) *Shear induced dilatancy*, referred to the formation of gaps by plastic shear (in the form of slip) carried out along a rough shear surface, as shown in Fig. 2.14(a). Mechanistically, it can be described by a plastic variation of volumetric strain, $d^p \epsilon_v$, induced by plastic shear $d^p \gamma$.

(2) *Cavitation induced dilatancy*, referred to the formation of voids by the application of hydrostatic stress, as shown in Fig. 2.14(b). Mechanistically, it can be described by a plastic variation of volumetric strain, $d^p \epsilon_v$, induced by hydrostatic stress τ_{kk}.

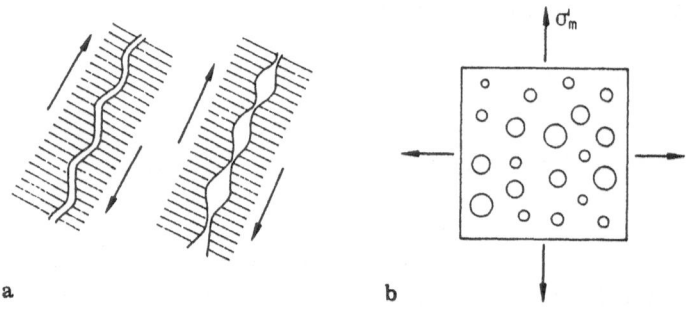

Fig. 2.14. Two typical plastic dilatancy phenomena. **a** Plastic dilatancy induced by slip along rough surfaces with the formation of gaps; **b** plastic dilatancy caused by voids formation and growth

The mechanics formulation of the second case will be postponed to section 8.2 as one essential ingredient of meso-damage theory. For the first case, a linear law is usually proposed to correlate the plastic volumetric change $d^p \epsilon_v$ and plastic shear $d^p \gamma$ for the simple case of Fig. 2.14(a)

$$d^p \epsilon_v = \beta d^p \gamma. \tag{2.146}$$

Under the participation of friction, the plastic shear increment relates to the Columb driving stress increment, $d\tau + \mu d\sigma$, by a plastic tangential shear modulus G_p. Thus

$$d^p \gamma = (d\tau + \mu d\sigma)/G_p, \quad d^p \epsilon_v = (\beta/G_p)(d\tau + \mu d\sigma). \tag{2.147}$$

With the further engagement of elastic deformation, the corresponding shear and volumetric increments are

$$d\gamma = d\tau/G + (d\tau + \mu d\sigma)/G_p,$$
$$d\epsilon_v = d\sigma/K + (\beta/G_p)(d\tau + \mu d\sigma). \tag{2.148}$$

The above one dimensional results are then generalized to the three dimensional case. The generalization is carried out by changing τ to $\bar{\sigma}/\sqrt{3}$, σ to $\tau_{kk}/3$, $d\epsilon_v$

to D_{kk}, $d\tau$ to $\check{S}$, and $d\gamma$ to $\mathbf{D'}$ (the deviatoric part of deformation rate tensor). Furthermore, it is frequently assumed that the deviatoric plastic deformation is coaxial with the stress deviator, and the hydrostatic stress τ_{kk} only influences the magnitude of the deviatoric plastic deformation but leaves its direction unaltered. Accordingly, we have

$$\mathbf{D'} = \check{S}/2G + (\sqrt{3}/2G_p\bar{\sigma})(\mathbf{Q} : \dot{\tau})\,\mathbf{S}$$
$$D_{kk} = \dot{\tau}_{kk}/3K + (\beta/G_p)\,\mathbf{Q} : \dot{\tau}. \tag{2.149}$$

Summing them together, one obtains the constitutive equation of the following *Rudnicki-Rice form*[38]

$$\mathbf{D} = [\mathbf{C} + (1/G_p)\,\mathbf{PQ}] : \dot{\tau} \tag{2.150}$$

where $\mathbf{C}$ is the elastic compliance tensor and assumes the specific form of

$$\mathbf{C} = (1/9K)\,\mathbf{1}\,\mathbf{1} + (1/2G)\,\mathbf{I_4} \tag{2.151}$$

for the case of elastic isotropy. The second rank symmetric tensor $\mathbf{P}$ in (2.150) denotes the direction of plastic strain rate and is described by

$$\mathbf{P} = (\sqrt{3}/2\bar{\sigma})\,\mathbf{S} + (\beta/3)\,\mathbf{1}. \tag{2.152}$$

A comparison between (2.143) and (2.152) reveals that the plastic strain rate $\mathbf{D}^p$ will be parallel to yield surface normal $\mathbf{Q}$ if and only if the Columb friction coefficient μ happens to be identical to the plastic dilatancy ratio β, otherwise the normality structure will be lost, as well as the contradiction to the principle of maximum plastic work. The experimental data of μ and β for rocks and high strength Martinsite steels are typically

Rocks: $\mu = 0.4$ to 0.9
                 (some data even ranges from 0.9 to 1.3)
                 $\beta = 0.2$ to 0.4

Martinsite steels: $\mu = 0.05$
                     $\beta = \mu/15$.

In conclusion, the friction coefficient μ is usually larger than the plastic dilatational ratio β, and consequently the loss of normality. The terminology of *non-associated flow rule* is invoked to address this situation. The non-associated flow rule motivates the search for new (physical or thermodynamic) basic postulate beyond the unsuitable PMPW as its theoretical foundation, and efficient computing algorithms to handle the non-symmetric mathematical formulation as exhibited in (2.150).

2.10 Limitations of Macroplasticity

During the preceding exploration about the general scope of macroplasticity, the readers must be fascinated by its mathematical elegance and capability to solve practical problems. Above-documented achievements undoubtedly compose a large proportion on human's triumph in the field of plasticity. As the study on plasticity goes deeper, however, some fundamental issues emerge to underline the limitations of macroplasticity, as well as the need of a different, yet complementary, approach based on mesoplasticity. These limitations, as we intend to enumerate below, do not disgrace the merit of macroplasticity, but rather indicate the fundamental need for the cooperation between classical continuum mechanics methodology and the expertise from material and physical sciences. Some of the limitations of macroplasticity may be caused by the limitation of contemporary knowledge on continuum mechanics; some of them, however, are intrinsic in the sense that they cannot be resolved by macroplasticity alone. The latter aspect promotes an intensive interaction between solid mechanists and material scientists, and results in significant advances in plasticity theory, as we shall address in the chapters to follow. At present moment, however, the painful practices on displaying the major limitations of macroplasticity has to be exercised as follows :

(1) *Detached from actual plastic deformation mechanisms.* The source of plastic deformation in a macroplasticity formulation is either assumed devoid of physical explanation or observed phenomenologically through macroscopic data. In macroplasticity, plastic deformation is defined by concept rather than through actual physical image. One of the unfavorable consequences for this hypothetical approach lies in the inability to distinguish features presented by a variety of plastic deformation mechanisms, traced back in a microstructural level from dislocation glide, phase transformation to microcracking damage. This lack of distinction to different plastic deformation mechanism exhibited by the macroscopic approach is in contrast to the closeness of mesoplasticity to the physical nature of plastic deformation as will be unfolded mechanism by mechanism in chapter 4. Another related consequence of macroplasticity, as we mentioned repeatedly throughout this chapter, is the lack of any *direct* (mechanical and experimental) means to detect current plastic deformation without the help of a hypothetical (and sometimes unjustified) process of elastic unloading. This inadequacy compares disadvantageous with the mesoplasticity approach where the current plastic deformation can be, at least in principle, predicted by Orowan-type theoretical model or measured from direct (microscopic) account of experimental data such as dislocation densities. The vision provided by mesoplasticity approach represents the actual events happening in the meso and/or microscopic level, and consequently more convincing and appealing than the phenomenological reasoning in macroplasticity. Furthermore, the possibility of overlooking or mistreating the plasticity behavior uniquely related to a special class of plastic deformation mechanism can be greatly reduced by careful mesoplasticity examination. The necessity of a complementary mesoplastic approach is then furnished from philosophical and psychological viewpoints as to lay down a sound physical background of plasticity theory and to provide some critical information missed in continuum hypothesis.

(2) *Empirical assumptions on material responses.* The basic practice in continuum mechanics approach relies on a priori, empirical assumptions for different ma-

terial responses, under the help of the general guidelines deduced axiomatically from the basic postulates and limited empirical testing data. Whereas a meso-plasticity analysis is linked with detailed material specification and microstruc-tural parameters of the material system under consideration. The significance of this distinction is highlighted by the ability (for mesoplasticity approach) or limited ability (for macroplasticity) to get material insights into constitutive for-mulation. This issue assumes increasing importance in the modern era when material design and mechanical property control becomes closely looped in the advances of new structural materials. With the advent of engineered multi-phase materials, most notably fiber composites and interface engineering, the prospect of designing material microstructure with specific properties presents a new opportunity as well as a new challenge for both solid mechanicians and material scientists, while macroplasticity approach seems inadequate than ever to guide the scientific and technological progresses in this direction. Intervened by the modern computer technology, the design and property prediction of high strength, high toughness structural materials have been integrated into a com-bined enterprise, with mesoplasticity taking the center-stage in the toughness predictions and the processing capability of the new material systems.

(3) *Ambiguities in the essential structures of macroplasticity framework.* Although new results on the essential structures of plasticity formulation have been con-stantly emerged from study on macroplasticity aspect, several vitally important issues are still elusive and probably cannot be fully resolved from macroplas-ticity knowledge alone. Those issues include the basic physical postulate which gives rise to the structure of flow rule, the selection of corotational rate, the cause and evolution of anisotropic hardening in accompany with material tex-ture development, the resolution (both physical and mathematical) of elastic and plastic deformation at large strain, the geometry of yield surface espe-cially the vertex formation, etc.. As the readers will find in the subsequent content that tremendous understanding toward these issues can be provided by a mesoplasticity exploration, and that new understanding also serves to drive the study of macroplasticity into many new frontiers..

(4) *Difficulties in the description of microstructural sensitive phenomena.* Phenom-ena such as phase transformation, flow localization, ductile fracture and ma-terial damage are extremely sensitive to the microstructural details, especially the inhomogeneities scattered inside the material systems. The homogeneous continuum devoid of any internal structures is assumed as the cornerstone in macroplasticity. However, it would undoubtedly cause barriers in the char-acterization of microstructural sensitive phenomena, rendering their accurate description intrinsically difficult if approached from macroplasticity method-ology alone. On the other hands, substantial progresses have been reported from mesoplasticity study which enable the scientists to demonstrate vividly the development of cup-cone failure, shear band and ductile fracture paths to the finest details in resembling the experimental predictions. The development of *computational micromechanics* in the past ten years is so enormous that it is upgraded into a status of quantitative simulation on the material responses and

even quantitative predictions of various material parameters based on more fundamental and independently measured physical quantities. Needleless to say, a much closer collaboration between solid mechanics community and material science community is now neccessary in order to describe quantitatively different classes of microstructures, to understand how a particular class of microstructure responds to the loads and how it fails, to formulate appropriate constitutive relations and damage evolution laws, and to develop the processing neccessary to obtain the desired microstructure.

(5) *Unable to handle applications of microplastic natures.* As we remarked earlier, a great deal of applications can be handled by the knowledge of macroplasticity. However, there are some practical applications which are intrinsically mesoplastic. Examples for these applications include ultra-precision machining, texture control of superconductive alloys, surface finishing improvement of mechanical processing, superplastic manufacturing, etc., not to mention the applications related to the material failures. These applications are closely related to the evolution of microstructures and the stress-strain history recorded in the individual grains.

The above limitations of macroplasticity stimulate rapid development of mesoplasticity in the past twenty years and attracts more and more interests from scientists and engineers. As readers may find in the following content, this subject has now transformed to a quantitative science discipline encompassing both systematic theoretical framework and application feasibility. To enjoy this accomplishment, the readers should take their ride to the following chapters and make their own comparisons and justifications from time to time.

2.11 References

1. Carroll, M.M., Appl. Mech. Rev., 38(1985), p.1301.
2. Pipkin, A.C. and Rivlin, R.S., ZAMP, 16(1965), p.313.
3. Valanis, K.C., Int. J. Solids Structures, 17(1981), p.249.
4. Mróz, Z., Acta Mechanica, 7(1969), p.199.
5. Bishop, J.F.W. and Hill, R., Phil. Mag., 42(1951), p.414 and p.1298.
6. Lee, E.H., J. Appl. Mech., 36(1969), p.1.
7. Hwang, K.C., Nonlinear Continuum Mechanics, Tsinghua Univ. Press, (1989), in Chinese.
8. Hutchinson, J.W., Finite strain analysis of elastic-plastic solids and structures, in Numerical Solution of Nonlinear Structural Problems, Hartung, R.F. Ed., ASME, (1973), p.17.
9. Lee, E.H., Mallet, R.L. and Wertheimer, T.B., J. Appl. Mech., 50(1983), p.554.
10. Dienes, J.K., Acta Mech., 32(1979), p.217.
11. Fardshisheh, F. and Onat, E.T., Proc. Symp. on Foundation of Plasticity, Warsaw, Noordhoff, (1974), p.89.
12. Agah-Tehrani, A. et al., J. Mech. Phys. Solids, 35(1987), p.517.
13. Hwang, K.C. and Cheng, L., Acta Mechanica Sinica, Special Issue on Plasticity, (1989), P.3, (in Chinese).
14. Hill, R., Aspects of Invariance in Solid Mechanics, Adv. in Applied Mechanics, 18(1978), p.1.
15. Drucker, D.C. and Palgen, L., J. Appl. Mech., 48(1981), p.479.
16. Yang, W., Brown, M.W. and Miller, K.J., Biaxial plastic analysis for cylindrical specimen, Report to Royal Society of London, (1988).

17. Mear, M.E. and Hutchinson, J.W., Mech. Materials, 4(1985), p.395.
18. Becker, R., Needleman, A., Richmond, O. and Tvergaard, V., J. Mech. Phys. Solids, 36(1988), p.317.
19. Hill, R., Mathematical Theory of Plasticity, Clarendon Press, Oxford, (1950).
20. Boehler, J.P., Applications of Tensor Functions in Solid Mechanics, (1987).
21. Kadaschevich, Yu.N. and Novozhilov, B.B., Appl. Math. & Mech., No.1 (1959), p.78, (in Russian).
22. Ziegler, H., Quart. Appl. Math., 17(1959), p.55.
23. Yang, W., Cheng, L. and Hwang, K.C., Objective corotational rates and shear ocscillation, Inter. J. Plasticity, 8(1992), p.653.
24. Neale, K.W., Solid Mechanics Archives, 6(1981), p.79.
25. Budiansky, B., (1959), J. Appl. Mech., 26(1959), p. 259.
26. Batdorf, S.B. and Budiansky, B., NASA TN 1871, (1949).
27. Christoffersen, J. and Hutchinson, J.W., J. Mech. Phys. Solids, 27(1979), p.465.
28. Pan, J. and Rice, J.R., Int. J. Solids Struct., 19(1983), p.973.
29. Coleman, B.D. and Gurtin, M.E., J. Chem. Phys., 47(1967), p.597.
30. Rice, J.R., J. Mech. Phys. Solids, 19(1971), p.433.
31. Hill, R. and Rice, J.R., SIAM J. Appl. Math., 25(1973), p.448.
32. Rice, J.R., Continuum mechanics and thermodynamics of plasticity in relation to microscale deformation mechanism, in Constitutive Equations in Plasticity, Argon, A.S. ed., The MIT Press, (1975), p.23.
33. Nemat-Nasser, S., J. Appl. Mech., 50(1983), p.1114.
34. Buchdahl, H.A., The Concepts of Classical Thermodynamics, Cambridge Univ. Press, Cambridge, (1966).
35. Il'yushin, A.A., Prikl. Mat. Mekh., 25(1961), p.503.
36. Drucker, D.C., in Proc. 1st U.S. Nat'l Congr. Appl. Mech., ASME, p.487.
37. Hill, R., J. Mech. Phys. Solids, 16(1968), p.229 and p.315.
38. Rudnicki, J.W. and Rice, J.R., J. Mech. Phys. Solids, 23(1975), p.371.

3 Introduction to Material Structures

3.1 Introduction

A central problem in mechanical science is the understanding of the behaviour of solids under external loads. The object under study can be a group of atoms or molecules, to the higher structure of metals, alloys, plastics, ceramics and rocks, and to the even larger structure such as machine or geological features which are built from these materials. While useful knowledge of material behaviour will only be gained at the right scale of observation (i.e., radioactivity at the sub-atomic scale and the optical interference at the lattice scale), very often a multi-scale observation is needed to a deeper understanding of the problem and the removal of superficially contradicting phenomenon which owns its clues on a more fundamental level. A good example in plasticity problems is the shear banding phenomenon at large deformation. The first observation of shear bands in optical microscope was reported by Adcock[1] in 1922 in a copper-nickel alloy, almost twelve years before the birth of the dislocation theory of crystals which tried to understand the atomic process involved in slip in crystals. Shear banding was thought to be so complicated by that time that further research interests on the formation of these shear bands did not reappear until the seventies with the wide spread use of electron microscope and X-ray diffraction.

Traditionally, many descriptions of the mechanical behaviour are based on the average properties of materials which are treated as continuum. This is opposed to the practice in solid state physics where the macroscopic properties have to be derived from the statistical average of those of the micro-components of the system being studied. The word structure refers to the architecture of the basic unit (at the length scale of observation) which builds up the system or sub-system. In material physics, the task of handling structural problems is much simplified as the architecture at the sub-micron level at which physicists are concerned is the arrays and distributions of atoms in a perfect lattice of the crystal. Many of the physical properties are *structure-insensitive* i.e., the specific heat or the melting point does not change much with the presence of lattice defects. On the other hand, all the plastic properties such as the yield stress, creep and fatigue strength are *structure-sensitive*. The methods of modern solid state physics allow us to calculate the elastic properties of metals, but we still have no convincing way of calculating their plastic properties.

The major framework of continuum mechanics is developed at a time modern analytical technique was not available for characterization of material structures. With the rapid development in modern analytical instrumentation and computing power, advanced quantitative models capable of describing many aspects of structures such as the spatial distribution of lattice phase and orientation become available. This provides for the very first time the possibility and opportunities for studying plasticity problems built on a continuum based microstructure formulation. The following sections attempt to give readers an appreciation of the modern description and modelling of material structures.

3.2 The Crystalline State of Matter

3.2.1 Crystalline versus Amorphous State

As far as atomic arrangement is concerned there are two known solid state structures: the *crystalline* state (with long range order) and the *amorphous* state (with absence of long range order or presence of short range order). An ordered arrangement of the atoms in a repetitive three dimensional pattern is the prime characteristics of the ideal crystalline state. This regularity of the internal arrangement is often reflected in an external regularity of form, and the term crystal refers to fragments of matter showing this feature which is important in the study of mineralogy. The development of perfect regularity corresponds to a minimization of energy of the solid at the state in which it is formed. Quantum mechanics treatments of the interactions of atoms in close proximity can predict quantitatively the minimal energy configurations for some simple systems of single atoms. Apart from the many crystalline materials in which the atomic arrays approximate closely to the crystalline state, there are a number of materials which do not develop this long range order. Some never develop this regularity more than a few atomic distance and approximate to an ideal amorphous state, which may be compared to the freezing of a liquid structure. The perfect amorphous state is really one extreme limit of the structural arrangements that are to be found in solid matters. A number of computer algorithms had been developed to investigate the properties of these random packed structures. While the study of amorphous state is becoming important in the development of many new advanced materials, it will not be dealt with here and readers who are interested in this subject can refer to more specialized works in amorphous physics.

The degree of crystallinity of materials depends on the conditions of their solidification and subsequent thermomechanical treatment. The great majority of all solid materials are crystalline. Most metals and minerals are crystalline under the normal conditions of processing. However they can be rendered amorphous by quenching so fast (up to $100°K$ per second) from the liquid state that the crystallization transformation has insufficient time to occur. This is the industrial basis for the production of glassy metals which possess good wear resistance and magnetic properties. On the other hand, polymers are considered either amorphous or crystalline with various degrees. Polymers are macromolecules formed by joining a large number of small molecules, or monomers, in a chain. The resulting large molecule may be in the form of a long linear chain, a chain with side branches or interconnected to form a three dimensional network. Not all polymers give rise to crystallinity. Branched polymer chains are not easy to pack together in a regular manner while cross-linked polymers cannot be arranged due to the links between the chains so that crystallinity is not possible.

Recently the development of a new class of materials called *nanocrystalline materials* (NCM) claims to exhibit an atomic structure differs from the two known solid structures. Nanocrystalline materials are single or multi-phase polycrystals, the crystal size of which is of the order of a few (typically 1-10) nanometers so that about 50% volume of the materials consists of grain or interphase boundaries (see Sec. 3.2 on grain boundary structures). Solids with some degree of crystallinity which although less than perfect are still sufficient to allow their properties to be investigated in crystallographic terms. Crystallographic studies now overflows into organic, inorganic, metallic, physical, and biological fields in ways that were inconceivable decades ago. Techniques available for probing into the atomic arrangement include X-ray, electron and neutron diffraction, small angle X-ray scattering and neutron scattering, Extended X-ray Absorption Fine Structure (EXAFS) measurements, field ion emission microscopy, laser Reimannn spectrum, and nuclear magnetic resonance.

3.2.2 Basic Crystallographic Geometry

The regular shape of crystals suggests that the atomic building units are regularly arranged. Crystallography is geometric by its nature and deals with the analysis of repeating patterns, the description and classification of crystal lattice. A *lattice* is a geometric concept that can be defined as "an infinite regular arrays of points in space, each of which has identical surroundings". Such an infinite discrete point set is called a *discontinuum* which is physically more real than the concept of *continuum* which does not exist at the atomic scale of observation. Some general properties of the discrete point sets applied to crystallography have been discussed by Engel[2].

A crystal is an excellent example of a pattern which repeats itself in three dimensions. Each basic repeating unit which is either single atom or group of atoms is known as the *motif*. A *crystal structure* is determined by the type of, number and the arrangement of atoms in the motif and by the relative positions of the motifs as defined by the space lattice. Crystal having the space lattice is described by three primitive lattice translation vectors in three dimensions, i.e.

$$\mathbf{r} = u\mathbf{a} + v\mathbf{b} + w\mathbf{c} \tag{3.1}$$

where u, v, w are the resolved scalar components of the lattice translation vector $\mathbf{r}$ on the three primitive vectors $\mathbf{a}$, $\mathbf{b}$ and $\mathbf{c}$. The displacement $\mathbf{r}$ is called a trans-

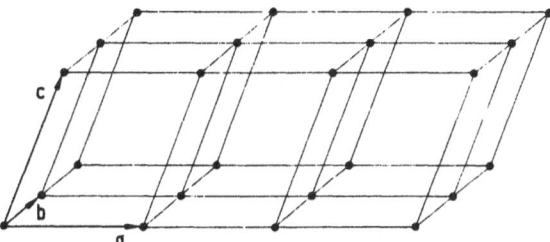

Fig. 3.1. A lattice cell described by 3 primitive vectors **a**, **b** and **c**

lation operation. A lattice *unit cell* of volume $\mathbf{a}.[\mathbf{b} \times \mathbf{c}]$ is obtained by drawing a parallelepiped with the three primitive vectors as three of its edges (Fig. 3.1).

A unit cell of a crystal has the following properties:

(i)   The complete set of parallelepipeds generated from it by the above translation operations covers all points in space and

(ii) each cell contains the equivalent of a lattice point (i.e., only one lattice point is included within the volume $\mathbf{a}.[\mathbf{b} \times \mathbf{c}]$)

If a lattice can be moved in a certain way but still appears exactly the same, the movements which leave the pattern of atomic arrangement unchanged are known as *symmetry operations* and different types of symmetry operations are called *symmetry elements*. A particular lattice may be described in any number of ways by different choices of the primitive vectors **a**, **b**, and **c** which are inclined to each other at angles α, β and γ. In addition the lattice could possess the following symmetry:

(i)   rotation of $2\pi/n$ about axis through the origin, where $n = 1$, 2, 3, 4, or 6. (as crystal lattice can have only those axes that are compatible with a space filling array)

(ii) reflections in planes containing the origin.

(iii) a centre of inversion which takes **r** into $-\mathbf{r}$.

The symmetry elements restrict the shape a lattice unit cell can take. On the basis of symmetry axis present, crystals can be divided into seven main groups as listed in Table 3.1.

Among the crystal systems, the triclinic has no symmetry and all three angles α, β and γ have to be specified. On the other hand, the cubic system has the highest symmetry of all. The three crystallographic axes are orthogonal and equivalent, being related by secondary three-fold axes running parallel to the body of the diagonals of the unit cell cube. Many technologically important metals and alloys belong to the cubic systems. An understanding of the symmetrical properties is not only important in the mathematically description of the geometry of crystals, but also of many physical and mechanical properties of polycrystalline materials as well.

To treat the various changes that evolve in crystal structures such as occurring in deformation and phase transformation, we often need to describe precisely the

Table 3.1. The seven crystal systems

System	Unit cell dimensions	
Triclinic	$a \neq b \neq c,$	$\alpha \neq \beta \neq \gamma \neq 90°$
Monoclinic	$a \neq b \neq c,$	$\alpha = \gamma = 90° \neq \beta$
Orthorhombic	$a \neq b \neq c,$	$\alpha = \beta = \gamma = 90°$
Tetragonal	$a = b \neq c,$	$\alpha = \beta = \gamma = 90°$
Hexagonal	$a = b \neq c,$	$\alpha = \beta = 90°, \gamma = 120°$
Trigonal	$a = b = c,$	$\alpha = \beta = \gamma < 120°$ and $\neq 90°$
Cubic	$a = b = c,$	$\alpha = \beta = \gamma = 90°$

relative locations of atoms and of planes of atoms. Two approaches are often used: the vector method for specifying crystallographic directions and planes, and the stereographic projections for treating crystal orientations in three dimensions. The former will be dealt with here while the crystallographic orientation will be discussed in Sec.3.4. A system of describing crystal planes and directions was developed by W.H.Miller based on three-figure groups known as *Miller indices*.

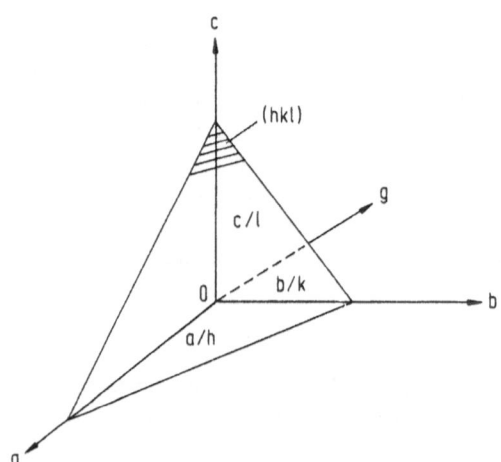

Fig. 3.2. The Miller indices of a plane with intercepts a/h, b/k, and c/l on the three axes

If the unit cell of a crystal is defined by the vectors **a**, **b**, and **c**, then any crystal plane which intercepts the axes in lengths proportional to $a/h, b/k, c/l$ respectively is denoted by its Miller indices (hkl) (Fig. 3.2). Fractional intercept distances from the unit cell have to be inverted (i.e. reciprocals taken) before they can be expressed as a ratio. Planes which run parallel to one of the crystallographic axes intersect it at infinity are given a value zero at that index to avoid the appearance of infinity in the indices. If a plane cuts any axis on the negative side of the origin, the index will be negative and is indicated by placing a minus sign above the index, such as $(\bar{h}\bar{k}\bar{l})$ which is pronounced as bar h, bar k and bar l. Examination of many crystals shows that the indices of both natural and cleavage faces are small numbers and these observations were summed up by crystallographer A.R.J.Hauy as the "the law of rational indices". To specify a direction of the vector **r** in the lattice, the three resolved components are read on the Cartesian coordinate axes and divided by their highest common factor. They are enclosed in square brackets. Some examples of directional indices and Miller indices of planes and directions in cubic crystals are shown in Fig. 3.3.

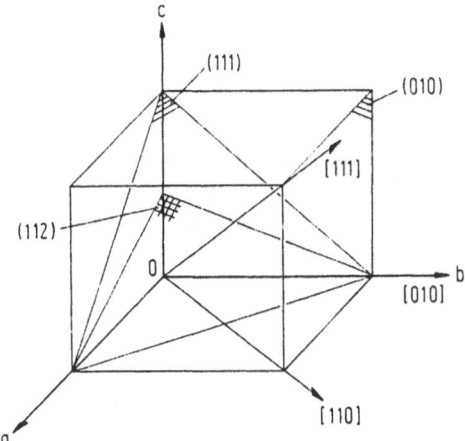

Fig. 3.3. Miller indices of (a) planes and (b) directions in cubic crystals

The possession of symmetry elements by a lattice renders particular sets of directions equivalent. $<uvw>$ with pointed brackets indicates $[uvw]$ and all the similar directions which are related to it by symmetry. For example, in the cubic system, $<100>$ means the six directions $[100]$, $[\bar{1}00]$, $[010]$, $[0\bar{1}0]$, $[001]$ and $[00\bar{1}]$. $<110>$ means twelve directions. As in the case of equivalent directions, the number of planes of one form depends on the symmetry of the lattice and increases as the symmetry increases. In the cubic system, all possible permutations of (hkl) of three non-zero different indices give 48 planes. Planes of (hkl) and $(\bar{h}\bar{k}\bar{l})$ are parallel and are identical . The family of planes reduces to 24. $\{100\}$ means the 3 sets (100), (010), (001). $\{110\}$ means the 6 sets (110), (101), (011), ($\bar{1}$10), ($\bar{1}$01), (0$\bar{1}$1). $\{111\}$ means the 4 sets (111), ($\bar{1}$11), (1$\bar{1}$1), (11$\bar{1}$) while (123) means the 24 sets.

Some basic relations between the directions and planes will be illustrated here for a cubic system with the use of matrices and vectors. A group of planes with a common line of intersection is called a "zone" of planes, The line of intersection is called the zone axis. Three sets of planes $(h_1\ k_1\ l_1)$, $(h_2\ k_2\ l_2)$, $(h_3\ k_3\ l_3)$ belong to one zone if the determinant

$$\begin{vmatrix} h_1 & k_1 & l_1 \\ h_2 & k_2 & l_2 \\ h_3 & k_3 & l_3 \end{vmatrix} = 0 \tag{3.2}$$

and a direction $[uvw]$ will lie parallel to a set of planes (hkl) provided $hu+kv+lw = 0$. The direction $[uvw]$ along which two sets of planes $(h_1k_1l_1)$ and $(h_2k_2l_2)$ intersect is given by

$$\begin{aligned} u &= k_1 l_2 &- k_2 l_1 \\ -v &= h_1 l_2 &- h_2 l_1 \\ w &= h_1 k_2 &- h_2 k_1 \end{aligned} \tag{3.3}$$

Then angle Θ between two directions $[u_1 v_1 w_1]$ and $[u_2 v_2 w_2]$ is

$$\cos \Theta = (u_1 u_2 + v_1 v_2 + w_1 w_2)/(u_1^2 + v_1^2 + w_1^2)^{-1/2}(u_2^2 + v_2^2 + w_2^2)^{-1/2}. \tag{3.4}$$

A direction $[xyz]$ is perpendicular to the plane (xyz) so that the equation (3.4) used to find the angle between the two directions applies equally to find out the angle between the two planes. The interplanar spacing d is particularly important in describing the diffraction from crystals. For a cubic crystal,

$$d = a(h^2 + k^2 + l^2)^{-1/2} \tag{3.5}$$

where a is the lattice constant.

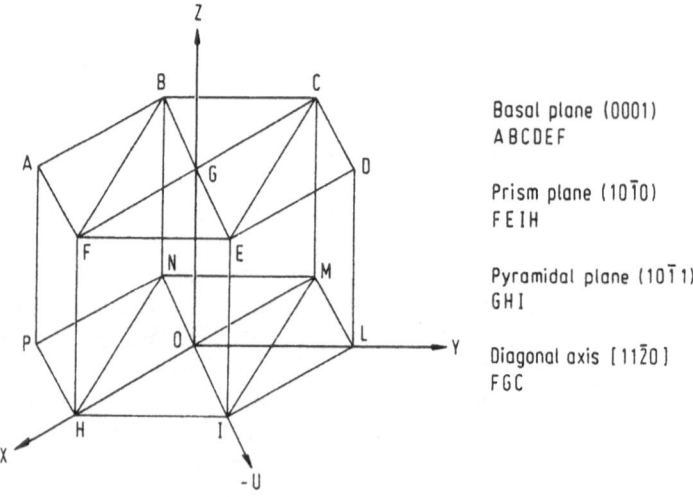

Basal plane (0001)
ABCDEF

Prism plane $(10\bar{1}0)$
FEIH

Pyramidal plane $(10\bar{1}1)$
GHI

Diagonal axis $[11\bar{2}0]$
FGC

Fig. 3.4. The Miller-Bravais indices for HCP structures

In the trigonal or hexagonal systems, the rotation axes imply the existence of three equivalent axes at 120° intervals. A four-index system (hkil) based on a non-primitive hexagonal cell is used. Three axes OX, OY and OU are taken in the basal plane along the three close-packed directions at 120° to each other, and a fourth axis OZ is set normal to the basal plane (Fig. 3.4). Intersections of crystal planes are then measured in unit lengths along the four axes in turn. The *Miller-Bravais* index of the plane is presented as (hkil) where i corresponds to the intercept on the u axis. The first three face indices are not independent since $h + k + i = 0$.

3.2.3 Miller Indices and the Reciprocal Lattice

The orientation and spacing of a set of crystallographic planes can be fully defined by the vector **g** (as shown in Fig. 3.2) which is perpendicular to the set of planes and of length proportional to the interplanar spacing d of equation (3.5). The *reciprocal lattice* leads to simplified geometric expressions for the spacing of crystal planes. To construct the reciprocal lattice, a vector **g*** is defined which is parallel to **g** but with a length proportional to the reciprocal of d. Secondly, the **g*** vectors corresponding to all the sets of planes in the lattice are drawn from the same origin. The tips of all the vectors will themselves form a lattice as shown in Fig. 3.5.

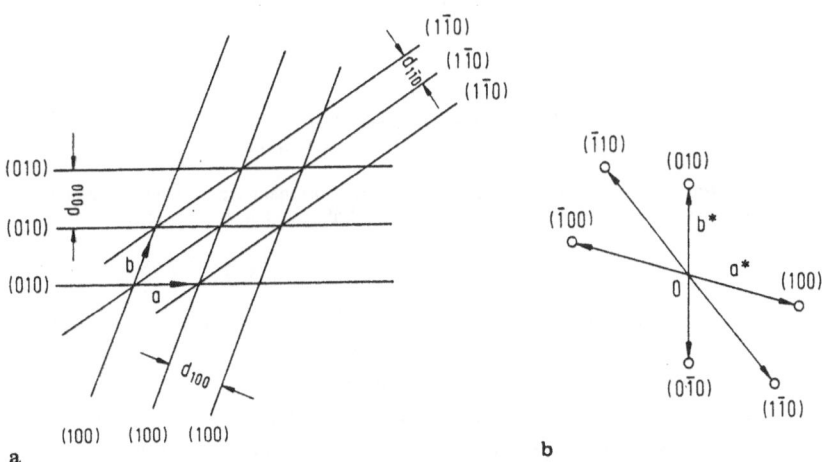

Fig. 3.5. Relationship between the (a) crystal base vectors and (b) the reciprocal base vectors

The translations of a primitive space lattice are denoted by **a**, **b** and **c** and the vector **p** to any lattice point is given by

$$p = ua + vb + wc. \tag{3.6}$$

The definition of the reciprocal lattice is that the basic vectors $\mathbf{a}^*$, $\mathbf{b}^*$, $\mathbf{c}^*$ are such that

$$\mathbf{a}.\mathbf{a}^* = \mathbf{b}.\mathbf{b}^* = \mathbf{c}.\mathbf{c}^* = 1 \qquad (3.7)$$

and

$$\mathbf{a}.\mathbf{b}^* = \mathbf{b}.\mathbf{c}^* = \mathbf{c}.\mathbf{a}^* = 0. \qquad (3.8)$$

Two important properties of the reciprocal lattice are listed here. Firstly, the vector $\mathbf{g}^*$ defined by $h\mathbf{a}^* + k\mathbf{b}^* + l\mathbf{c}^*$ is normal to the plane of Miller indices (hkl) in the real lattice. Secondly the magnitude g^* of this vector is the reciprocal of the spacing of (hkl) in the primary lattice, i.e. $d = 1/g^*$.

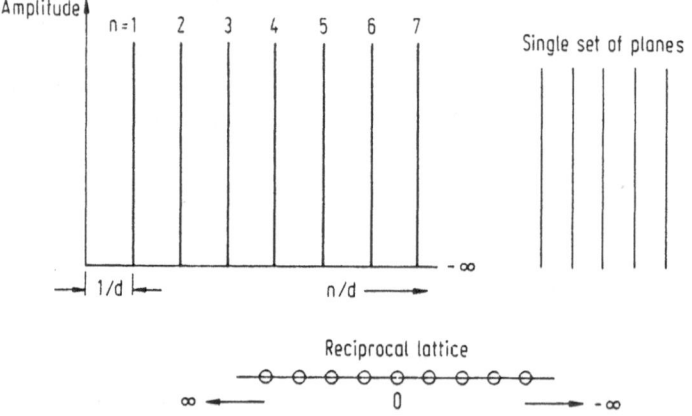

Fig. 3.6. Fourier transform of a single set of planes as one-dimensional reciprocal lattice (After Windle, A.H., A First Course in Crystallography, G.Bells and Sons Ltd., London, 1977, p.152)

Mathematically the reciprocal lattice is the Fourier transform of the real lattice. The translation from the origin of the reciprocal lattice to any lattice point is described by the vector $\mathbf{g}$ which has a magnitude equal to the reciprocal of the wavelength of one of the Fourier components of the real lattice. A single set of planes is replaced by a series of points and its Fourier transform drawn as a one-dimensional reciprocal lattice is shown in Fig. 3.6. For example the (100) set of planes will give a row of reciprocal lattice points with co-ordinates $\pm 000, 100, 200, 300 \ldots$ expressed in terms of the reciprocal lattice vectors $\mathbf{a}^*$. If all various sets of plane are considered, their Fourier transforms will build up an infinite two-dimensional reciprocal lattice.

The reciprocal lattice is a simplified geometric expression for the spacing of crystal planes. This in itself is useful in studies of crystals using the techniques of X-ray, electron or neutron diffraction. It can be seen that the relation between the reciprocal lattice and the electron diffraction pattern is a particularly close one, so close that the pattern is identical to a net of the reciprocal lattice [3]. The conditions for diffraction to occur is governed by the well known Bragg's equation given by

$$n\lambda = 2d \sin \Theta \tag{3.9}$$

where λ is the wavelength of the incident radiation, d is the spacing of the planes which makes an angle of Θ with the incident electron beam and n is a whole number. An equivalent way of defining the conditions under which a diffracted beam will be reflected is by the Ewald sphere construction (Fig. 3.7).

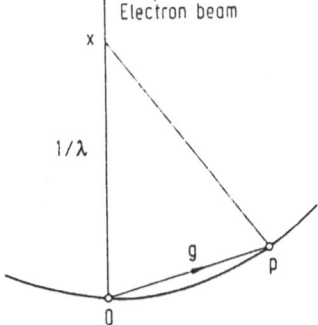

Fig. 3.7. Ewald sphere showing the geometrical condition for Bragg's reflection from a set of plane with reciprocal lattice vector g

A line is drawn through O, the origin of the reciprocal lattice and of length $1/\lambda$ to the point X. A sphere, known as Ewald sphere is now inscribed with radius $1/\lambda$. A strong diffraction beam parallel to XP will form if a reciprocal lattice point P lies on this sphere. Bragg's law will be satisfied if

$$\sin \Theta = g\lambda/2 \tag{3.10}$$

where $g = 1/d$. The electron diffraction pattern represents a planar section through the origin of the reciprocal lattice perpendicular to the beam direction. Some typical electron diffraction patterns for a FCC crystal is shown in Fig. 3.8. Electron diffraction patterns of FCC metals have indices all odd or all even whereas $h+k+l$ is always even on the pattern from BCC crystals. The solving of diffraction patterns takes time and this can be expedited with computer-generated patterns displayed directly on screen for immediate comparison with experimentally determined patterns.

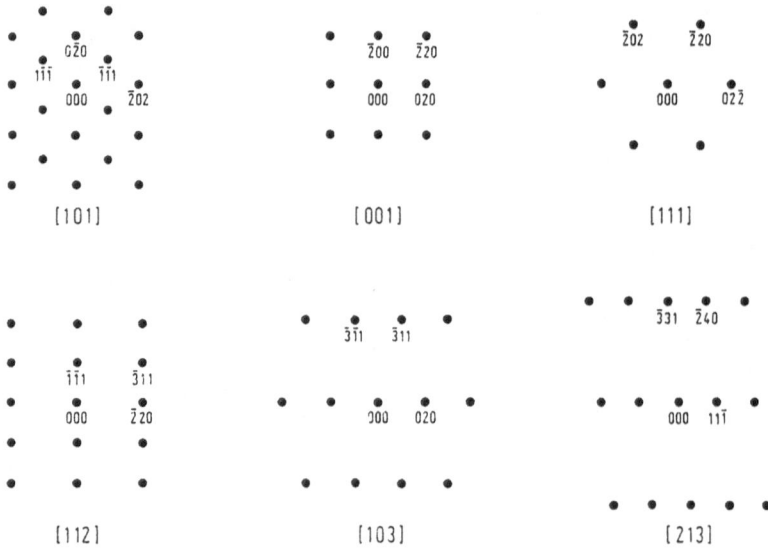

Fig. 3.8. Some common diffraction patterns in FCC lattice

3.2.4 Crystal Structures

A primitive unit cell contains only a lattice point per unit cell. The unit cells obtained from the seven crystal systems contain only one lattice point per cell. For example the lattice point at the eight corners of a simple cube is shared between eight adjacent cells. From the seven crystal systems as listed in Table 3.1, it is always possible to select non-primitive unit cells by placing extra points within the simple cells to make body-centered unit cells, or at the centres of the faces of the unit cells to create face-centred cells. Altogether there are fourteen possible combinations of the space lattices and the unit cells. The resulting arrays are known as the *Bravais lattices* named after the French crystallographer Auguste Bravais in the nineteenth century.

The fundamental feature of a crystalline solid is its periodicity of structure. If we substitute atoms, group of atoms, molecules, for the points on the fourteen Braves lattices we obtain the actual crystal structures of which the number can be infinite by making minor alterations to the atomic patterns on the point lattice. Crystals having same space lattice can have different motif in the lattice point (i.e. NaCl and LiCl). There are many ways of classifying crystal structures. These can be based on the type, number and arrangements of atoms in the motif and its relative positions defined by the lattice as discussed above. Other criteria used are the types of atomic bonding and the degree of close packing in the crystal. According to the types of bonds in a crystal structure, we can have *metallic* structures, *ionic* structures, *covalent* structures and *molecular* structures. Based on the density of

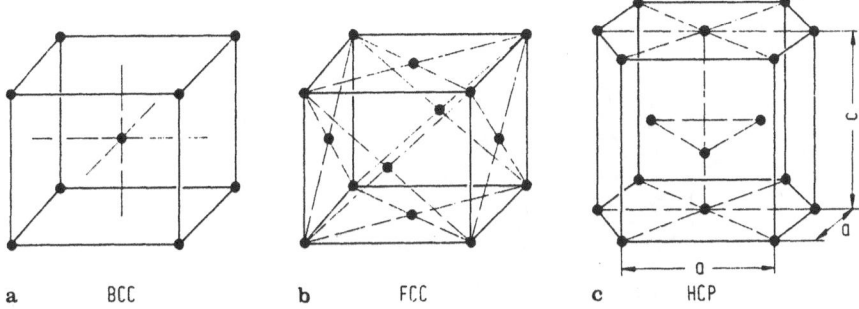

Fig. 3.9. Unit cells of the (a) BCC, (b) FCC and (c) HCP structures

packing of atoms in the crystal, we have the *body centre cubic* structures (BCC), the *face-centre cubic* structure (FCC) and the *hexagonal close-packed* structure (HCP) as shown in Fig. 3.9.

If we think an atom as a ball of finite size as in the so-called rigid ball model, the atoms are most closely packed if each is centered on a point of a planar hexagonal lattice which is called a close-packed plane. A close-packed plane contains three close-pack directions. The most densely packed three-dimensional structure is obtained by stacking close-pack planes. Both the FCC and HCP structures are *closest packed structures* obtained by stacking close packed planes in different sequence as shown in Fig. 3.10. The positional sequence ABCABCABC... gives rise to the FCC structure whereas the HCP structure corresponds to the stacking sequence ABABAB... (or ACACAC... or BCBCBC...). The BCC structure is not as closely packed as the other two. These three crystal structures occur with about equal frequency amongst the metals. In alloys where different atoms are present, a suitable combination of large and small atoms can give rise to a *topologically close-packed phases (TCP)* (e.g. σ phase (FeCr) in stainless steel) which has higher space utilization than the FCC or HCP structure with the same atomic species.

Ionic, metallic and most intermetallic bonds are non-directional and tend to form densely packed structures. The covalent bond is directional and will be less densely packed. Many materials have more than one type of bonding in the solid and complex structures may exist as crystalline, semicrystalline or amorphous. Some solids show reversible changes of crystal structures at certain temperatures without any change of chemical composition and this phenomenon is known as *allotropy* when referred to the elements and *polymorphism* when referred to chemical compounds.

The packing discussed so far concerns with the packing of identical spheres. If element of similar atomic size is added to a structure, it will substitute randomly for the host atoms producing a *substitutional solid solutions*. If there is a preference for having like or unlike nearest neighbours an ordered substitutional solid solutions will result (i.e. ß-brass of 50% Cu atoms and 50% Zn atoms at room temperature).

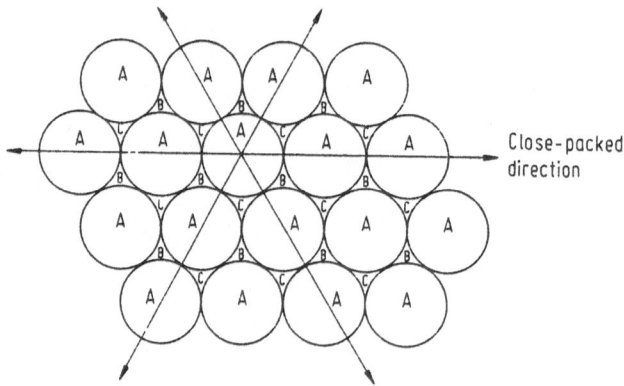

Fig. 3.10. Atom centres of different layers to form closed-packed structure. The closed packed directions are also indicated

Perfectly ordered alloys form a new lattice that is larger than the crystal lattice of the disordered alloy. Such a new lattice is called a *superlattice*. When the alloying atom is much smaller than the host atom, it will occupy the interstices of the parent crystal lattice without appreciable distortion of the host lattice, and an *interstitial solid solution* may result (i.e. α-iron).

The major differences among metals, ceramics and polymers lies in the manners their atoms are organized in the crystalline state or non-crystalline state. However considerable overlapping may occur among the different groups especially when new composite materials and micro-alloying materials are concerned. Materials which do not crystallize on being cooled to low temperature form a rigid structure known as *glass*. Metals, ceramics and polymers illustrate qualitative differences in this regard. Metallic structures usually consist of only a few atoms and therefore crystallize easily. Ceramic crystals are generally more complex. They usually contain both metallic and non-metallic elements with ionic or covalent bonds. Frequently occurring chemical formulas in ceramics are: AX, AX_2, A_2X, ABX_3, A_2X_3, and AB_2X_4, where A and B are metals and X is a nonmetal. They include a board range of silicates, metallic oxides and combinations of silicates and metal oxides. Elements such as carbon, silicon, certain carbides, borides are also considered as ceramics. Crystallization occurs readily in ceramics although glass formation is also common. Glass formation is often the rule in polymers due to the difficulty in the rearrangement of the long molecular chains. The mechanical properties are strongly affected by the type of bonding. Atoms can transfer a bond from one neighbour to another fair easily and this gives rise to the high ductility associated with metals. In ceramics the combination of ionic and covalent bonds leads to high strength and brittle fracture. In thermoplastic polymers the covalent bonds within the molecules and the van der Waals forces between molecules lead to the phenomenal elongation, while in thermosetting polymers the network structure leads to higher strength and low elongation.

3.3 Defects in Crystalline Materials

In real crystalline materials, the arrangements of atoms are far from the ideal lattice configurations. These deviations from the perfect lattice arrangement are called *crystal defects*, which can arise during crystal growth, thermal agitation, plastic deformation or irradiation. According to their geometric characteristics, crystal defects can be classified into three groups as follows:

(1) *point defects*: such as vacancies, interstitial atoms or substitutional atoms.

(2) *line defects*: mainly referred to dislocations and

(3) *planar defects*: such as grain boundaries, twin boundaries and stacking faults.

A schematic illustration of the various crystal defects in a polycrystalline materials is shown in Fig. 3.11.

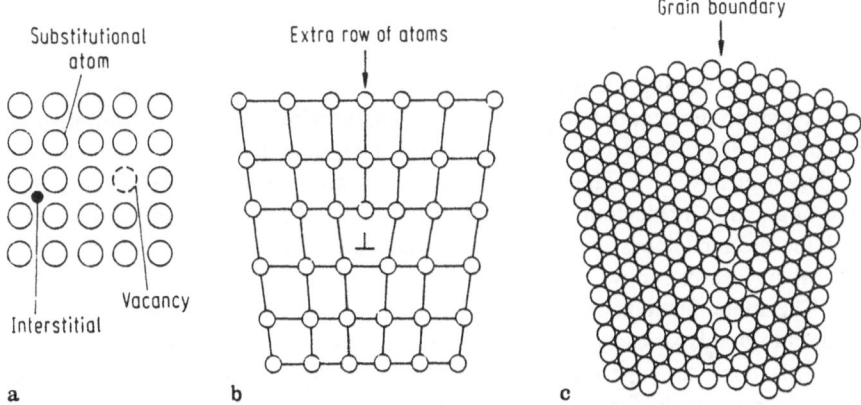

Fig. 3.11. Schematic drawing of various crystal defects. **a** Point defect, **b** line defect, **c** planar defect

These defects not only perturb locally the translational invariance and point symmetry inherent in the crystal lattice, but also trigger non-uniform plastic deformation at a stress level far below the theoretical shear strength of the crystals. The generation, annihilation, and interaction of these defects constitute the physical basis for the plastic deformation of materials. In this section, only major features of defects in crystalline materials are introduced and a quantitative treatment of these defects involved in the plastic deformation process will be discussed in Chapter 4 and 5 of this book.

3.3.1 Point, Line and Planar Defects

Vacancy (an empty lattice position) and interstitial (extra atom not linked to the lattice) arise during quenching, plastic deformation or irradiation. When an atom is displaced from its equilibrium position, it can either migrate to the crystal surface or move to the interstitial space in the lattice. The atoms which take up the interstitial space can be of the same species of atoms and not necessarily the small atoms frequently found in the solid solution of alloys. Either the vacancy or the interstitial atom will distort the local lattice and cause an elastic strain field around the defect. While substitutional atoms (atoms of different chemical species) can be eliminated by higher purity materials, vacancies and interstitial atoms will exist at a finite temperature. The free energy of a crystal has a minimum in correspondence to some concentration of vacancies and interstitial atoms, due to their entropic contribution. Each type of defect is associated with a characteristic diffusion coefficient. Lattice vacancy is the cause of the diffusion of many important transport phenomena such as surface treatment, annealing, sintering of powders, and semiconductor doping. Vacancy also plays an important role in the yield point and creep phenomenon in metals.

The most important two dimensional or line defect is the *dislocation* which can arise during crystal growth, quenching of crystals from high temperature, and plastic deformation. Dislocations refer to the type of crystal defects that atoms are dislocated or misaligned from their otherwise perfect lattice positions. There are two types of dislocations: *edge* and *screw dislocations*. The atomic arrangement around the edge dislocation is shown in Fig. 3.11b and that for a screw dislocation is shown in Fig. 3.12. The distortion around a screw dislocation is different from that of an edge dislocation. There is no extra plane of atoms in a screw dislocation and the atomic planes are arranged around the dislocation line in a continuous helical surface or screw.

The main function of a dislocation is to permit easy breaking and forming of atomic bonds so as to facilitate plastic deformation. Under the influence of an external stress, an edge dislocation can glide or slip in a direction perpendicular to its length, while for a screw dislocation the slip direction is parallel to the dislocation line. The energy requires to move an edge dislocation on its *slip plane* is small as the process involves the re-arrangement of a few atoms around the dislocation core. No transfer of matters are involved and such a movement is called *conservative*. However for an edge dislocation to *climb* upwards or downwards, it will be necessary to remove or add extra atoms on the extra half plane. This type of movement is *non-conservative* as climb requires thermal activation and is favoured only at high temperatures. A screw dislocation does not has a preferred slip plane to glide and its motion is less restricted than an edge dislocation. It can *cross slip* (move from a slip plane into another adjacent slip plane) with a common slip direction. Fig. 3.13 shows the *slip band* (e.g. at A) and cross slip (e.g. at B) in an aluminium crystal. The slip band consists of cluster of slip steps on closely spaced parallel sip planes.

Generally, the slip plane is a plane of highest packing density and the slip direction is the closest-packed direction. In FCC metals, the slip planes are the {111} planes and the slip directions are <110>. There are 12 slip systems in the FCC lattice. The BCC lattice is not a closed-packed structure as the FCC, and higher shearing stress are required to cause slip. Slip is found to occur on the

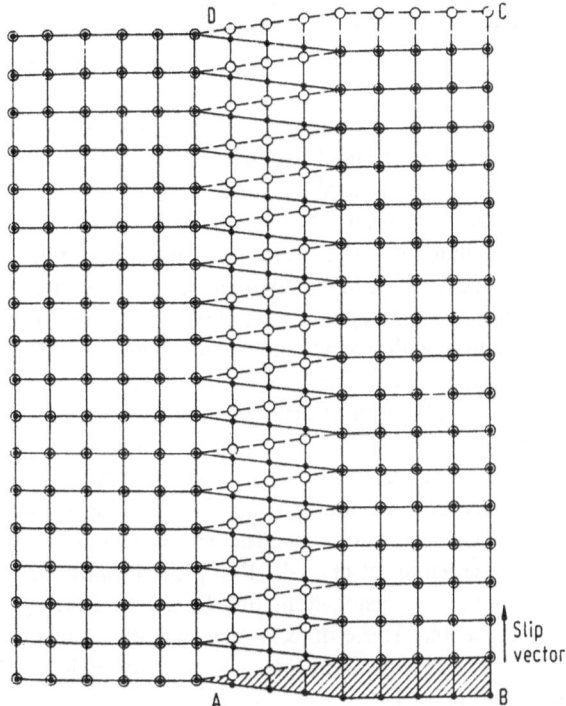

Fig. 3.12. Atomic arrangement around the screw dislocation. Solid circles indicates atoms below the slip plane and open circle indicates atoms above slip plane. The dislocation line is indicated by AD. (From Read, W.T, Dislocations in Crystals, McGraw-Hill, New York, 1953, p.17.)

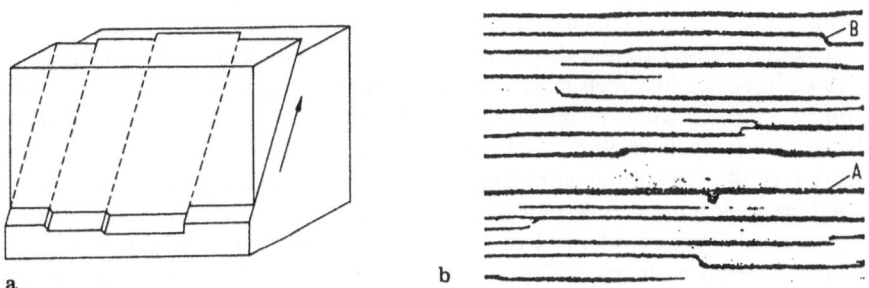

Fig. 3.13. Slip band and cross slip in aluminium. Micrograph in (b) corresponds to top face of model shown in (a). (From Cahn, R.W., J.Inst.Metals, 79, 1951, p.129)

{110}, {112} and {123} planes and along <111> directions. There are 48 such slip systems in BCC crystals. Since slip occurs on several planes, slip lines in BCC metals have a wavy appearance as dislocations cross slip often to another plane. In the HCP metals, the slip occurs on the {0001} basal plane in the <$11\bar{2}0$> direction. The limited number of slip systems (i.e. 3) account for the low ductility and extreme orientation dependence of plastic deformation of HCP metals.

Dislocations in covalent solids such as diamond, germanium and silicon are immobile at low temperatures as the atoms are held together by strong, directional covalent bonds. Dislocation glide is only possible at elevated temperatures and this explains partly the brittleness of these materials. In organic polymers, the long molecular chains unfold in the crystalline region or draw out of amorphous region, giving a large shape change. The primary bonds in the polymer chain are not broken during plastic deformation, and vacancies and dislocations play only a minor role.

The density of dislocation increases with plastic deformation. There is a direct proportionality between dislocation density and the shear strain found in a crystal. *Dislocation density* is defined as the total length of dislocation per unit volume or number of dislocations intersecting a unit area measured in cm^{-2}. In an annealed metal the dislocation density is between 10^5 to 10^8 cm^{-2} which may increase to 10^{12} cm^{-2} during plastic deformation. The multiplication of dislocations and interactions of dislocation with point defects and other dislocations generate more complex defects and lead to work-hardening of crystals. Fig. 3.14 shows trails of defects and "debris" left behind by moving dislocations in a Fe-3% Si alloy after it had been deformed by about 1 percent. These defects may include dislocation dipoles, jogs, Lomer-Cottrell lock, forest dislocations, etc., and their formation will be discussed in Chapter 4 and 6 of this book.

Errors or faults in the stacking sequence (see Sec. 4.5.2) can arise in most metals during heat-treatment and plastic deformation. For the FCC structure, the stacking sequence is ABCABC, where ABC refer to the atom centers in the closed-packed planes in Fig. 3.10. When slip has occurred between an A and B layer, the stacking sequence is then altered. Depending on the stacking sequence, the stacking fault in FCC metals can be considered as a thin HCP region or submicroscopic twins. From dislocation theory, a stacking fault in a FCC metal is an *extended dislocation* caused by the dissociation of a perfect dislocation into *partial dislocations* (Sec 4.5.3) which leave a planar fault between them as they move apart (Fig. 3.15). The width of the stacking fault region is determined by the balance between the repulsive force between the two partial dislocations and the surface tension of the faulted region pulling them together. Seeger[4] has shown that the equilibrium spacing is inversely proportional to the *stacking-fault energy* (SFE) of the material. Typical values of SFE corresponding to room temperature for some common metals is shown in Table 3.2. The measurement of SFE has been reviewed by Gallagher[5] and is found to be sensitive to the chemical composition of the materials.

Transmission electron microscopy has been used to perform quantitative analysis of crystal defects such as dislocations, stacking faults, twin and grain boundaries in materials which can be produced in thin foils. Defects are observed due to scattering of electrons in the strained region around the defects. The contrast expected from a specimen depends on the diffraction conditions (i.e. on particular crystal

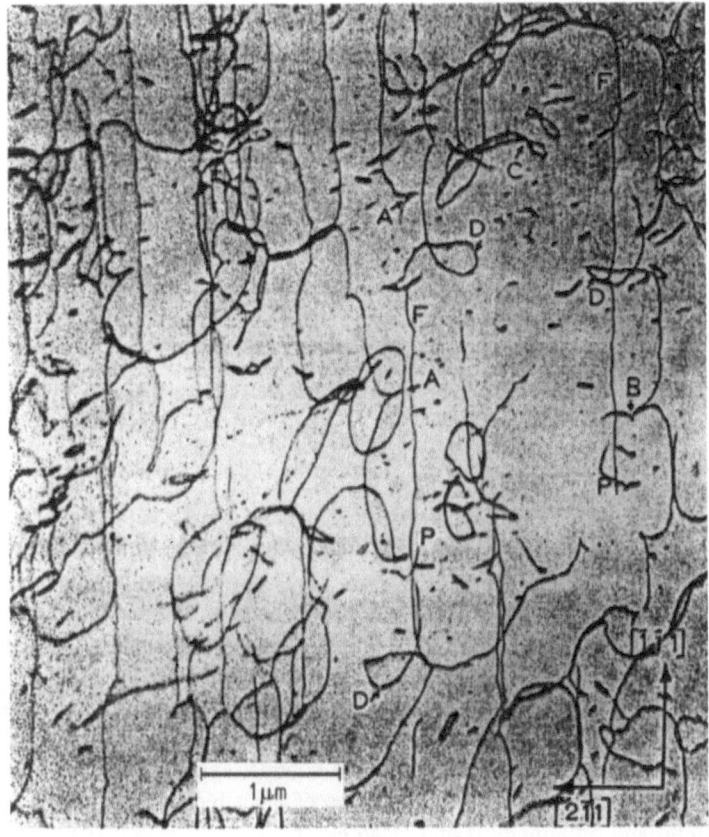

Fig. 3.14. Transmission electron micrograph showing trails of defects left behind by moving dislocations in a Fe-3% Si alloy. (From Low, J.R. and Turkalo, A.M., Acta.Metall.,10(1962), p.215.)

Table 3.2. Typical values of stacking-fault energy for some common metals and alloys (Abstracted from Murr,L.E., Interfacial Phenomenon in Metals and Alloys, Addison Wesley, 1975)

Metal	SFE mJm^{-2}
Brass	< 10
Austenitic stainless steel	21
Silver	22
Gold	45
Copper	78
Nickel	128
Aluminium	166

planes which are diffracting strongly and their proximity to Bragg's angle). Depending on the diffraction conditions, the image of a dislocation can be a single

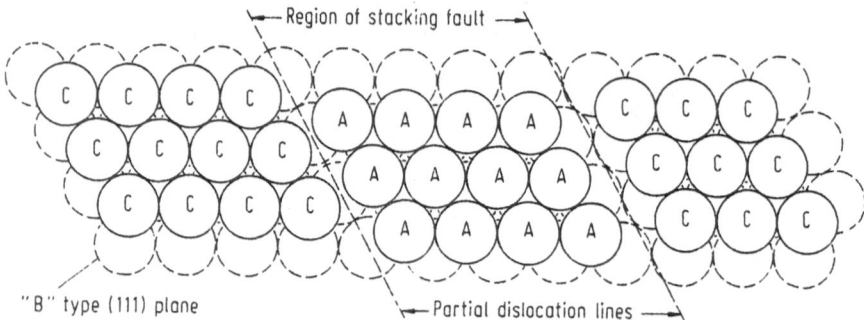

Fig. 3.15. Creation of a stacking fault bounded by two partials dislocations (After Guy, A.G., Intro-duction to Materials Science, McGraw-Hill, 1972, p.173)

dark line on one side of dislocation, two dark lines, wavy line or no change in contrast (invisible). The Burgers vector **b** (see Sec. 4.2.1) of a dislocation line can be determined by using the *invisibility criteria*. Perfect screw or edge dislocations may be completely invisible in the foil when $\mathbf{g}.\mathbf{b} = 0$, where the magnitude of $\mathbf{g}$ is the reciprocal of the spacing of (hkl) plane and a vector normal to the reflecting plane referred in Sec. 3.2.3. Stacking faults also produce a characteristic diffraction image. The development of computer simulation of experimental images provide a speedy technique to compute the electron image contrast expected for different assumed Burgers vectors. More detailed exploration of the subject can be found in the work of Lotretto and Smallman[3] and Head et. al.[6].

3.3.2 Grain and Twin Boundaries

In a polycrystalline material, each grain is surrounded by a grain boundary which exists because the grains on either side are either different in composition, structure or orientation. The interfaces between grains of different compositions are called *interphase boundaries*, and those between crystals of the same composition are called *self-boundaries*. Recently the emphasis in the study of mechanical properties have moved away from the processes which occur inside the individual grains to those which are governed by the boundaries between the grains. In this section only self-boundaries are introduced while interphase boundaries will be discussed in Sec. 3.3.3.

Fig. 3.16 shows the self-boundaries in an annealed α-brass. A self-boundary accommodates the orientation change from one grain to the other. In general, there are five independent crystallographic parameters to characterize the orientation relationship of two neighboring grains. Two degrees of freedom define the axis of misorientation, d, and one is required to specify the angle of misorientation, Θ.

Fig. 3.16. Grain and twin boundaries (self-boundaries) in an annealed α-brass

If the first grain is rotated through Θ about d, by definition the orientation of the second grain is obtained. The remaining two degrees of freedom are required to define the orientation of the boundary plane. Any experimental analysis of grain boundary structure is based on measurement of at least the first three of these parameters[7].

Grain boundaries are usually divided into "low angle" or "high angle" according to the misorientation between the crystal on either side. Numerous models of the structure of grain boundary have been put forward. The dislocation model[8] for a *tilt boundary* was considered as composed of an array of perfect edge dislocations parallel to a tilt axis which defines the misorientation Θ for one degree of freedom (see Fig. 5.12 in Sec. 5.3.2). If the lattice are rotated about an axis normal to the boundary plane, a *twist boundary* results. For small angle boundaries, the geometrical description in terms of dislocation is physically valid. The properties of such boundaries can be computed from dislocation theory and reasonable agreement has been found [9]

Fig. 3.17 shows a high resolution electron micrograph of a pure tilt boundary in a molybdenum bicrystal whose angle is 14° around the common [001] axis. The dislocation structure is clearly evidenced. As the angle of mismatch increases, the dislocations are so close as to loose their individual identities. The mismatch angle at which this happens ($10° - 15°$) serves as a dividing line to categorize grain boundaries as either low angle or high angle boundaries described in the literatures. For symmetrical tilt angle boundary, Glieter[10] has shown that geometrically the boundaries can be described as a uniform array and a superimposed non-uniform array of dislocations. For random high angle boundaries, it is difficult to postulate a suitable dislocation model.

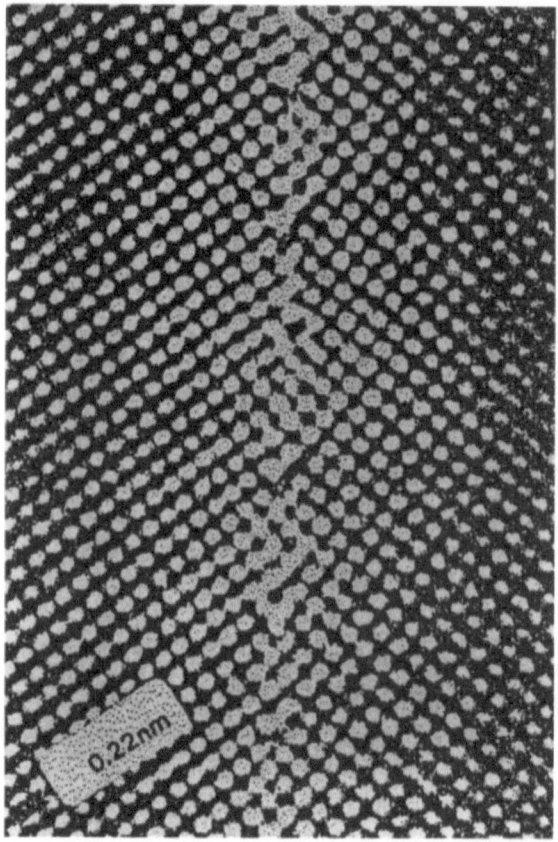

Fig. 3.17. Atomic imaging of a tilt boundary in a where x and y are integers representing the Cartesian molybdenum bicrystal (From Bourret, A. and Penisson, J-M, Jeol News 25E, No.1, 1987, p.2.)

High angle boundaries were originally thought to be amorphous films. However, not all high angle boundaries have an open disordered structure. There are some *special high angle boundaries* which have significantly high boundary mobilities and the low interfacial energy than the random boundaries. Major special high angle boundary models include coincidence site lattice (CSL) models[11, 12], O-lattice models[13], and plane matching model[14]. In the coincidence site lattice boundary (also referred to as *Kronberg-Wilson* (K-W) boundaries), two crystal lattices related by a rotation Θ about an axis <hkl> have certain lattice sites in common. The angle of rotation is

$$\Theta = 2\tan^{-1}(Ny/x) \qquad (3.11)$$

where x and y are integers representing the Cartesian coordinates of the lattice point joined to an origin and is related to the reciprocal of the density of common lattice sites Σ by

$$\Sigma = x^2 + N^2 y^2 \tag{3.12}$$

and $N = (h^2 + k^2 + l^2)^{-1/2}$ in cubic crystals. If Σ yields a even number, it is divided repeatedly by two to give an odd number. An example of the geometrical relationships giving rise to a $\Sigma = 5$ CSL is shown in Fig. 3.18

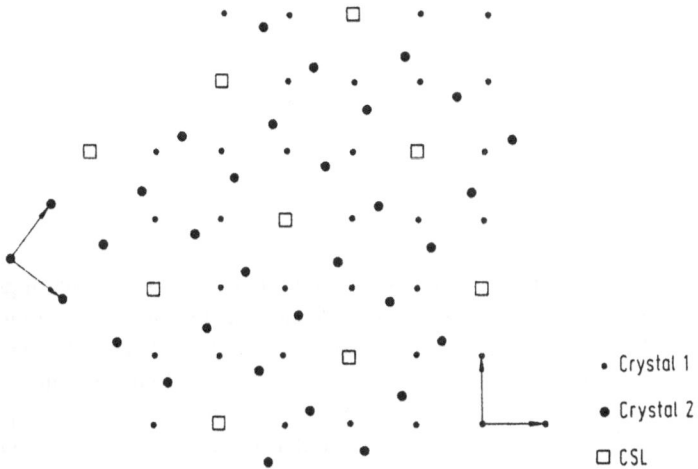

Fig. 3.18. Geometrical relationship of a $\Sigma = 5$ CSL grain boundary (From Randle,V. and Ralph, B., Revue Phys.Appl., 23(1988), p.501.)

A CSL exists only for precise values of the rotation axis and misorientation between the two grains. Some Σ-values for coincidence boundaries in cubic system with <100>, <110> and <111> rotation axis are shown in Table 3.3 If there is an angular deviation from an exact high density (low Σ) CSL, the CSL structure can be maintained by arrays of secondary intrinsic grain boundary dislocations in a manner analogous to the misorientation across a low angle boundary being conserved by primary grain boundary dislocations. If high values of Σ are allowed, any boundary may be described in terms of a CSL although the physical significance is small. The O-lattice model constitutes a generalization of the CSL model while the plane-matching model is more descriptive and based on optimum fit of closed packed planes. The grain boundary structure is a complex subject. Further references can be found in the work of Bolloman[15] and Gleiter and Chalmer[16].

In FCC crystals, a twin is connected to its origin by reflection in a {111} twining plane. Due to crystal symmetry of cubic crystals, this rotation relationship is equivalent to 60° <111> misorientation with $\Sigma = 3$ in the CSL model of grain boundary structure, although other misorientations different from $\Sigma = 3$ have also

Table 3.3 Some axis-angle pairs for CSL in the cubic lattice system. (Abstracted from Pumphrey, P.H. and Bowkett, K.M., Script.Met., 5, 1971, p.365.)

<100> axis		<110> axis		<111> axis	
Θ	Σ	Θ	Σ	Θ	Σ
22.62	13a	26.53	19a	27.80	13b
28.07	17a	38.94	9	38.21	7
36.87	5	50.48	11	46.83	19b
53.13	5	70.53	3	60.00	3
61.93	17a	86.63	17b	73.17	19b
67.38	13a	93.37	17b	81.79	7
112.62	13a	109.47	3	92.20	13b
118.07	17a	129.52	11	147.80	13b
126.87	5	141.06	9	158.21	7
143.13	5	153.47	19a	166.83	19b
151.93	17a			180.00	3
157.38	13a				

("a" and "b" distinguish pairs of different CSL's with the same Σ-values.)

been observed by Kopezky et al. in copper[17]. A twin boundary is a special high angle boundary with a structure similar to an intrinsic stacking fault. An atomic model of annealing twin formation was advanced by Gleiter[18]. It was observed that {111} planes of migrating grain boundary contain steps. Atoms are emitted from steps of the disappearing grains and absorbed out of the boundary into steps of the growing grains. This atomic movement is similar to the crystal growth from the vapour state.

Fullman and Fisher[19] suggested that the formation of twins lowered the total interfacial energy of the boundaries. The density of annealing twins (generally determined as the number of coherent twin interface per grain) has been found to vary with the stacking fault and prior cold deformation. The frequency of annealing twins increases with decreasing stacking fault energy. Increasing the amount of prior cold work increases the twin density. As there is no obvious reasons why stacking fault energy should decrease with increasing plastic flow, Form et al.[20] was of the opinion that the reoriented matrix in the twin provides a more rapid reduction in the dislocation density during recrystallization annealing and not because of the twinning process may lead to a reduction in interfacial energy. Annealing twins and deformation twins must be clearly distinguished. Deformation twins are produced by shearing operations on {111} planes and will be further discussed in Sec. 5.1, as one of the basic plastic deformation mechanism.

3.3.3 Coherency of Interphase Boundaries

Interphase boundaries are different from grain or twin boundaries as they separate regions of different compositions or crystal structures. Like a grain boundary, the boundary between two different phases can be specified by describing the orientation relationship between the lattices of the two crystals, and the orientation of the boundary itself. Furthermore, the interfaces may have a certain degree of coherence which determines the matching of lattices. Interphase boundaries can be divided based on their degree of atomic matching into three classes: *coherent, semicoherent* and *noncoherent*. There are two meanings of coherency. Firstly, coherency implies coincidence of the interface with a crystallographic plane or the matching of orientation relationship between the two phases such that

$$\{h_1k_1l_1\}_\alpha || \{h_2k_2l_2\}_\beta \quad \text{and} \quad [u_1v_1w_1]_\alpha || [u_2v_2w_2]_\beta$$

where α and β denote two phases separated by the interface. The second meaning of coherency refers to the elastic accommodation of lattice mismatch across the interface by straining one or both of the two lattices.

A schematic drawing of a strain-free and strained coherent interface is shown in Fig. 3.19. Strain-free coherent interfaces (Fig. 3.19a) arise when the two phases match perfectly at the interface plane. They are also called fully coherent interfaces. When the distance between the atoms in the interface is not identical it is still possible to maintain coherency by elastic stretching of the bonds as shown in Fig. 3.19b.

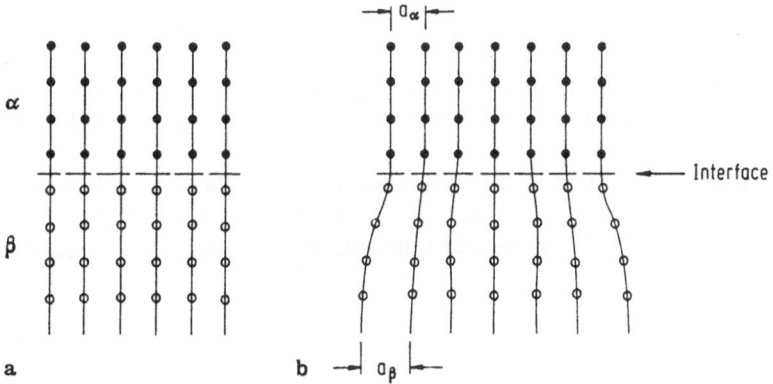

Fig. 3.19. Coherent interface with (a) no misfit strain and (b) coherency strain

Examples of coherent interface can be found in
(i) Matching of the close-packed planes of an FCC and HCP crystal such as the (0001) plane of the HCP silicon-rich κ phase and the (111) plane of the FCC copper-rich α-phase in Cu-Si alloys.
(ii) Formation of Ag precipitate in a supersaturated solid solution of Ag in Al, sharing a continuous FCC lattice with the matrix.

(iii) Precipitation of $MgFe_2O_4$ particles within MgO. The crystal structures of both phases differ but possess a common oxygen ion sub-lattice and an interface passing through the oxygen ions.

Usually the matching of the corresponding planes of two specially related crystals is only approximate. The misfit δ between two lattices α and β may be defined as

$$\delta = (a_\beta - a_\alpha)/a_\alpha \qquad (3.13)$$

where a_α and a_β are the lattice parameters of unstressed α and β phase respectively. The coherency interface produces a strain field in the matrix and hence an increase in the hardness of the system. This forms the industrial basis of precipitation hardening of Al-Cu alloy (see Sec. 6.3).

The energy of the boundary increases with the misfit parameter. When the misfit is large, energetically it will be favoured to replace the coherent interface with a semicoherent interface in which the deviations from coherency is taken up by the formation of dislocations to accommodate the misfit as shown in Fig. 3.20. The lattice misfit can be completely accommodated without any long-range strain fields by a set of edge dislocations. The number of extra planes N to be accommodated in the upper crystal in unit distance is

$$N = 1/a_\alpha - 1/a_\beta. \qquad (3.14)$$

The dislocation spacing $D = N^{-1}$, and

$$D = a_\alpha \cdot a_\beta/(a_\beta - a_\alpha) = a_\beta/\delta. \qquad (3.15)$$

When the misfit is 0.1, the misfit dislocations are separated by about ten atom spacing. The matching in the interface is almost perfect except around the dislocation core.

Noncoherent interfaces are those where perfect plane matching is lacking or no coincidence relationship exists (Fig. 3.20). Very little is known about the atomic structure of noncoherent interfaces which have features in common with random grain boundaries.

3.4 Quantitative Description of Microstructures

The constituents of materials determined with an optical or an electron microscope are called microstructures which reveal the following information: (1) the crystal parameters such as the nature of each phase, crystal defects, and boundaries, (2) the relative amounts of each phase, and (3) the morphology which includes the geometry, shape, size, and orientation distributions of the various phases. Metallographic aspects of crystal structures and defects have been mentioned in Sec. 3.2 and Sec. 3.3. In this section only the morphological features of the microstructures are discussed and crystallographic orientations are elaborated in Sec. 3.5.

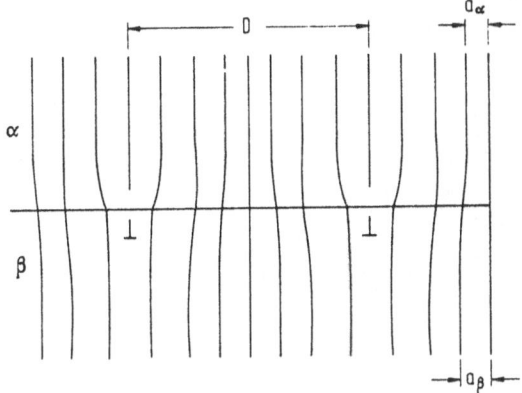

Fig. 3.20. A semicoherent interface with edge dislocations to accommodate the lattice misfit

The parameters of structure are three-dimensional. However, three-dimensional parameters are difficult to measure and the best way to obtain a quantitative description of microstructures is from two-dimensional parameters which are derived from three-dimensional models. An abstract description of geometrical parameters in two dimensions is given in terms of mathematical morphology[21]. Some basic topological features of microstructures are discussed below.

3.4.1 Basic Topological Features

A sketch of the single phase isotropic polycrystalline structure is shown in Fig. 3.21. The topological features of interest are

Table 3.4. Dimensionality for various topological features

Feature	Dimensions
Grain volume	3
Grain face	2
Grain edge	1
Grain vertex	0

Three-dimensional surfaces and edges appear as lines and points respectively in the two dimensional section, i.e. their dimensionality decreases by one. For a specimen having n_3 grains, n_2 grain faces, n_1 grain edges, and n_0 grain vertex, the topological features must satisfy the Euler-Poincare relationship,

$$n_0 - n_1 + n_2 - n_3 + \ldots (-1)^k n_k = 1 \tag{3.16}$$

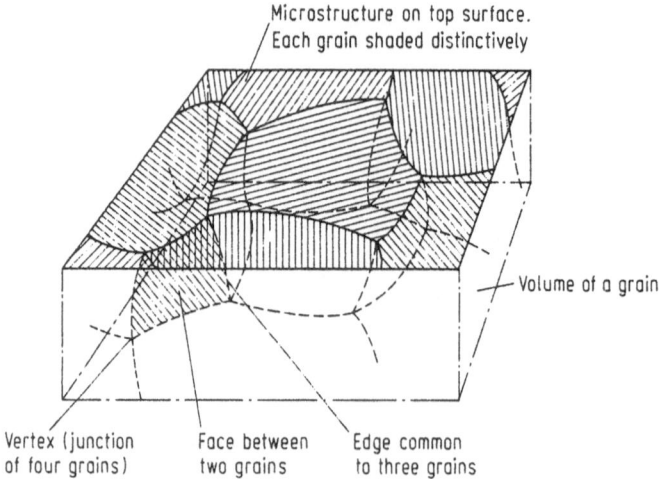

Microstructure on top surface.
Each grain shaded distinctively

Volume of a grain

Vertex (junction
of four grains)

Face between
two grains

Edge common
to three grains

Fig. 3.21. Topological features of a grain structure (From Guy, A.G., Introduction to Materials Science, McGraw-Hill, 1972, p.130)

where k is the dimensionality. The distribution of grain shape is characterized by the topological parameters such as the number of faces per grain or the number of edges per grain face in two dimensional sections. The shape correlation between neighbouring grains takes the form

$$\sum_{r=3}^{9}(6 - r)n_2(r) = n_3 + 1 \tag{3.17}$$

where r is the number of edges of a face of a polyhedron. The average number of edges per grain face in a typical alloy should thus be less than 6. Good correlation exists between the number of edges of a grain and the average number of sides of its neighbour.

Some basic parameters relating the features of a three-dimensional microstructures are volume fraction V_v and surface area per unit volume S_v. These parameters have been used to correlate the mechanical properties of materials and are easily derived from two-dimensional measurements. Two fundamental equations derived by Dehoff and Rhines[22] are as follows:

$$L_v = 2P_a \tag{3.18}$$

$$S_v = 2N_L \tag{3.19}$$

where

L_v = density of lines per unit volume
P_a = density of line intersection per unit area
S_v = density of surfaces per unit volume
N_L = density of surface intersections per unit length of the test line.

If structural features such as grains are elongated or otherwise non-uniform in various directions, care must be taken to collect test data on a sufficient number of randomly oriented test planes. On the other hand, the non-uniformity of the microstructure can be assessed with separate set of lines parallel to the x , y or z direction of the specimen. While the surface density S_v and the line density L_v can be obtained by measurements of their respective densities on two dimensional sections. The density of grains in a space-filling microstructure N_v has remained elusive. Much controversy has existed regarding whether grain size should be measured using average grain area on random two-dimensional section or the average intercept length of a random test line with grain surfaces. Various microstructure modelling techniques[23] have been proposed to correlate N_v from measurable two-dimensional parameters L_v and S_v.

3.4.2 Quantitative Metallography

In quantitative metallography or stereometric microscopy, informations on the spatial structure are obtained by analyzing two-dimensional sections. Lineal analysis or point counting rather than direct measurement of areas is generally used. The determination of the volume fraction of constituents in a two-phase alloy is illustrated in Fig. 3.22. Let A_i be the area of the β phase on a test plane, i.e. on one photomicrograph. The volume V_i of β phase in a layer of space of $(\Delta_x)_i$ thick is

$$V_i = A_i(\Delta_x)_i. \tag{3.20}$$

The volume fraction of the β phase V_v is given by

$$V_V = V_i/L^3 = \int_{x=0}^{x=L} (A_x/L^2)(dx/L) \tag{3.21}$$

where L represents the size of a cube volume being considered. In the expression, Ax/L^2 is the area fraction of the β phase at height x. The integral of (3.21) denotes the average value of the area fraction A_A, and hence, $V_V = A_A$. Similarly, the following expression relating the volume fraction V_V, area fraction A_A, lineal fraction L_L, and fraction of test points P_P can be derived:

$$P_P = L_L = A_A = V_V \tag{3.22}$$

In a single phase material, L_L then gives the average intercept grain size. For grains or interphase boundaries, the usual measure is S_v (surface area per unit volume). It will be easier to measure N_L (number of intercepts per unit length) which is related to S_v by (3.19). Other features such as grain edges are specified by L_v (length in unit volume) which is determined by P_a (points per unit area) in (3.18). A summary of the quantities that are available for measurement by means of quantitative microscopy and for which exact measurement relationships have been offered with mathematical proof is listed in Table 3.5. These topological properties

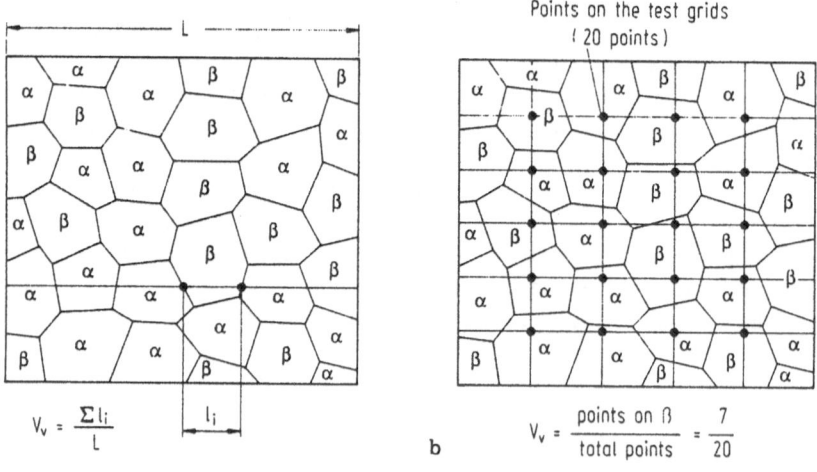

Fig. 3.22. Determination of volume fraction of β phase on a plane section. **a** Lineal analysis, **b** point counting method

Table 3.5. Fundamental Relationships of Quantitative Microscopy (From Rhines, F.N., Metal Progress, Aug., 1977, p.62.)

Geometric property	Symbol	Measuring Relationship	Nature of Sampling Device	Units
Length of line	L_V	$L_V = 2N_A$	N_A = point intercepts in unit area	cm^{-2}
Area of surface	S_V	$S_V = 2N_L$	N_L = line intercepts with unit length	cm^{-1}
Volume fraction	V_V	$V_V = A_A = L_L = P_P$	A_A = area in unit area L_L = length in unit length P_P = points per point	cm^{0}
Curvature of line	C_V	$C_V = \pi T_V$	T_V = planar tangencies in unit area	cm^{-3}
Torsion of line	R_V	$R_V = I_V$	I_V = inflections in unit volume	cm^{-3}
Curvature of surface	M_V	$M_V = T_A$	T_A = lineal tangencies in unit area	cm^{-3}
Number	N_V	Direct counting	Serial sections	cm^{-3}
Connectedness	G_V	$G_V = B_V - k_V + 1$	B_V = branches K_V = nodes in unit volume	cm^{-3}

of an irregular continuum are all structure-insensitive parameters and hence they can be used to correlate with properties of the material being studied.

The sampling techniques in quantitative microscopy are often time consuming and laborious. Sometimes approximate feature determinations are obtained by comparison with standard charts such as quantity of inclusions in steel and grain size measurement. A rapid and accurate measurement of complex parameters, especially size distribution often requires the use of *quantitative image analysis* (QIA). A quantitative image analyzer consists of a microscope, a scanner, a computer and a monitor. The principle of detection is illustrated in Fig. 3.23. The scanner converts the optical image of the prepared sample formed by the microscope into an electrical signal whose magnitude depends upon the brightness or darkness of the segment of the field of view it is scanning. The detection of features (such as second phase particles , porosity, inclusions or fracture surfaces) is based upon the video level which corresponds to the grey value of the features. These video levels are compared with an operator selected video level called a threshold. The analyzer can be adjusted to select the features brighter or darker than the threshold level.

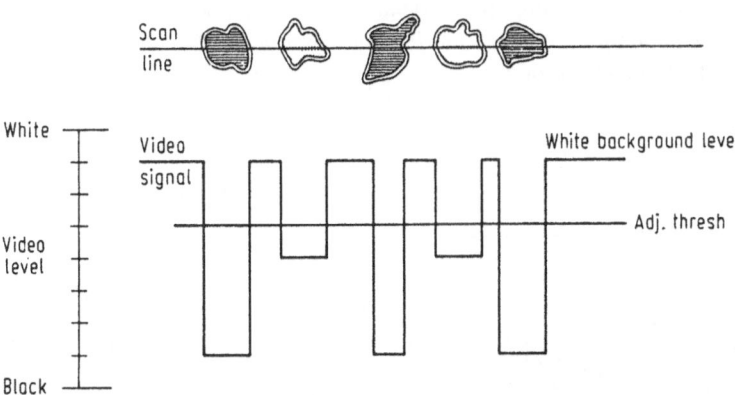

Fig. 3.23. Principle of detection of features in quantitative image analysis (After Coppenfeim, J.C., Microstructural Science, 9(1981), p.163.)

Different phases can also be separated by adding the video signal and the measurement of energy dispersive X-ray signals using a scanning electron microscope. The actual measurement of the detected features can be in various forms such as total area, percentage of area, projected length, or feature count. Some typical applications of QIA technique include grain size measurement, porosity content in sintered products, particle size distribution in dispersion hardened alloy, and percentage of graphite area in ductile iron. The hardware and software power of modern image processing systems such as high speed and high resolution image storages, complex image enhancement procedures and advanced detection algorithms have been developed to a high efficiency which has led to their intensive use in quantitative metallography.

3.5 Crystallographic Textures

3.5.1 Introduction

The word texture has different meanings in different scientific disciplines. In metrology texture refers to the degree of roughness or smoothness of a surface while in textile industry it refers to the woven arrangement of fibres or yarns. The meaning of texture is wider in geological studies which includes the size of the component of a crystals in a rock, their shape, distribution and orientation. Texture in many deformed rocks are expressed as mesoscopic planar or linear structures which allow the rock to be split more easily in some directions than others. The word crystallographic texture used here refers specifically to the orientation distribution of crystallities in a polycrystalline material which can be natural solids (e.g. minerals and rocks) or technological materials (e.g. metals, ceramics and crystalline polymers). In a polycrystalline material, each of the crystalline phase has an orientation that differs from those of its neighbours. It is unusual for the crystallities to have a random distribution of crystallographic orientations. The non-random distributions that occurred are called *preferred orientations* or *crystallographic textures* (hereby abbreviated as textures in the book).

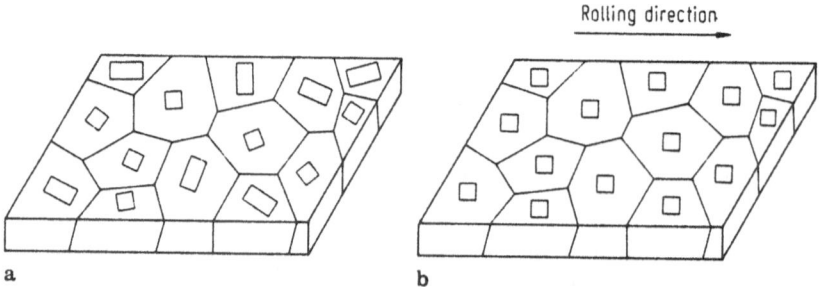

Fig. 3.24. Schematic diagram of a weak texture (a) and a strong texture (b)

A schematic diagram of a weak texture and a strong texture in sheet metal is illustrated in Fig. 3.24. The orientations of the crystallographic planes lying parallel to the rolling plane are indicated for each of the grain. Fig. 3.24b shows a very strong sheet texture with all grains similarly oriented (e.g. the cube face of each grain lies parallel to the sheet plane and the cube edge points in the rolling direction). Such a sheet would behave almost like a single crystal.

Textures are developed during solidification, plastic deformation, annealing and phase transformation of a crystalline material and are very sensitive to the solid state processes. A major source of anisotropy of the physical and mechanical properties of a material is caused by the presence of textures. The anisotropy is desirable in the case of grain orientated magnetic materials or undesirable in the formation of earing in deep drawn cups. Texture control has become a much investigated area in the processing and manufacturing of many technological important materials such as sheets metals for automobiles, electrical steels for power transmission and polycrystalline superconductors.

Grain orientation can be determined by a variety of methods which include optical etch-pit techniques and diffraction by X-rays, electrons or neutrons. The X-ray method is most commonly used in the determination of textures of sheet materials and is based on the back reflection technique[24] or the transmission technique[25]. Readers who are interested in the experimental determination of grain orientations and textures should consult the text of Cullity[26], and Hatherly and Hutchinson[27].

3.5.2 Representation of Grain Orientation

To specify the orientation g of a grain in a polycrystalline material, a crystal coordinate system K_C (i.e. the local coordinate system) has to be chosen in each crystal. In cubic crystals, the three cubic axes are usually chosen. A sample coordinate system (i.e global coordinate system) has to be defined consisting of three prominent sample directions. In a rolled sheet, the reference directions are the rolling direction (RD), transverse direction (TD) and the normal direction (ND). The grain orientation is then defined by the rotation which transforms the sample coordinate system K_S into the crystal coordinate system K_C (Fig. 3.25).

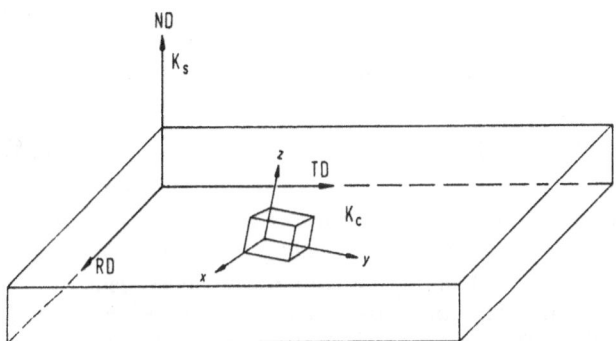

Fig. 3.25. Sample coordinate system K_s and crystal coordinate system K_c

The orientation of the crystal with respect to the sample axes can be characterized in many different ways. One of the most frequently used classical representations of an orientation is to specify the Miller indices (hkl) of the crystallographic plane parallel to the rolling plane and the indices $[uvw]$ of the crystal direction

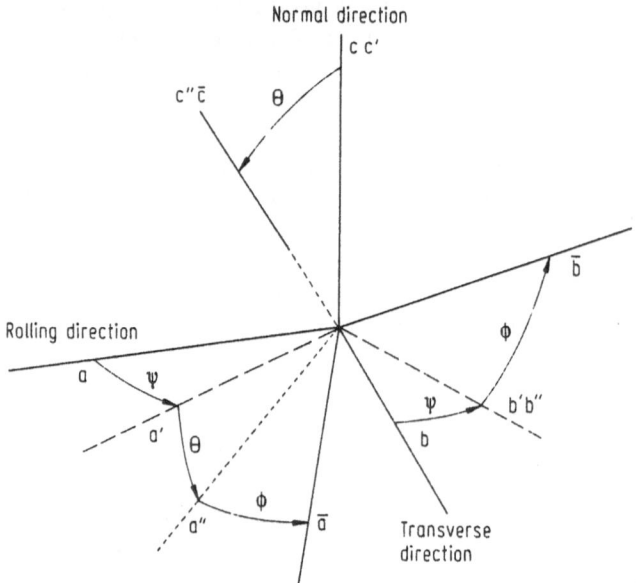

Fig. 3.26. The successive rotations through the Euler angles which relate the crystal axes to the sheet reference axes. (From Davies, G.J., et al. J. Appl.Cryst., 49, 1971, p.67.)

parallel to the rolling direction, i.e.

$$g = (hkl)[uvw].$$ (3.23)

The orientation g can also be denoted by a matrix

$$g = [g_{ij}], \quad g_{ij} = \cos <x_i, x_j'>.$$ (3.24)

The orientation matrix consists of the direction cosines of the axes x_i of the crystal coordinate system K_C with respect to the axes x_j' of the sample coordinate system K_S. Of the nine components in (3.24), only three are linearly independent.

Another way of representing the orientation is to use the Euler angles ψ, θ and ϕ as adopted by Roe[28]. These angles specify the orientation of the crystal coordinate system with respect to the sample coordinate system. The angles θ and ψ define the orientation of the crystallite c axis in the sample space and ϕ specifies the rotation of the crystallite around its own c axis. Axes a, b, c of the crystal coordinate system coincident with the sheet reference axes (Fig. 3.26) are rotated successively by ψ about c to a', b', c'; by θ about b' to a'', b'', c''; and by ϕ about c'' to $\bar{a}, \bar{b}, \bar{c}$ which represent the final position of the crystal axes. Bunge and Haessner[29] on the other hand have taken the θ rotation about the crystallographic

a-axis. The two sets of Euler angles are related by:

$$\psi = \varphi_1 - \pi/2$$
$$\theta = \Phi$$
$$\phi = \varphi_2 + \pi/2. \tag{3.25}$$

The Euler angles (ψ, θ, ϕ) specified here are equivalent to β, Θ and ϕ defined in Fig. 7.7 of Chapter 7.

In cubic crystals, the relation between the traditional descriptions (hkl) $[uvw]$ and the Euler angles are as follows:

$$h = -\sin\theta\cos\phi, \quad k = \sin\theta\sin\phi, \quad l = \cos\theta,$$
$$u = \cos\psi\cos\theta\cos\psi - \sin\psi\sin\phi$$
$$v = -\cos\psi\cos\theta\sin\phi - \sin\psi\cos\phi$$
$$w = \cos\psi\sin\theta. \tag{3.26}$$

The Euler angles can be plotted as three rectangular coordinates in a three dimensional space which may be called the *orientation space* or *Euler space*.

3.5.3 Pole Figures and Inverse Pole Figures

Textures are usually described by means of *pole figures*, *inverse pole figures*, and the *orientation distribution function (ODF)*. Pole figures are stereographic projections which show the statistical distribution of particular crystallographic directions of grains in an polycrystalline aggregate relative to the sample coordinate system K_S. In the stereographic projection, a sphere is imagined to surround the crystal and the normal to each face of the crystal projects from the centre of the sphere. The point at which each normal touches the sphere is projected back through the equatorial plane towards the south pole of the sphere. The equatorial plane is called the plane of projection and the points where these lines cut the equatorial plane are the *stereographic poles* of the corresponding crystal faces.

The orientation of a single grain can be represented by projecting its $\{100\}$ (or $\{110\}$ or $\{111\}$) poles on to the equatorial plane at the appropriate angular positions relative to the reference directions (Fig. 3.27a). When the $\{100\}$ poles are plotted, a $\{100\}$ pole figure is resulted (Fig. 3.27b). The clustering of poles shows the existence of a texture in a polycrystalline sample (Fig. 3.27c) and is plotted as density contours on the pole figure (Fig. 3.27d). Contours level greater than $1\times$ random imply a concentration of poles and those less than 1 times random a depletion in the poles concerned. If the resulting poles are distributed uniformly over the area of projection, there is no preferred orientation and the specimen is said to have a random texture which is rare in real materials.

There are two approaches to the description of pole figures. In the first description, regions of high pole density are matched with the single crystal orientation which is considered representative of the sheet texture. In a rolled sheet, this ideal orientation is represented by the Miller index $\{hkl\}$ which is parallel to the sheet

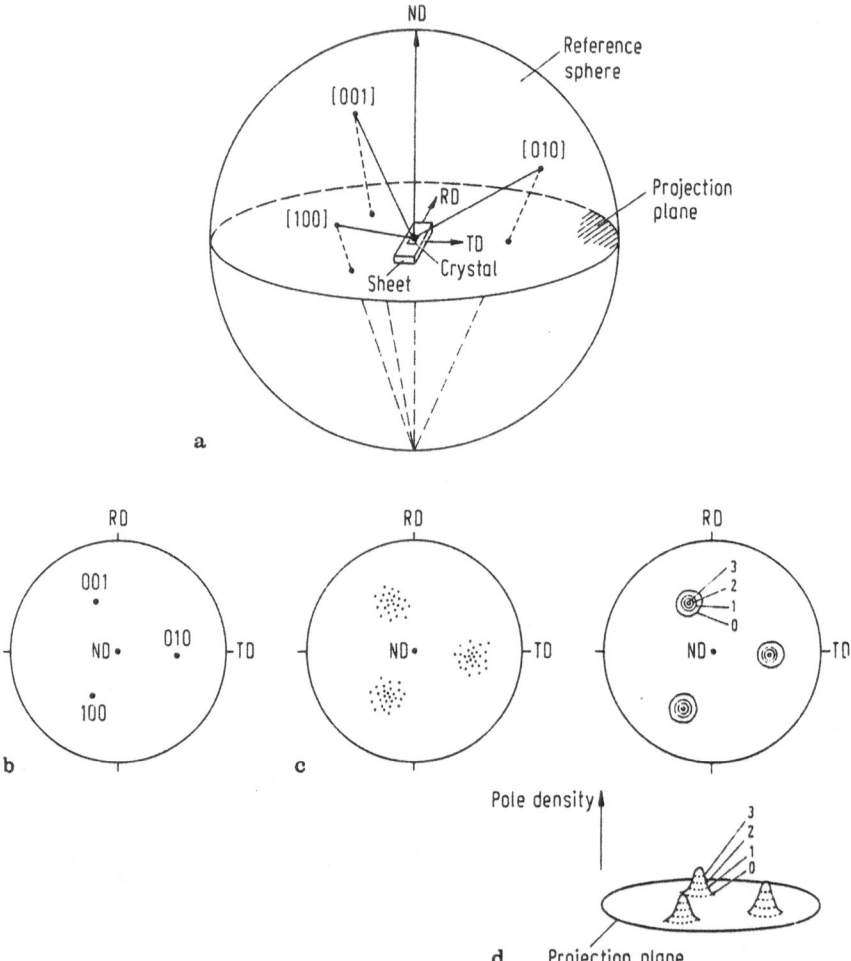

Fig. 3.27. **a** Projection sphere and reference direction; **b** stereographic projection of the <1 0 0 > poles of a single grain; **c** the clustering of {100} poles in a textured material; **d** contour map of pole density. (After Hatherly, M. and Hutchinson, W.B., An Introduction to Textures in Metals, The Institution of Metallurgists, Monograph 5, 1979)

normal and the other, $<uvw>$ which is parallel to the rolling direction. Depending on the symmetry of the crystallographic orientations involved, the number of equivalent orientations present in the pole figures of rolled sheet may be one, two or four. For orientations of maximum symmetry, a single set of poles will be needed whereas in orientations of minimum symmetry four equivalent sets will be required. The {111} pole figure of some common orientations are shown in Fig. 3.28a where only one single set is shown, and in Fig. 3.28b where the complementary equivalent orientations are all shown. In rolling, the texture has two reflection planes

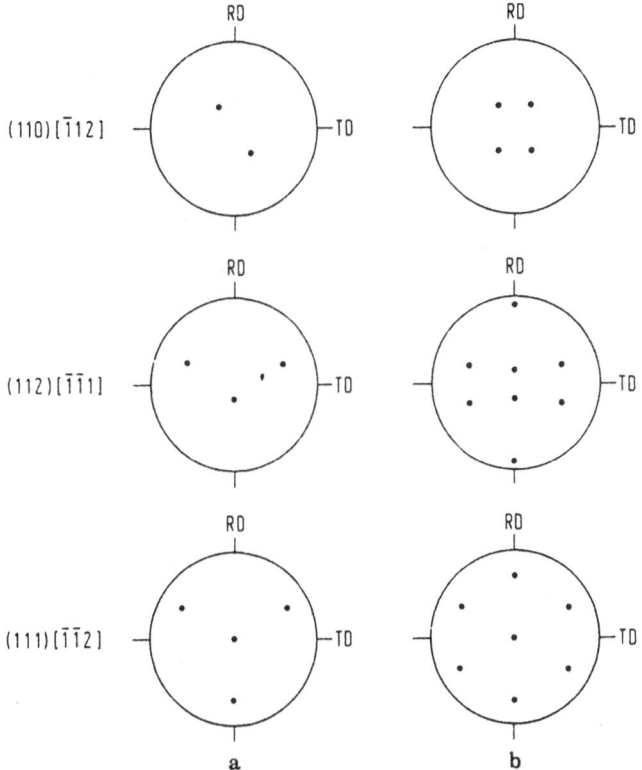

Fig. 3.28. {111} pole figures of some common orientations in cubic lattice. Showing a single set (a) and the symmetrical sets (b)

of symmetry where the four quadrants of the pole figures are equivalent. In wire drawing, the wire texture would exhibit a rotational symmetry.

In another approach advocated by Grewen and Wasserman[30], the pole figure can be represented by fibre axes and textures are defined by rotation about their respective fibre axis. The textures produced by axially symmetric deformation processes such as wire drawing usually exhibited rotational symmetry and such textures can be best described by defining the crystallographic directions $<uvw>$ which is parallel to the deformation axis. All possible orientations that can be developed by rotation about the fibre axis will be equally represented. Such textures are called *fibre textures*. When the pole figure is complex, the texture is often described by more than one set of ideal orientations or by mixtures of different fibre textures.

Besides pole figure data, inverse pole figure can also be obtained as experimental data of textures. In a pole figure, the orientation of some crystal directions relative to the specimen coordinate system in stereographic projection is shown, whereas in an inverse pole figure the orientation of the specimen axes relative to the crystal coordinate axes is shown. The inverse pole figure uses a unit stereo-graphic triangle as a reference frame with contour lines to show the frequency with which the various directions in the crystal coincide with the specimen axis under consideration. For sheet material, it is a common practice to determine inverse pole figures for specimen directions such as ND, RD, and TD of the rolling geometry. The (111) pole figure and the inverse pole figures for an age-hardened and cross-rolled Al-3%Cu alloy are shown in Fig. 3.29 for comparison. The major texture component present is the {011} <111> orientation.

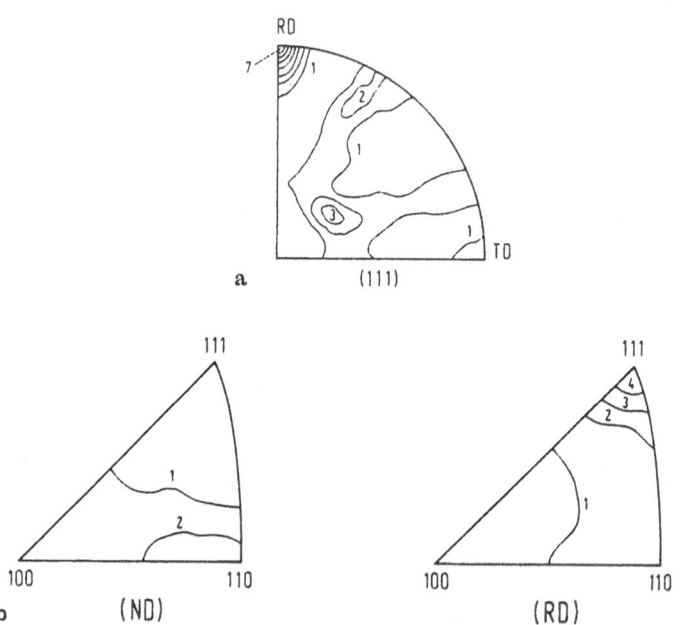

Fig. 3.29. Texture of an aged hardened Al-3%Cu alloy cross-rolled 80% in thickness as represented by (a) {111} pole figure and (b) inverse pole figure. (Lee, W.B., unpublished work)

3.5.4 Crystallite Orientation Distribution Function

Pole figures are a method of reducing the complex distributions of three dimensional orientations to a two dimensional orientation distribution. Ambiguities arise in the interpretation of pole figures if peaks of different orientations are not well separated in the projection plane. The three dimensional orientation distribution function $w(g)$ which is a full description of a texture represents the probability of a crystallite having an orientation described by three Euler angles. Each orientation of the crystallite is represented uniquely by a point in the Euler space. The texture of a polycrystalline sample then appears as a "cloud" of orientation points from which a continuous orientation function has to be constructed. If $\Delta V(g)$ is the volume of all the crystallites having orientations in the range from g to Δg and V is the volume of the whole sample then the orientation distribution $w(g)$ is given by

$$w(g)dg = \Delta V(g)/V. \tag{3.27}$$

The orientation distribution function is expressed as $w(\psi, \theta, \phi)$ in the Roe notation or as $f(\varphi_1, \Phi, \varphi_2)$ in Bunge notation. In the case of a random orientation distribution,

$$w_{random} = 1/8\pi^2 \tag{3.28}$$

and

$$f_{random} = 1. \tag{3.29}$$

The normalization factors are different and the corresponding integrals of the two functions over the orientation space are

$$\iiint w(\psi, \theta, \phi) \sin\theta \, d\theta \, d\psi \, d\phi = 1 \tag{3.30}$$

$$\iiint f(\varphi_1, \Phi, \varphi_2) \sin\Phi \, d\Phi \, d\varphi_1 \, \phi\varphi_2 = 8\pi^2. \tag{3.31}$$

To construct an orientation distribution function, the orientation of a large number of individual crystallites has to be measured by various methods such as electron diffraction. A more common method of determining the ODF is from the X-ray pole figure measurement. The measured intensity of the reflected beam is related to the (hkl) pole density q_{khl} which is an integral over the orientation distribution function. The mathematical methods of arriving at the orientation distribution function have been developed by Williams[31] based on vector method (also known as the biaxial pole figure method) and Bunge[32] and Roe[33] based on the series expansion method. In the vector method, the pole densities P_i are considered at discrete points in the experimental pole figure and the pole density is approximated by a linear expression of the form

$$P_i = \sum_{j=1}^{j} c_{ij} f_j \tag{3.32}$$

where f_j are unknown functional values corresponding to certain orientation points, and c_{ij} are purely mathematical coefficients. (3.24) leads to very large system of linear equations to be solved.

The second method which is based on series expansion is more widely used. In this method, the pole density data $q_{j(\chi,\eta)}$ from the experimental $(hkl)_j$ pole figure are expanded in a series of spherical harmonics as

$$q_j(\chi,\eta) = \sum_{L=0}^{\infty} \sum_{m=-L}^{L} Q_{Lm}^j \cdot P_L^m(\cos\chi) \cdot e^{-im\eta} \qquad (3.33)$$

where the angles χ and η are the polar coordinates which specify the orientation of the normal direction of the lattice plane $(hkl)_j$ with respect to the sample coordinate system, Q_{Lm}^j are the coefficients and $P_L^m(\cos\chi)$ is the normalized associated Legendre function. The orientation distribution function $w(\psi,\theta,\phi)$ is decomposed into a mathematically known function of the Euler angles but with unknown coefficients W_{lmn} of the form

$$w(\psi,\theta,\phi) = \sum_{L=0}^{\infty} \sum_{m=-L}^{l.} \sum_{n=-L}^{L} W_{Lmn} \cdot Z_{Lmn}(\cos\theta) \cdot e^{-im\psi} \cdot e^{-in\phi} \qquad (3.34)$$

where Z_{Lmn} is a generalization of the associated Legendre function. The relation between the ODF and the pole density data is then replaced by a system of linear equations between the corresponding coefficients W_{Lmn} and Q_{Lm}^j. Based on the generalization of the Legendre addition theorem[34] the following relation is obtained,

$$Q_{Lm}^j = 2\pi \left[2/(2L+1)\right]^{1/2} \sum_{n=-L}^{L} W_{Lmn} \cdot P_L^n(\cos\Theta_j) \cdot e^{-in\Phi_j} \qquad (3.35)$$

where the angles Θ_j and Φ_j are the polar coordinates of the normal direction of the lattice plane $(hkl)_j$ with respect to the crystal coordinate system. The coefficients Q_{Lm}^j are determined from the inversion of (3.33), i.e.

$$Q_{Lm}^j = 1/2\pi \int_0^{2\pi} \int_0^{\pi} q_j(\chi,\eta) \cdot P_L^m(\cos\chi) \cdot e^{im\eta} \cdot \sin\chi \, d\chi d\eta \qquad (3.36)$$

and are calculated for every value of the indices L and m and for every measured $(hkl)_j$ pole figure. In (3.34), the number of unknown W_{Lmn} depends on the different values of index n. If the harmonics series is expanded up to L, and n runs from $-L$ to L, there will be $2L+1$ values of n and W_{Lmn}. This would require $2L+1$ pole figures to solve the same number of equations. Due to crystal symmetry, the number of pole figures required is much less. The orientation density is finally synthesized for every required orientation (ψ,θ,ϕ) according to (3.34). Instead of the infinite series as shown in (3.34), all series expansion method must be terminated at a finite value of L. For a good approximation of the ODF (i.e. error $< 15\%$) the series may be expanded up to $L = 22$. The error introduced by truncation of the series (i.e. reduction in the height of maxima in the recalculated pole figure of the ODF) can be estimated by calculating the square deviation integral between the measured pole density and the recalculated pole density.

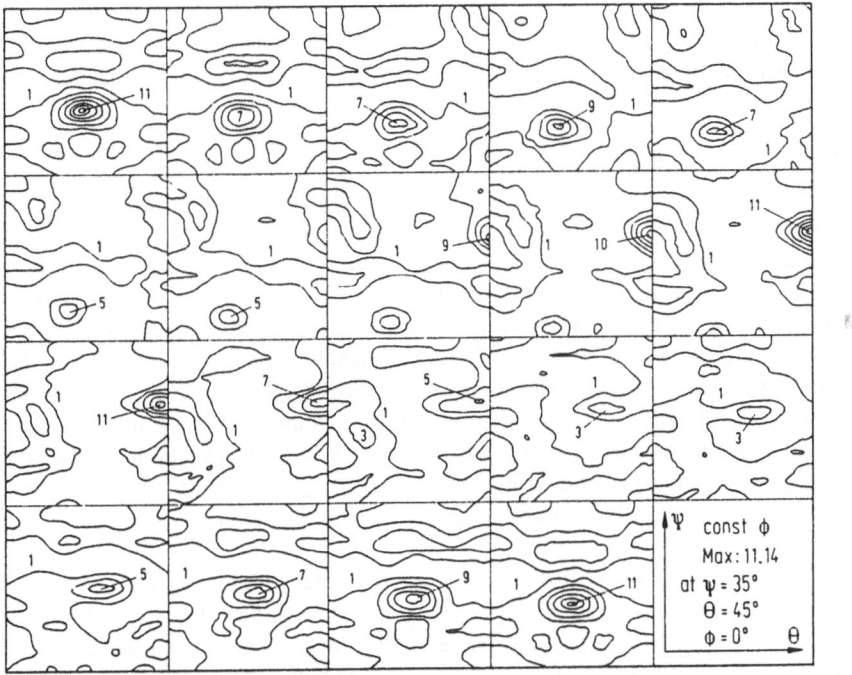

Fig. 3.30. Orientation distribution for an aged-hardened and cross-rolled Al-3% Cu alloy. (From Lee, W.B. unpublished work)

Results of ODF analysis are normally represented by plotting the probabilities in Euler space in 5° steps for the angular variables (ψ, θ, ϕ) and taking constant sections of one of the Euler angles, usually constant ϕ sections. The orientation distribution for the same aged-hardened and cross-rolled Al-3%Cu alloy (as illustrated in Fig. 3.29) calculated from three pole figures is depicted in Fig. 3.30 which shows a maximum intensity at $\psi = 35°$. $\theta = 45°$, and $\phi = 0°$ (i.e. at $\{011\}<111>$ orientation). By integration of the distribution function over a given region, the volume fraction of crystallites having an orientation within that region of the Euler space can be determined. The calculation of the ODF from experimental pole figures requires a large amount of calculation and the mathematical techniques are often uncommon to engineers who want to use the method. However, the basic mathematics has been translated into computer programmes and commercial softwares are available for routine laboratory analysis of textures of engineering materials.

A detailed description of texture analysis in materials science can be found in the work of H.J.Bunge[35].

3.6 Evolution of Deformed Microstructures

A basic description of the plastic behaviour of materials is given by its plastic flow curve. At small strain this behaviour is usually modelled by a simple parabolic equation such as

$$\sigma = K \epsilon^n \qquad (3.37)$$

where σ is the equivalent stress and ϵ is the equivalent strain. The assumption of a unique equivalent flow curve can be expressed as the assumption of a unique continuous series of "deformed state". Macroplasticity uses such equivalence criteria derived from generalizations of the observed macroscopic plastic behaviour of materials at small strains. At large strains, these equations are not adequate and the modelling cannot be done by the phenomenological descriptions alone. The complexities are caused by the interactions between the microstructure and the plastic flow in which the pattern of plastic flow is controlled by the microstructure and the deformation can rearrange the microstructure. In this section, a qualitative account of the evolution of cold-worked structure is given. The deformation mechanism responsible for the evolved deformation structures will be discussed in later chapters.

Plastic deformation is highly inhomogeneous not only on the specimen scale but also on the grain scale. Comparison in different zones of a single grain often reveals strain heterogeneities. The evolution of microstructures during plastic deformation of crystalline materials has been extensively studied over the last decades. Much of the earlier structural investigations were limited to low strains ($\epsilon < 0.5$) and foils for transmission electron microscopy (TEM) studies were from the rolling plane sections only. The interpretation of heavily deformed structure is difficult when viewed through the rolling plane section as the thickness of the deformation sub-structure is often smaller than the thickness of the foil. It was not until quite recently that a better understanding of the microstructural changes of materials at large strain was achieved by TEM investigation on planes normal to the rolling plane on heavily rolled samples.

The cold-rolled microstructure of metals has been extensively studied, the evolution of which is influenced by the stacking-fault energy (SFE), the temperature and rate of deformation, and the initial grain size. The deformation structures of high to medium SFE FCC metals have been reviewed in details by Gil.Sevillano et al.[36]. During deformation, dislocations interact and tend to cluster into arrangements of high dislocation density which are separated by regions of low density. In medium to high SFE metals rolled to low thickness reduction ($< 10\%$), the structure consists of an array of *dislocation cells* as shown in Fig. 3.31. The structure of a cell is that of a fairly dislocation free-interior surrounded by walls of dislocations. As straining proceeds, the grain interior can be delineated by *dense dislocation walls* (DDW) which extend across a significant fraction of the grain. The average misorientation across such DDWs is larger than across the ordinary cell walls formed within the volumes delineated by the DDWs.

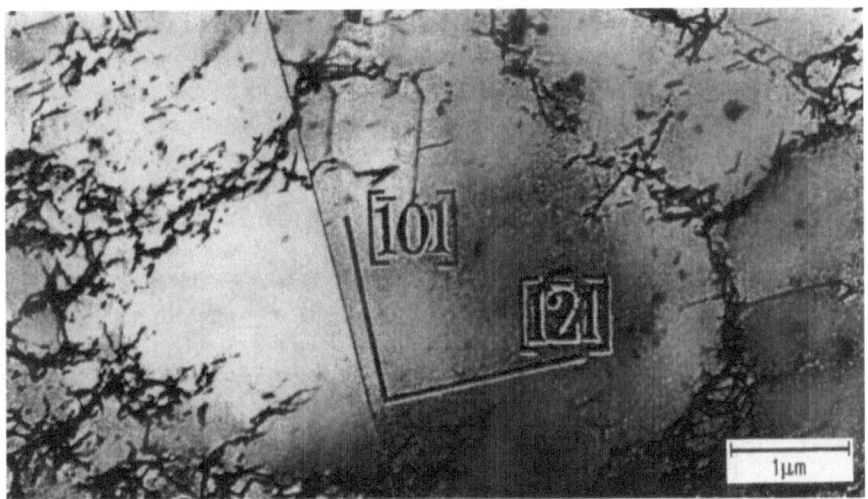

Fig. 3.31. Dislocation walls and dislocation free region in deformed copper (From Steeds, J.W., Proc. R.Soc., A295, 1966, p.343.)

Cells could form by a relaxation process[37] when dislocations are free to arrange themselves. Cell wall thickness decreases with strain while cell interiors appear progressively cleaner. The progressive sharpening of the cellular walls leads to the creation of subgrains which are cells that have undergone some sort of recovery. Cells form most readily when screw dislocation are able to cross slip easily. The tendency to form cellular structure thus increases with the SFE of metals, and the cell size decreases with strain but reaches a limiting value at a reduction around 30%[38]. At low strains, the misorientations between neighbouring cells are of order of 1° and rise to about 10° at higher strains. Once a cell structure is developed the pattern of flow becomes influenced by the microstructure, and the material would deform like composites as the cell walls block the cross-slip that creates the dislocation cell. The details of cell formation are still obscure, and the complex dislocation configurations cannot in general be predicted quantitatively on the basis of the minimum energy principle alone.

For reductions between 10% to 65%, the slip processes that lead to the development of the elementary structure cease. Another feature known as a microband (MB) is developed[39]. Microbands are elongated plate-like regions of material bounded by dislocation boundaries, and contain a high dislocation density. The appearance of MBs corresponds to "slip bands" observed by optical microscopy and to particular strain markings in metallographic specimens. Their formation involves the cooperative movement of dislocations on planes that are much more closely spaced ($< 0.008\mu$m). These MBs appear to develop from DDWs and lie parallel on $\{111\}$ planes. With increasing strain, the number of DDWs and MBs increases and intersecting sets of DDWs and MBs can form a parallelogram structure. The subdivision of grains into regions delineated by DDWs and MBs has been related by Hansen[40] to the slip pattern in grains where individual grains deform by fewer than five independent slip system, whereas the formation of equiaxed cells is favoured at early stage of homogeneous slip (Fig. 3.32).

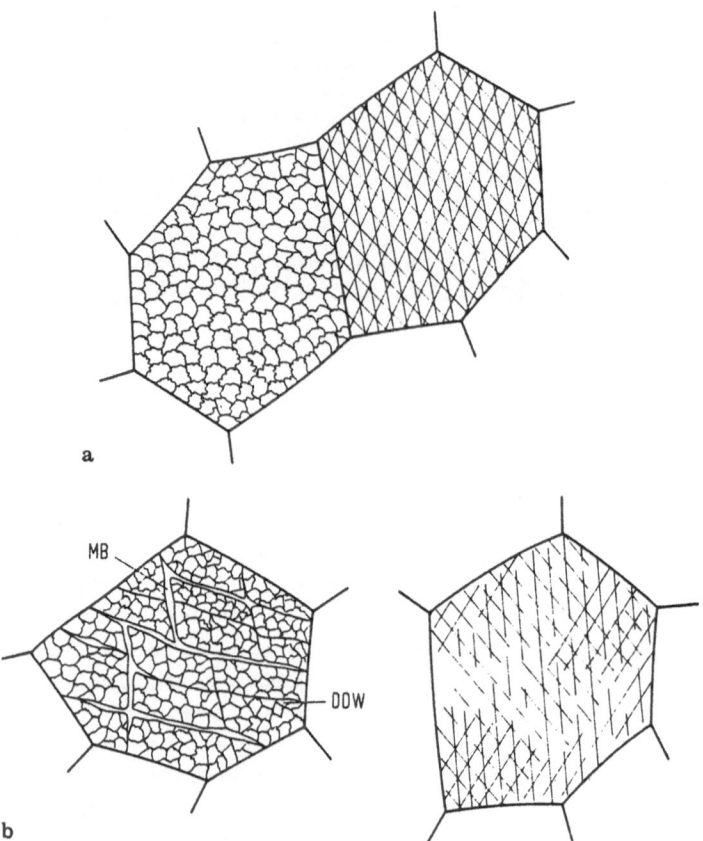

Fig. 3.32. Dislocation structure and slip line pattern in grains. **a** Homogeneous slip. **b** inhomogeneous slip (From Hansen, N., Materials Science and Technology, 6(1990), p.1039.)

In both polycrystals and single crystals, subdivision of grains into relative homogeneous regions separated by *transition bands*[41] is also observed. The formation of transition bands resembles to certain extent the formation of microbands although on a large scale. A grain can split in zones with different crystallographic orientations. These different oriented zones are separated by "transition bands" which consists of layers of elongated subgrains. The mutually misoriented fragments of the original grain are called *deformation bands*. Deformation bands arises when two zones of a grain rotate towards different orientations as result of local difference in stress state or different combination of slip systems to yield the imposed strain state. The tendency for formation of deformation bands is greater in polycrystalline specimen due to the constraints imposed by neighbouring grains.

Cellular structure is not observed in metals of low SFE such as brass. Dislocations completely dissociate and cross slip is inhibited. From the early beginning, the deformation substructure shows a banded and linear arrays of dislocations, and bundles of faulted materials. Mechanical twinning (see Sec. 5.1) becomes the major mode at moderate strains. Many of the grains contain bands of fine twins that are distributed in a faulted structure of high dislocation density. According to Duggan et al.[42] the difficulty of slip in other than the twin composition plane leads to overshooting, and both the parent and twin structure remain coupled and rotate until the {111} twin is parallel to the rolling plane. A TEM micrograph of a heavily twinned region in a cold-rolled α-brass is shown in Fig. 3.33. The alignment of twin boundaries to the rolling plane is almost perfect at high rolling reductions.

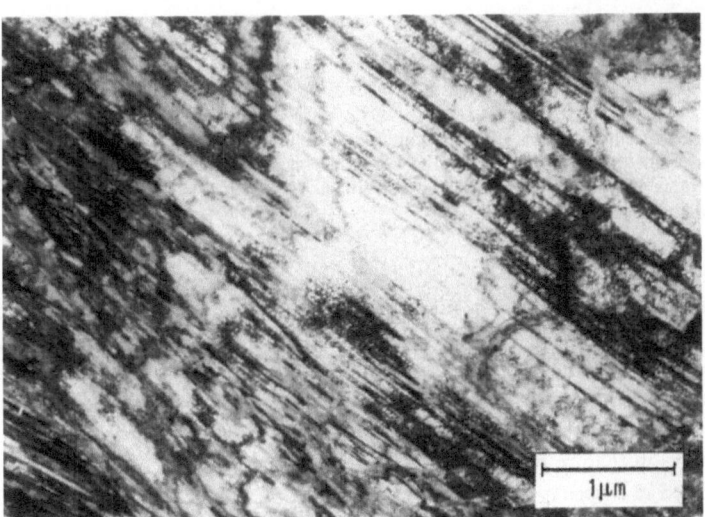

Fig. 3.33. TEM micrograph of a heavily twinned region in α-brass cold-rolled 84% in thickness

The resolved shear stress in the fine twin lamellae is not sufficient for slip and flow localization in the form of a shear band then develops. By 65% reduction, shear bands appear as dark etching lines under optical microscope and are confined within each grain (Fig. 3.34). The volume of shear bands increases with rolling strain and beyond 92% reduction, the microstructure is dominated by a profuse array of shear bands dividing the matrix into a series of isolated rhomboidal prisms. These bands are flat sheet-like structures composed of elongated crystallites from 0.02 to 0.1 μm in width, and usually inclined at some angles to the rolling plane. The frequently occurring shear angles are about 35° but other shear band angles have also been reported. Not all grains form shear bands. For example, grains with orientation near {110}<112> or {110}<110> do not show any large-scale organization[43]. The evolution of the deformation structure is highly orientation dependent.

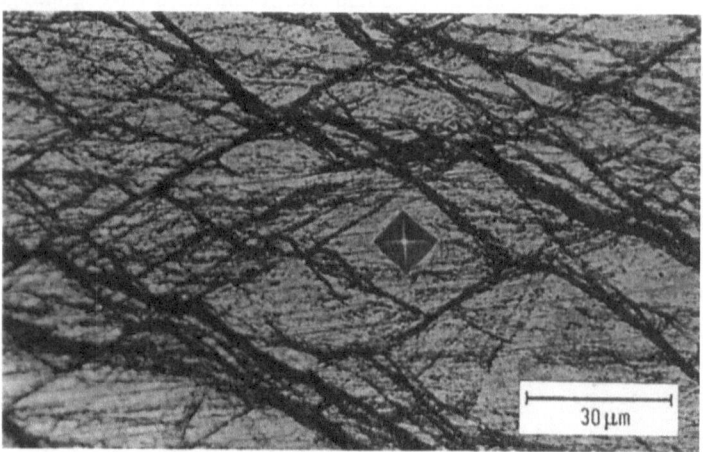

Fig. 3.34. Optical micrograph of α-brass after 65% cold rolling reduction showing shear bands (coarse-dark etching bands at about 35° to the rolling direction when measured in longitudinal plane) and microhardness indent in twinned volume. (From Lee, W.B. and Duggan, B.J., Metals Technology, March, 1983, p.85.)

At rolling reductions greater than 90%, the twinned volume is largely replaced by shear band material and new bands cease to develop. Homogeneous slip is thought to resume in the fine substructure and the microstructure transforms to a rough cellular structure similar to that of pure copper at the same strain. In medium to high SFE materials, shear banding also replaces microband formation as the dominant deformation feature but only at very high reductions[44]. Shear bands are frequently observed in a variety of metals and alloys deformed by rolling, extrusion, and forging. Further discussion on the mechanism of shear band formation is given in Sec. 5.2 of Chapter 5.

At large rolling deformation beyond about 60%, the original grain boundaries are no longer discernible and the structures scarcely distinguishable at the optical level. It becomes physically impossible for the original grains to maintain a shape imposed upon them by the macroscopic deformation process alone. The reduction in the average grain thickness may be less than the macroscopic value.

Such discrepancy is accounted for by the formation of new grains during cold deformation[45]. The initial high work-hardening modulus (as determined by the ratio between the ultimate tensile stress and the rolling strain) is associated with a high dislocation density and cell formation. Low work-hardening moduli are found when new grains are nucleated at regions of high dislocation density. The high density of vacancies generated during cold work after high strains may enhance the recrystallization rate but it seems difficult to establish as to the static or dynamic origin of the observed nuclei.

3.7 References

1. Adock, F., J.Inst. Metals, 27(1922), p.73.
2. Engel, P., Geometric Crystallography, D.Reidel Publishing Co., Holland, 1896.
3. Loretto, M.H. and Smallman, R.E., Defects in Electron Microscopy , Chapman and Hall, London, 1975.
4. Seeger, A., Dislocations and Mechanical Properties of Crystals (Edited by Fisher et al.), John Wiley, (1957).
5. Gallagher, P.C.J., Met. Trans., 1 (1970), p.2429.
6. Head, A.K., Humble,P., Clarebrough, L.M., Morton, A.J. and Forwood, C.T., Computer Electron Micrographs and Defect Identification, North Holland, 1972.
7. Randle, V. and Ralph, B., J.Mat.Sci., 21(1986), p.3823.
8. Read,W.T. and Shockley, Phys.Rev., 78(1950), p.275.
9. Gifkin, R.C., Mat. Sci.and Eng., 2 (1967), p.181.
10. Gleiter, H., Phys.Stat.Sol.(b), 45(1971), p.9.
11. Kronberg, M.L. and Wilson, F.H., Trans.Met.Soc.A.I.M.E., 85(1949), p.501.
12. Brandon, D.G., Ralph, B., Ranganathan, S. Wald, M.S Acta.Metall., 12(1964), p.813.
13. Bollowman, W., Phil.Mag., 16(1967), p.12.
14. Pumphrey, P.H., Grain Boundary Structure and Properties, Academic Press, (1976).
15. Bollman, W., Crystal Defects and Crystalline Interfaces, Springer-Verlag, New York, (1970).
16. Gleiter, H. and Chalmers, B., Prog.Material.Sci., 16(1972), p.1.
17. Kopezky, Ch.V., Noviko, V.Yu, Flionova, L.K. and Bolshakova, N.A., Acta.Metall., 33(1985), p.1985.
18. Gleiter, H., Acta.Metall.,17(1969), p.1421.
19. Fullman, R.L. and Fisher, J.C., J.Appl.Phys., 22(1951), p.1350.
20. Form, W., Gindraux, G. and Mlyncar, V., Mat.Sci., Jan.(1980), p.16.
21. Serra, J., Image Analysis and Mathematical Morphology, Academic Press, (1980).
22. Dehoff, R.T. and Rhines, F.N., Quantitative Microscopy, McGraw-Hill, New York, 1968.
23. Hansen, K.L., Acta.Metall., 27(1979), p.515.
24. Schulz, L.G, J.App. Phys., 20 (1949), p.1030.
25. Decker, B.F., Asp, E.T., and Harker, D., J.App.Phys. 19 (1948), p.388.
26. Cullity, B.D., Elements of X-ray Diffraction, Addison Wesley, (1967).
27. Hatherly, M. and Hutchinson, W.B., An Introduction to Textures in Metals, The Institution of Metallurgists, Monograph 5, (1979).
28. Roe, R.J, J.Appl.Phy., 37(1966), p.2069.
29. Bunge, H.J. and Haessner, J.Appl.Phys., 39(1968), p.5503.
30. Grewen, J. and Wassermann, G., Acta.Metall., 3(1955),p.154.
31. Williams, R.O., Trans.A.I.M.E, 242(1968), p.104.
32. Bunge, H.J., Z.Metallkd., 56(1965), p.872.
33. Roe, R.J., J.Appl.Phys., 36(1965), p.2024.
34. Rose, M.E., Elementary Theory of Angular Momentum, John Wiley and Sons, New York, 1957.
35. Bunge, H.J., Texture Analysis in Materials Science, Butterworths, 1982.
36. Gil Sevillano, J., van Houtte, P. and Aernoudt, E., Progress in Materials Science, 25(1981), p.69.

37. Maser, S., Seeger, A. and Theieringer, H., Phil. Mag., 22, 1963, p.515.
38. Swann, P.R., Electron Microscopy and Strength of crystals, Interscience, New York, (1963), p.131.
39. Malin, A.S. and Hatherly, M., Metal Science, August, 1979, p.403.
40. Hansen, N., Materials Science and Technology , 6(1990), p.1039.
41. Walter. J.L. and Koch, E.F., Acta.Metall., 10(1962), 10, p.1059.
42. Duggan, B.J., Hatherly, M., Hutchinson, W.B., and Wakerfield, P.T., Metal Science, August, 1978, p.343.
43. Duggan, B.J. and Lee, W.B., Proccedings of the 7th International Symposium on Metallurgy and Materials Science, 8-12 Sept., 1986, p.297.
44. Malin, A.S. and Hatherly, M., Metal Science, August, 1979, p.463.
45. Nutall, J. and Nutting, J., Metal Science, Sept.(1978), p.431.

4 Deformation Mechanisms I: Dislocations

4.1 Basic Mechanisms for Plastic Deformations

Plastic deformation, although often appears to be macroscopically homogeneous, is intrinsically heterogeneous when viewed from a meso and/or micro scale. Occasionally, *localized plastic flow* can even be observed in a macroscopic level. The examples include *shear banding*, localized *necking* and the localized plastic deformation near the *stress concentration zone* such as the crack tips. The shear banding mentioned above can take different appearances such as macroscopic slant shear band in plane strain tension, segregated Lüders bands in uniaxial tension and cruciform intensive shear bands occurred during metal upsetting, as depicted in Fig. 4.1(a), (b) and (c), respectively. For a round bar subject to uniaxial tension, the plastic deformation is concentrated within a portion of the specimen, and necking occurs under certain flow localization condition.

Even for macroscopically uniform plastic deformation, discrete slip with specific crystallography preference usually highlights the deformation process. The slip phenomenon, no matter occurred in a macro or meso scale, is dominated by the dislocation glide mechanism. The importance and the historical merit of the dislocation theory endorse our devotion of a systematic account to various aspects of the dislocation theory, encompassing from section 4.2 to section 4.5. The quantitative framework developed would also benefit the subsequent discussions on the other deformation mechanisms addressed in Chapter 5, as well as the hardening mechanisms in Chapter 6 and the plastic constitutive relations in Chapter 7.

Beside dislocations, plastic deformation can proceed via other mechanisms such as diffusion controlled creeping process, grain rotation and grain boundary sliding, twinning and phase transformation. The *creep* process becomes pronounced in the high temperature regime where the material particles is highly mobilized and the characteristic time of diffusion is greatly reduced. *Twinning* is another intergranular plastic deformation mechanism different from the dislocation mechanism. Its kinematic description and mechanics modelling will be addressed in section 5.1. Twinning is also one of the key issues in the coincidence grain boundary structure, as well as in the meso-arrangement of transformed phase in *transformation induced plasticity* (TRIP). Both dislocations and twinning attribute to the formation of shear bands, which will be dealt with extensively in section 5.2. *Grain rotation and grain*

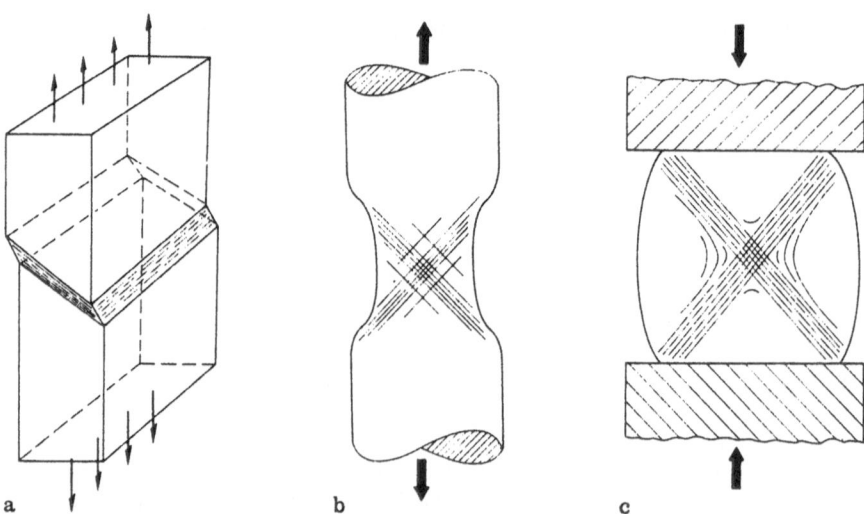

Fig. 4.1. Plastic deformation by macroscopic shear banding. a Macroscopic slant shear band formed in plane strain tension; b segregated Lüders bands formed in uniaxial tension; c cruciform intensified shear zone in metal upsetting

boundary sliding also play an important role in the plastic deformation mechanisms, especially for the so-called superplasticity behavior which attracts a great deal of recent attention and provides remarkable application values in the metal forming practices. This aspect of plastic deformation will be explored to some extent in section 5.3. Various features of the phase transformation plasticity, in combination with its potential applications in material toughening, will be examined in section 5.4, due to the ever-growing research activities in this area during the recent years.

4.2 Introduction to Dislocations

4.2.1 Origin of Dislocation Theory

In as early as the end of the nineteenth century, metallurgists observed that plastic deformation is caused by slip at certain crystallographic planes. The experimental observations at that time also indicated the existence of *macroscopic slip bands* (called as the *Lüders bands* by metallurgists). Subsequent calculations by Frenkel and his successors, however, revealed that the critical applied shear stress estimated for a *uniform* plastic slip along a specific crystallographic plane is two or even three magnitudes higher than the shear stress actually measured from the experiments. This extraordinary disagreement between the early uniform deformation model of plasticity and experimental measurements is largely due to the fact that the atomic bonds across the slip plane do not disrupt *simultaneously*. In reality, a variety of defects exist in the actual crystalline solids. Those defects not only perturb locally the translational invariance and point symmetry inherent in the crystal lattice, but also trigger *non-uniform* plastic deformation at a stress level far below the theoretical shear strength.

According to their geometric characteristics, lattice defects can be classified into three groups. Namely, (1) point defects: such as vacancies and interstitials; (2) line defects: mainly referred to dislocations; and (3) planar defects: stacking faults, grain boundaries, etc.. Among defects listed above, the line defects plays a unique role in the crystal plasticity theory.

In the eventful year of 1934, dislocation theory was proposed independently by Orowan, Polanyi and Taylor. The birth of dislocation theory brought hope to the stagnancy of crystal strength and deformation theory, and laid down the essential framework for the subsequent rapid development of mesoplasticity in the past half century. Taylor's dislocation model[1] is illustrated in Fig. 4.2. It closely resembles the dislocation configuration people observed twenty years later. The solid circles in Fig. 4.2(a) represent an array of atoms in an idealized simple cubic lattice, linked by solid lines signifying the bonds formed between adjacent particles. Under the application of an external shear stress τ, the atoms in the leftmost upper-half column (marked by A_1B_1) would shift slightly to the right, then its engagement with the lower-half column changes from the previous column C_1D_1 to the next column C_2D_2, as shown by the hollow circles and the dash lines in Fig. 4.2(a). In doing so, an extra upper-half atom column A_2B_2 is created and the particle re-organization is clearly illustrated by solid lines and solid circles in Fig. 4.2(b). The inserting upper-half column A_2B_2 cuts in like a knife with its edge touching down the slip plane denoted by the dot-dash lines in Fig. 4.2(a), (b) and (c). This *edge dislocation* is represented symbolically by ⊥, while the thick vertical stroke symbolizes the extra upper-half column of atoms and the slim horizontal bar denotes the slip plane. The above-mentioned bonding switch can proceed successively from the left to the right until the free surface on the right hand side is reached (as shown in Fig. 4.2(c)) to finalize a complete offset by plastic slip. As an interesting analogy, the process of dislocation movement is quite similar to the creeping of earthworm.

The above kinematic description on Taylor's dislocation model is associated with edge dislocations. The essential features of an edge dislocation are:

(1) In every step of dislocation motion, only a few atomic bonds need to be broken apart. The whole process of dislocation motion could be accomplished in a Domino fashion, rendering the low stress level required for dislocation movement.

(2) Dislocation is a line defect, the *dislocation line* divides a slip plane into slipped and unslipped part.

(3) After sweeping through the crystal lattice, one dislocation line produces a specific amount of plastic slip defined by a vector **b**.

(4) The **b** vector for an edge dislocation is perpendicular to the dislocation line.

Five years after the invention of dislocation theory, Burgers[2], a Dutch metallurgist, proposed another type of dislocation mechanisms called *screw dislocation*,

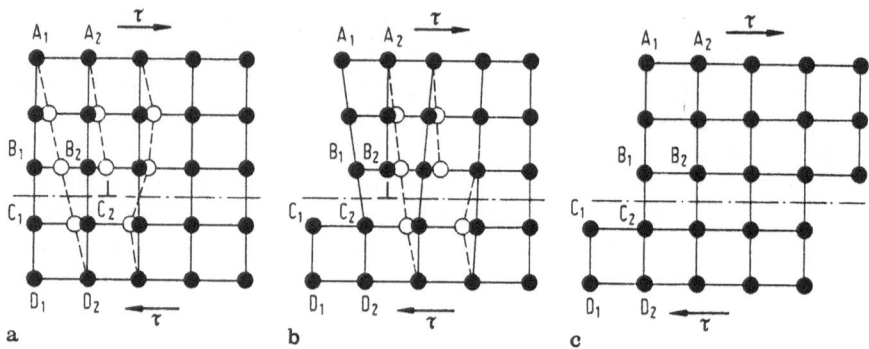

Fig. 4.2. Edge dislocation model of Taylor. **a** Bonding switch starts from the left to form an edge dislocation; **b** dislocation moves into the crystal lattice; **c** dislocation emerges from the surface on the right to create a complete offset

as exhibited in Fig. 4.3. The **b** vector for a screw dislocation (named after Burgers as the *Burgers vector*) is parallel to the dislocation line, and the atoms are packed along a helix spiraled around the pencil axis of the dislocation line.

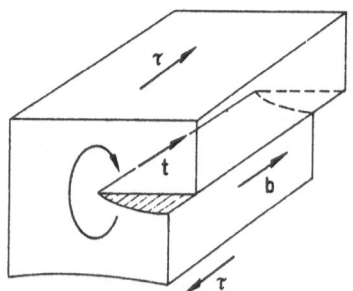

Fig. 4.3. Schematic representation of a screw dislocation

The invention of dislocation model is an excellent example demonstrating that human's intellectual thinking surpasses the experimental observations. The concrete, direct confirmation for the existence of dislocations was delayed to until early fifty's. However, such direct experimental observations upgrade the dislocation theory to a convincing science discipline. Major experimental techniques to observe dislocations are :

(1) surface etch pitting,
(2) decoration technique (decorated by impurity atoms such as AgCl to make dislocations visible under optical microscopy),
(3) direct observation under transmission electronic microscopy (TEM),

(4) X-ray diffraction technique,
(5) field-ion or field-emission microscopy.

4.2.2 Dislocation Kinematics

Tangential vector. The dislocation configuration introduced in the previous sub-section can be visualized as a smooth dislocation line L of arbitrary shape or a smooth dislocation segment L in the case of a dislocation network. Discussion on topological restrictions to a dislocation line will be deferred until later. A unit *tangential vector* **t** can be defined at any point of L and varies continuously along L as shown in Fig. 4.4. It is important to maintain a consistent positive direction of **t** along L, which must not be altered as soon as we fix it.

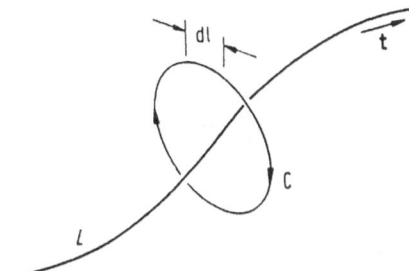

Fig. 4.4. Dislocation line and tangential vector

Burgers vector. Two different definitions of Burgers vector have been employed in crystallography. They would be examined here by the example of an edge dislocation. The positive direction of **t** is set to be perpendicular to the paper plane and directed inward.

(1) *FS/RH convention (or Frank convention).* According to FS/RH (finish to start / right hand) convention, a closed circuit is first drawn clockwise in the (distorted) crystal lattice containing the dislocation. Then, a corresponding circuit passing through the same lattice points is drawn in a reference idealized (non-distorted) crystal lattice. The latter circuit is probably unclosed. The Burgers vector **b** is defined as the vector closing the circuit in the reference lattice from the finishing point to the starting point, as shown in Fig. 4.5(a). The advantage for this definition lies on the fact that **b** so-defined is an exact lattice vector of the idealized crystal. But it suffers by the disadvantage that **b** cannot be defined in the crystal lattice which actually contains the dislocation. This convention is consistent with a continuum mechanics approach based on the reference configuration.

(2) *SF/RH convention (or Hirth-Lothe convention [3]).* According to SF/RH (start to finish / right hand) convention, a closed circuit is first drawn clockwise in

a reference, idealized crystal lattice. Then, a corresponding (unclosed) circuit passing through the same lattice point is drawn in the crystal lattice containing the dislocation. The closing vector $\hat{b}$ for the open circuit situated in the actual crystal lattice is termed *local Burgers vector*, as shown in Fig. 4.5(b). To make $\hat{b}$ so-defined have the similar direction as the b defined by FS/RH convention, we purposely choose the positive direction of $\hat{b}$ as from the starting point to the finishing point of the unclosed circuit. Though $\hat{b}$ can be obtained directly from the deformed crystal, it is usually not an exact lattice vector due to the lattice distortion near the dislocation line, and slight difference will occur for different circuits chosen. However, if the circuit chosen (termed *Burgers circuit* in the sequel) is far away from the dislocation core, one approximately has

$$\hat{b} \approx b \qquad\qquad (4.1)$$

The convention so-adopted is consistent to a continuum mechanics approach based on the current configuration.

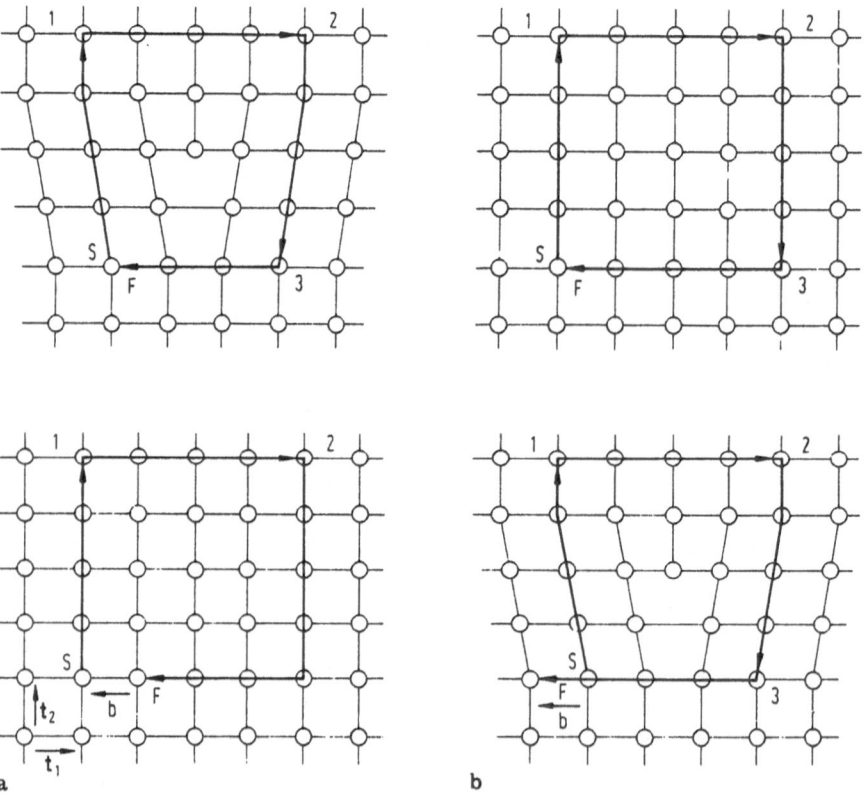

Fig. 4.5. Two definitions of Burgers vector

For the convenience of mechanics description, we will assume SF/RH convention, as well as the approximated equation (4.1) in the sequel. Such a definition can be expressed explicitly by the following mathematical formula

$$\mathbf{b} = \oint_C \partial \mathbf{u}/\partial l \, dl \tag{4.2}$$

where $\mathbf{u}$ the displacement vector, C an arbitrary closed circuit around L, and l the arc length measurement along circuit C, see Fig. 4.4 for an illustration. When following the SF/RH convention and equation (4.1), the sense of rotation for Burgers circuit C, and consequently the direction of tangential vector $\mathbf{t}$, obey the right hand convention. And this circuit should be taken outside the core-distortion zone. It then follows from Weingarten theorem[4] that the path independency implied in the integral of (4.2) will be justified provided the Burgers circuit C transverses only along the "good region" in the crystal lattice.

The above definition on Burgers vector is obviously relevant to the orientation of tangential vector $\mathbf{t}$, for the latter specifies the positive rotation sense of Burgers circuit. It is easy to prove that

$$\mathbf{t} \longrightarrow -\mathbf{t} \Rightarrow \mathbf{b} \longrightarrow -\mathbf{b}. \tag{4.3}$$

That is, the direction of the Burgers vector will be switched if we alter the positive sense of a dislocation line.

Curved dislocations. We now move on to the general situation of a *curved dislocation* as shown in Fig. 4.6. The slip plane with outward normal $\mathbf{m}$ is divided by the dislocation line and the exterior boundary into the shaded slipped part A_{slip} and unshaded non-slipped part, as portrayed in Fig. 4.6(b). It was shown by Nabarro[5] that the dislocation line can terminate at free surfaces, other dislocations, grain boundaries or sites of other defects, but *not within the crystal*. This property makes a clear division between the slipped and unshipped parts topologically possible.

Due to the arbitrariness of Burgers circuit C in equation (4.2), the whole curvilinear dislocation line L only attaches to a unique Burgers vector $\mathbf{b}$. This constant-valued vector $\mathbf{b}$ is in general inclined to the $\mathbf{t}$ vector, and the inclination angle changes pointwise along the dislocation line L, as shown in Fig. 4.6 (b). The edge dislocation and screw dislocation shown in Fig. 4.2 and Fig. 4.3 respectively correspond to the following particular cases

$$\text{Edge dislocation:} \quad \mathbf{b} \cdot \mathbf{t} = 0, \tag{4.4}$$

i.e., the Burgers vector is perpendicular to the dislocation line as indicated in point C of Fig. 4.6(b).

$$\text{Screw dislocation:} \quad \mathbf{b} \cdot \mathbf{t} = \pm|\mathbf{b}| = \pm b \tag{4.5}$$

indicating the Burgers vector is parallel to the dislocation line, like the point A in Fig. 4.6(b). The positive or negative sign in (4.5) is associated with the right or

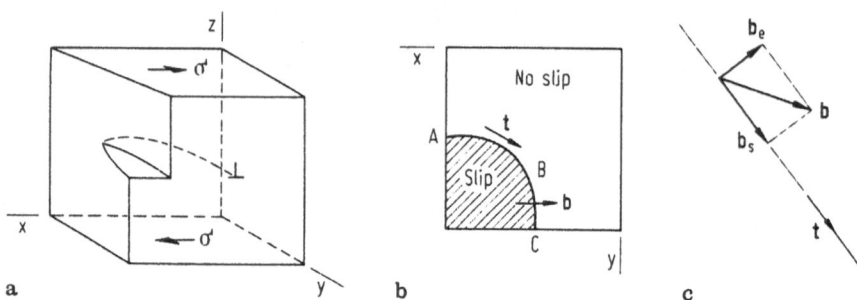

Fig. 4.6. Curvilinear dislocation (*b* is parallel to the *x*-axis in the figure). **a** A curved dislocation developed arround the corner of a cube; **b** projection on the slip plane; **c** decomposition of the Burgers vector

left screw dislocation, respectively. The readers can easily check that the screw dislocation illustrated in Fig. 4.3 is a right screw dislocation.

The dislocation characteristics at an arbitrary point B along L is featured by both edge and screw components, so comes the terminology of *mixed dislocation*. As demonstrated in Fig. 4.6(c), a generic Burgers vector **b** can be decomposed into a screw part $\mathbf{b}_s$ and an edge part $\mathbf{b}_e$

$$\mathbf{b} = \mathbf{b}_s + \mathbf{b}_e$$
$$\mathbf{b}_s = (\mathbf{b} \cdot \mathbf{t})\mathbf{t} \qquad \mathbf{b}_e = \mathbf{t} \times (\mathbf{b} \times \mathbf{t}) \tag{4.6}$$

where "×" denotes vectorial product.

Node. In certain circumstances, a single dislocation line can branch out into two lines. Alternatively, a triple joint of dislocation lines, referred to as *node*, will be formed as shown in Fig. 4.7. By means of the arbitrariness of Burgers circuit, computations of Burgers vector from two different Burgers circuits C and C' in Fig. 4.7 should lead to agreeable result. Consequently,

$$\mathbf{b}_1(\mathbf{t}_1) = \mathbf{b}_2(\mathbf{t}_2) + \mathbf{b}_3(\mathbf{t}_3). \tag{4.7}$$

Following the sign consistency convention of the tangential vectors from the preceding discussions, $\mathbf{t}_1$ in (4.6) directs toward node O, whereas $\mathbf{t}_2$ and $\mathbf{t}_3$ direct away from node O. If one redefines the positive directions of various tangential vectors appropriately, a *node theorem* of connected dislocation lines can be derived by the generalization of equations (4.3) and (4.7) :

If the positive directions of all tangential vectors $\mathbf{t}_i (i = 1, \ldots, N)$ *are assigned to point away from the dislocation node, then for a node linking N dislocation segments, the following relation holds,*

$$\sum_{i=1}^{N} \mathbf{b}_i(\mathbf{t}_i) = 0. \tag{4.8}$$

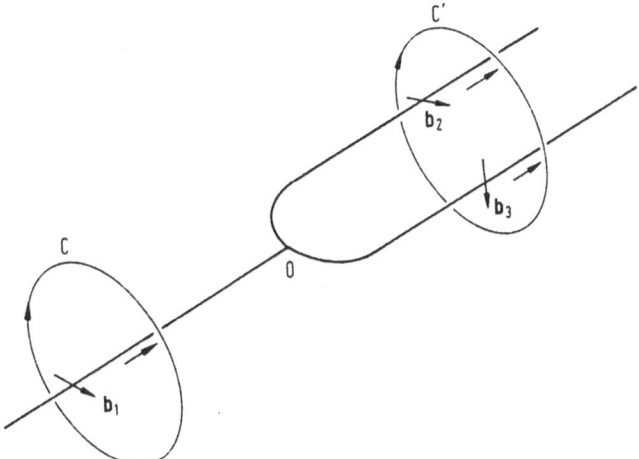

Fig. 4.7. Dislocation node

Similar to the Kirchhoff node theorem for an electric circuit, Burgers vector is conserved at a dislocation node.

Dislocation gliding plane. *Dislocation gliding plane* is defined as the plane spanned by the Burgers vector **b** and the tangential vector **t**. It is also called *slip plane* if massive dislocation glide occurs in that plane. The outward normal vector **m** characterizing this plane is

$$\mathbf{m} = (\mathbf{b} \times \mathbf{t})/|\mathbf{b} \times \mathbf{t}|, \tag{4.9}$$

where the positive direction of $\mathbf{b} \times \mathbf{t}$ is assigned by the right hand convention for counter-clockwise rotation from vector **b** to vector **t**. Equation (4.9) implies that the dislocation gliding plane can be uniquely determined only if the edge component of Burgers vector is non-zero. For the particular case of a pure screw dislocation, however, the dislocation gliding plane becomes non-unique. In fact, any planes containing **t** (and therefore **b**) could become the dislocation gliding planes. Such an indeterminate feature on the gliding behavior of a pure screw dislocation gives rise to the phenomenon of *cross-slip* as pictured in Fig. 4.8. For as long as the vectorial product of the slip plane normals $\mathbf{m}_1$ and $\mathbf{m}_2$ in Fig. 4.8 keeps parallel to **b**, the screw dislocation could cross-slip from the slip plane $\mathbf{m}_1$ to the slip plane $\mathbf{m}_2$. Cross-slip greatly promotes the mobility of dislocations.

Dislocation generated plastic deformation. Now we consider the plastic deformation caused by dislocations. An generic point on the slipped part A_{slip} of slip plane **m** is marked by **X**, as shown in Fig. 4.9. The displacements at **X** just above

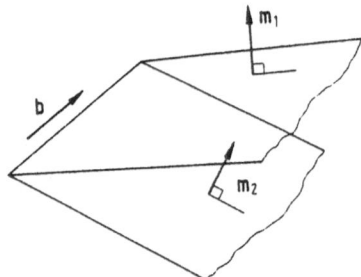

Fig. 4.8. Cross-slip

and beneath the slip plane are related by the following dislocation condition

$$\mathbf{u}(\mathbf{X}_+) - \mathbf{u}(\mathbf{X}_-) = \mathbf{b}, \tag{4.10}$$

where $\mathbf{X}_+$ or $\mathbf{X}_-$ represents point above or beneath A_{slip} which are infinitely close to $\mathbf{X}$. The above expression simply states that the displacement at $\mathbf{X}$ will dislocate a Burgers $\mathbf{b}$ when the point $\mathbf{X}$ is swept by a dislocation. If elastic lattice distortion is neglected, the above formula could be rewritten in terms of the deformation gradient $\mathbf{F}$ as

$$\int_{\mathbf{X}_-}^{\mathbf{X}_+} \mathbf{F} \cdot d\mathbf{X} = \mathbf{b} \qquad \forall \mathbf{X} \text{ in } A_{\text{slip}}. \tag{4.11}$$

where the integration path could be any circuit from $\mathbf{X}_-$ to $\mathbf{X}_+$. From (4.11), it is not difficult to show that the amount of plastic deformation generated by a mobile dislocation $\mathbf{b}$ on the slip plane $\mathbf{m}$ is, see reference [6]

$$\mathbf{F} = \frac{1 + \mathbf{b}\mathbf{m}\delta(\mathbf{m} \cdot \mathbf{X})}{1} \quad \text{if } \mathbf{X} \begin{array}{l} \text{in} \\ \text{not in} \end{array} A_{\text{slip}} \tag{4.12}$$

where $\delta(.)$ is the Dirac delta function and the origin of $\mathbf{X}$ is set at A_{slip}. Equation (4.12) describes the essential features of plastic deformation based on dislocation glide.

Dislocation generated slip deformation has the following features:
(1) Conservation of mass is observed, as a distinction from the climb motion of dislocation.
(2) The gliding is taken place in a specific slip system $(\mathbf{m}, \mathbf{b})$.
(3) The crystal lattice is preserved during dislocation glide.
(4) The dislocation glide in the slip plane is advanced *step by step*.
(5) Slip always starts from the inhomogeneities in the crystal lattice.

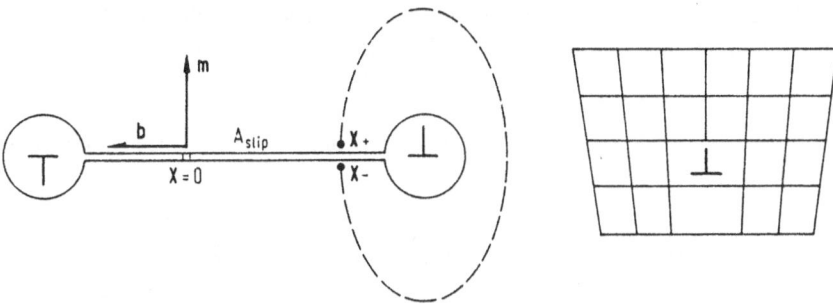

Fig. 4.9. Displacement above or beneath the slip plane

4.3 Elasticity Theory of Dislocation Fields

Far before the invention of dislocation theory, the linear elasticity field around a dislocations was solved by Timpe (1905) for an isotropic medium and elaborated later by Volterra (1907). A comprehensive account on the early works on elastic dislocation fields was also documented in the treatise of Love (1927). The research for dislocation fields in anisotropic crystal media was pioneered by Stroh (see reference [3]) through sextic formulation, and a systematic exploration on anisotropic dislocation fields was provided in reference [7].

4.3.1 Simple Dislocation Fields

Attention is first directed to the linear elastic field associated with a *straight dislocation* in an *isotropic* medium. Our discussion will be carried out from the pure screw and edge dislocations, and then generalized to the case of mixed dislocations.

Screw dislocation. Consider a screw dislocated cylinder of length L and radius R, and the dislocation line lies along the cylinder axis, as shown in Fig. 4.10. Cylindrical coordinates (r, Θ, z) are set up in the figure, with the dislocation line along the z-axis whose positive direction coincides with the tangential vector $\mathbf{t}$. The Burgers vector for the right screw dislocation in Fig. 4.10 is also along the positive direction of the z-axis. For such a dislocation configuration, the only nontrivial cylindrical displacement is the displacement in z-direction, and it can be represented by a helix surface

$$u_z = b\Theta/2\pi. \tag{4.13}$$

Equation (4.13) indicates the displacement in z-direction will increase an amount of b (the length of the Burgers vector $\mathbf{b}$) when the polar angle Θ rotates a complete circle (2π). One can easily identify the screw dislocation field to the anti-plane shear problem, and its only non-trivial cylindrical component of stress is

$$\sigma_{\Theta z} = Gb/2\pi r \tag{4.14}$$

From the above two formulae, the *line density of dislocation strain energy* (energy per unit length of dislocation line) becomes

$$W \equiv \Pi^L/L = \int_{r_o}^{R} (\sigma_{\Theta z}^2/2G)2\pi r dr = (Gb^2/4\pi)\ln(R/r_o) \tag{4.15}$$

where Π^L represents the self-energy of the dislocation line L. The symbol r_o in (4.15) denotes the radius of dislocation core. The above continuum mechanics analysis will break down for $r < r_o$, and the value of r_o is often comparable to the magnitude of b. The energy contained in the dislocation core should be calculated by particle kinetics or statistical mechanics approach, and this portion of energy is only estimated as from 10 to 15 percent of the total energy of a dislocation line. A description for such calculations will not be attempted here. The cutoff radius of dislocation field, R, is usually taken as half of the distance between the neighboring dislocations, where the stress fields arisen from the neighboring dislocations cancel each other out to realize the hypothetical analysis model shown in Fig. 4.10.

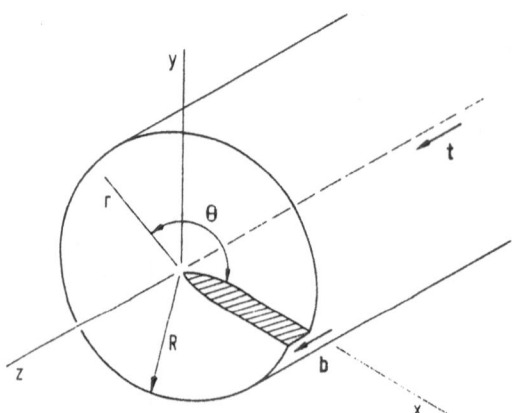

Fig. 4.10. Elastic field for a right screw dislocation

Equations from (4.13) to (4.15) reveal the following characteristics of the dislocation fields :
(1) The displacement fields are multi-valued.
(2) The strength of dislocation fields is proportional to b; while the line density of dislocation strain energy is proportional to Gb^2.

(3) Both stress and strain have the $1/r$ singularity.
(4) The strain energy is linear to a logarithmic factor $\ln(R/r_o)$. Finite length scales R and $r_o (R > r_o)$ are introduced to avoid unboundedness of the strain energy.

These features of screw dislocations will preserve for the general dislocation fields.

Edge dislocation. We next consider the same cylinder as shown in Fig. 4.10 but an edge dislocation with Burgers vector along the positive x-direction is embedded instead. The geometry of such a dislocated cylinder is drawn in Fig. 4.11 where the tangential vector **t** aligns along the positive z-direction. The solution for this plane strain problem was first attacked by Timpe in 1905, and its detailed solution procedure, as well as a complete documentation on the dislocation fields, can be found in the textbook *Theory of Elasticity* by Timoshenko and Goodier[8], pages 88 to 89. The self stress and displacement fields for edge dislocations under cylindrical coordinates are

$$\sigma_{rr} = \sigma_{\Theta\Theta} = -[Gb \sin\Theta]/[2\pi r(1-\nu)]$$
$$\sigma_{r\Theta} = [Gb \cos\Theta]/[2\pi r(1-\nu)]$$
$$\sigma_{zz} = \nu(\sigma_{rr} + \sigma_{\Theta\Theta})$$
$$\sigma_{rz} = \sigma_{\Theta z} = 0$$

(4.16)

and

$$u_r = (b/2\pi)\{-[(1-2\nu)/2(1-\nu)]\sin\Theta \ln r + \sin\Theta/[4(1-\nu)] \quad (4.17)$$
$$+ \Theta\cos\Theta\}$$
$$u_\Theta = (b/2\pi)\{-[(1-2\nu)/2(1-\nu)]\cos\Theta \ln r - \sin\Theta/[4(1-\nu)]$$
$$- \Theta\sin\Theta\}$$

for $r_o < r < R$.

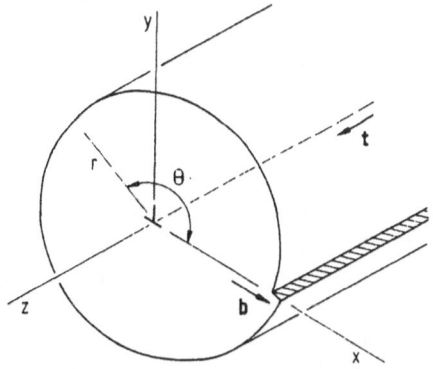

Fig. 4.11. Elastic field for an edge dislocation

Similar to the strain energy calculation for a screw dislocation, integration in the annular region of outer radius R and inner radius r_o can be performed to obtain the line density W for the strain energy associated with an edge dislocation

$$W = [Gb^2/4\pi(1-\nu)]\ln(R/r_o). \tag{4.18}$$

Comparing equations from (4.16) to (4.18) with their screw dislocation counterparts from (4.13) to (4.15), one finds
(1) The four basic characteristics summarized for screw dislocations also suit for edge dislocations.
(2) Beside the multi-valued terms independent of r, the displacements of edge dislocations contain additional single-valued terms. Among them, the terms proportional to ln$r$ usually dominate near the dislocation cores.

Mixed dislocation. The above results are generalized to arbitrary straight dislocations with Burgers vector inclined an angle β to the dislocation line, see Fig. 4.12. A Cartesian coordinate system (x,y,z) is set up there with z-axis directed along the positive direction of the tangential vector **t** of the mixed dislocation. The positive x-axis of the x-y plane (which is perpendicular to the z-axis) is chosen to coincide the edge component $\mathbf{b}_e$ of the Burgers vector. The lengths of the edge component $\mathbf{b}_e$ and the screw component $\mathbf{b}_s$ are respectively

$$|\mathbf{b}_e| = b_e = b\sin\beta, \quad |\mathbf{b}_s| = b_s = b\cos\beta. \tag{4.19}$$

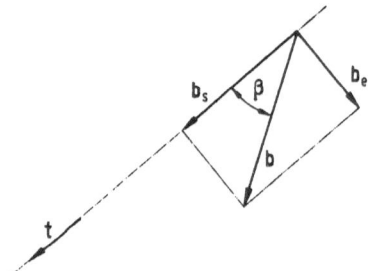

Fig. 4.12. Mixed dislocation and its decomposition into the edge and screw parts

The field variables for the mixed dislocation so-defined are functions of x and y only. For a linear elastic and isotropic medium, the displacement and stress fields for straight dislocations can be obtained by a superposition technique. That is, one simply substitutes the b_e in (4.19) into the places of b in (4.16) and (4.17); and substitutes the b_s in (4.19) into the places of b in (4.13) and (4.14). The interaction between screw and edge dislocations is absent in the above-mentioned stress and displacement fields, rendering a straightforward composition formula on the line density of strain energy associated with a mixed straight dislocation

$$W = E(\beta) \ln(R/r_o) \tag{4.20}$$

where $E(\beta)$ is called prelogarithmic energy coefficient. The $E(\beta)$ for an isotropic material is

$$E(\beta) = (Gb^2/4\pi)\{\cos^2 \beta + \sin^2 \beta/(1 - \nu)\} \tag{4.21}$$

and is proportional to the self inner-product of the Burgers vector. Formulae (4.20) and (4.21) are able to degenerate to formula (4.15) for a screw dislocation where $\beta = 0$ and to formula (4.18) for an edge dislocation where $\beta = \pi/2$.

4.3.2 Green Function for Linear Elasticity

The solutions derived in the previous subsection are only for straight dislocations in an isotropic medium. The dislocation fields in actual crystal lattice, however, reflect the following features :
(1) The actual dislocations are inevitably developed and moved in the *anisotropic* crystal grains, and the grain scale is considerably larger than the scale of dislocation-core.
(2) Dislocations inside crystal are usually taken the shape of loops.
     According to those demands, the elastic distortion fields for dislocation loops in general anisotropic media will be discussed in this and the next subsections. The solution for a straight dislocation can always be obtained from the dislocation loop solutions as a special case. The readers can also consult reference [7] for a more detailed account on this subject.
     Attention is focused on the equilibrium field in an infinite, linear elastic solid. The equilibrium and constitutive equations (presented in the index notation described in Chapter 2) of linear elasticity are

$$\sigma_{ij,j} + f_i = 0 \tag{4.22}$$
$$\sigma_{ij} = L_{ijkl}\epsilon_{kl} = L_{ijkl}u_{k,l}.$$

The fourth rank elasticity tensor L_{ijkl} in (4.22) has the following Voigt symmetry

$$L_{ijkl} = L_{jikl} = L_{ijlk} = L_{klij}. \tag{4.23}$$

The governing equation of linear elasticity (named as Navier equation) is obtained by combining the two sets of equations in (4.22)

$$L_{ijkl}u_{k,lj} + f_i = 0 \tag{4.24}$$

which is valid for any X_i in the three dimensional infinite space. If the body force term in (4.24) only consists of a concentrated unit force acted at $\mathbf{X}'$ and directed to X_m-axis, i.e.

$$f_i = \delta_{im}\delta(\mathbf{X} - \mathbf{X}') \tag{4.25}$$

where δ_{im} is the Kronecker delta, then the corresponding displacement field is termed the *Green function*, $G_{km}(\mathbf{X} - \mathbf{X}')$, of the linear elastic solid. Hence, the Green function is defined by the following equation

$$L_{ijkl} G_{km,lj} = -\delta_{im}\delta(\mathbf{X} - \mathbf{X}') \tag{4.26}$$

for any $\mathbf{X}$ and $\mathbf{X}'$ in the three dimensional infinite space, representing the displacement in the k-direction caused by a unit concentrated force acting at $\mathbf{X}'$ along the m-direction. For the isotropic case, equation (4.26) will lead to the famous Kelvin solution, see equation (4.39) in the sequel. For arbitrary anisotropic elastic solid, Equation (4.26) can be solved by various approaches listed below :
(1) Fourier transform,
(2) perturbation method for weakly anisotropic material,
(3) Fredholm method,
(4) Stroh formalism, see reference [3] for detailed description,
(5) Radon transform, as described in reference [7].

We now approach this problem by Fourier transform which readers are more likely to be acquainted with. The following relationships between a function $G(\mathbf{X})$ and its Fourier transformed phase function $g(\mathbf{K})$ are well-known

$$g(\mathbf{K}) = \int\!\!\!\int\!\!\!\int_{-\infty}^{\infty} G(\mathbf{X})\exp\{i\mathbf{K}\cdot\mathbf{X}\}\,d^3\mathbf{X}$$

$$G(\mathbf{X}) = (1/8\pi^3)\int\!\!\!\int\!\!\!\int_{-\infty}^{\infty} g(\mathbf{K})\exp\{-i\mathbf{K}\cdot\mathbf{X}\}\,d^3\mathbf{K} \tag{4.27}$$

where $\mathbf{K}$ stands for coordinates in the phase space. The volume elements $d^3\mathbf{X}$ in physical space and $d^3\mathbf{K}$ in the phase space are abbreviated for $dX_1 dX_2 dX_3$ and $dK_1 dK_2 dK_3$, respectively. The two formulae recorded in (4.27) are three dimensional forward and inverse Fourier transform formulae.

After these preparations, we are able to perform the Fourier transform to equation (4.26) and making use of the integration property of Dirac delta function to simplify the right hand side. The result turns out to be

$$\int\!\!\!\int\!\!\!\int_{-\infty}^{\infty} L_{ijkl}\, G_{km,jl}\exp\{i\mathbf{K}\cdot(\mathbf{X} - \mathbf{X}')\}\,d^3(\mathbf{X} - \mathbf{X}') = -\delta_{im}. \tag{4.28}$$

An algebraic equation in the phase space can be obtained if one carries out integration by parts twice on the left hand side of (4.28) and utilizes the asymptotically decaying property of G_{km} when away from $\mathbf{X}'$,

$$L_{ijkl}\, K_j K_l\, g_{km}(\mathbf{K}) = \delta_{im} \tag{4.29}$$

where g_{km} is the Fourier transform of Green function. It is observed from equation (4.29) that g_{km} is homogeneous in $\mathbf{K}$ of degree -2. Therefore, if a unit vector $\mathbf{z}$ parallel to $\mathbf{K}$ is introduced

$$z = \mathbf{K}/K \quad \text{where} \quad K = |\mathbf{K}| \tag{4.30}$$

then (4.29) can be rewritten as

$$M_{ik}\,g_{km} = \delta_{im}/K^2 \tag{4.31}$$

where

$$M_{ik} = L_{ijkl}\,z_j\,z_l \tag{4.32}$$

is the symmetric *Christoffel stiffness matrix*, also termed the *acoustic tensor*. By the positive-definiteness of elasticity tensor L_{ijkl} provided by thermodynamic argument, one can show that M_{ik} is also positive-definite. Which enables the inversion of (4.31) as

$$g_{km}(\mathbf{K}) = M_{km}^{-1}/K^2 \tag{4.33}$$

where M_{km}^{-1} is homogeneous in $\mathbf{K}$ of degree zero. The Green function can be calculated through inverse Fourier transform

$$G_{km}(\mathbf{X} - \mathbf{X'}) = (1/8\pi^3) \iiint\limits_{-\infty}^{\infty} (M_{km}^{-1}/K^2)\,\exp\{-i\mathbf{K}\cdot(\mathbf{X} - \mathbf{X'})\}\,d^3\,\mathbf{K}. \tag{4.34}$$

The above triple inverse integral can be simplified to a single definite integral as follows (whose derivation is illustrated in Appendix 4A)

$$G_{km}(\mathbf{X} - \mathbf{X'}) = [1/8\pi^2|\mathbf{X} - \mathbf{X'}|] \int\limits_{0}^{2\pi} M_{km}^{-1}(\pi/2, \phi).d\phi. \tag{4.35}$$

In the above formula, we write $M_{km}^{-1}(\mathbf{z})$ as $M_{km}^{-1}(\Theta, \phi)$ where Θ describes the angle between vector $\mathbf{z}$ and the straight line passing through $\mathbf{X} - \mathbf{X'}$. Thus, the integration route in (4.35) is along the unit circle lying on a plane perpendicular to the straight line running through $\mathbf{X} - \mathbf{X'}$, along which the rotation angle ϕ (defined with respect to a fixed datum line) varies from zero to 2π, as shown in the figure of Appendix 4A. From (4.35) and the symmetry of Christoffel stiffness matrix M_{ik}, the following symmetry properties of the Green function $G_{km}(\mathbf{X} - \mathbf{X'})$ can be established

$$G_{km} = G_{mk}; \quad G_{km}(\mathbf{X} - \mathbf{X'}) = G_{km}(\mathbf{X'} - \mathbf{X}) \tag{4.36}$$

i.e., the Green function is symmetric with respect to its indices and to the switch of its independent variables.

The solution (4.35) suits for arbitrary anisotropic elastic solid. For the special case of an isotropic medium, the elasticity tensor takes a simple form

$$L_{ijkl} = \lambda\delta_{ij}\delta_{kl} + G(\delta_{ik}\delta_{jl} + \delta_{il}\delta_{jk}). \tag{4.37}$$

It can be used to obtain a simple inversion expression of the Christoffel stiffness matrix

$$M_{ik}^{-1}(\mathbf{z}) = \{\delta_{ik} - z_i z_k/[2(1-\nu)]\}/G \tag{4.38}$$

which in turn yields an explicit formula for the Green function of isotropic material

$$G_{km} = (1/8\pi GR)\,\{2\delta_{km} - (\delta_{km} - T_k T_m)/[2(1-\nu)]\} \tag{4.39}$$

where $R = |\mathbf{X} - \mathbf{X'}|$, and $\mathbf{T} = (\mathbf{X} - \mathbf{X'})/R$ is a unit vector aligned with $\mathbf{X} - \mathbf{X'}$.

4.3.3 General Solution for Dislocation Loops

Attention is then focused on a spatial dislocation loop L situated in an arbitrary anisotropic medium, as depicted in Fig. 4.13(a). Owing to the continuity requirement of Burgers vector $\mathbf{b}$, it should be constant everywhere along the dislocation loop L. A smooth surface S_{cut} bounded by L but arbitrary otherwise can be attached to L, and it can be visualized as a hypothetical gliding plane. The jump value on the dislocation displacement $\mathbf{u}$ when across S_{cut} can be written down from (4.10) as

$$\mathbf{u}^+ - \mathbf{u}^- = \mathbf{b} \tag{4.40}$$

where $\mathbf{u}^+$ or $\mathbf{u}^-$ represents the displacement when approached from the upper or the lower side of S_{cut}, see Fig. 4.13. The stress and displacement fields associated with the dislocation loop L are denoted by σ_{ij} and u_i. To avoid the difficulty related to non-continuum dislocation core region, a domain V_{cut}, where the core-distortion region is cut off, is chosen and enclosed clockwise by surfaces S_*^-, S_{tube}, S_{cut}^-, S_{cut}^+, S_*^+ and S_{ext} as shown in Fig. 4.13(b). For any point $\mathbf{X}'$ in V_{cut}, the governing field equation is given by

$$L_{kpij}\, u_{i,j'p'}\,(\mathbf{X}') = 0 \tag{4.41}$$

where the subscripts with a prime such as j' and p' signify the differentiations with respect to X_j' and X_p'. This convention will be adopted hereafter.

For the same point $\mathbf{X}'$, its displacement in the i-direction caused by a unit concentrated force acting at $\mathbf{X}$ along the m-direction (i.e., the Green function) should satisfy the following equation

$$L_{kpij}\, G_{im,j'p'}\,(\mathbf{X} - \mathbf{X}') = -\delta_{km}\delta(\mathbf{X} - \mathbf{X}') \tag{4.42}$$

for any $\mathbf{X}$ and $\mathbf{X}'$ in the domain V_{cut}. Equations (4.41) and (4.42) provide us an apparatus to construct the displacement field u_i for dislocation loop from the Green function obtained in the previous subsection. Toward this end, equation (4.41) is first multiplied by $G_{km}(\mathbf{X} - \mathbf{X}')$, and then this product is subtracted by equation (4.42) multiplying by $u_k(\mathbf{X}')$. Which yields

$$L_{kpij}\left\{ u_{i,j'p'}(\mathbf{X}')\, G_{km}(\mathbf{X} - \mathbf{X}') - u_k(\mathbf{X})\, G_{im,j'p'}(\mathbf{X} - \mathbf{X}') \right\}$$
$$= u_m(\mathbf{X}')\delta(\mathbf{X} - \mathbf{X}'). \tag{4.43}$$

By means of the symmetry of L_{kpij} with respect to the index switch between (i,j) and (k,p) (as observed from the last equation of (4.23)), the left hand side of the above formula can be put into a gradient form. So one can proceed to integrate equation (4.43) in V_{cut} with respect to $\mathbf{X}'$. The integrals on the left and the right hand sides can be evaluated by the Gauss theorem and by the property of Dirac delta function, respectively. The displacement field for dislocation is consequently

$$u_m(\mathbf{X}) = \oint_{S'} L_{kpij}\, [u_{i,j'}(\mathbf{X}')G_{km}(\mathbf{X} - \mathbf{X}')$$
$$- u_k(\mathbf{X}')G_{im,j'}(\mathbf{X} - \mathbf{X}')]\, dS_p'. \tag{4.44}$$

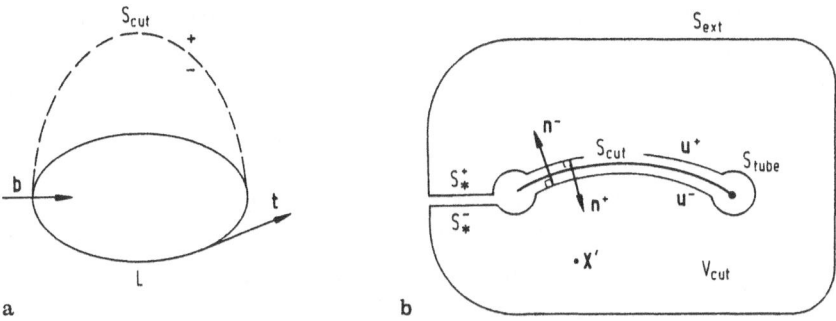

Fig. 4.13. Dislocation loop L, attached hypothetical slip plane S_{cut}, and integration surface enclosing V_{cut}

The closed surface S' (as the function of $\mathbf{X}'$) is composed of surfaces S_*^-, S_{tube}, S_{cut}^-, S_{cut}^+, S_*^+ and S_{ext}, as shown in Fig. 4.13(b).

It is noticed that the dislocation solution and the Green function are continuous across S_*, leading to the cancellation of the integrals along surfaces S_*^- and S_*^+. Furthermore, when S_{ext} expands to infinity and S_{tube} shrinks to the dislocation loop, the relevant integrals will vanish by the self-equilibrium nature and the asymptotic property of the dislocation solution (referred to subsection 4.3.1), as well as the asymptotic property of the Green function (as demonstrated in equation (4.35)). Therefore, only integrals along S_{cut}^- and S_{cut}^+ in the close-surface integration in (4.44) contribute to the displacement field. Consequently, (4.44) can be simplified to

$$u_m(\mathbf{X}) = -L_{kpij} \iint_{S_{cut}^+ + S_{cut}^-} u_k(\mathbf{X}')G_{im,j'}(\mathbf{X} - \mathbf{X}')dS_p' \qquad (4.45)$$

where the continuity of the first term in the integrand of (4.44) is also utilized to eliminate its contribution. (4.45) can be further simplified by the dislocation condition (4.40)

$$u_m(\mathbf{X}) = -L_{kpij}\, b_k \iint_{S_{cut}} G_{im,j'}(\mathbf{X} - \mathbf{X}')dS_p' \qquad (4.46)$$

where the Green function is provided by (4.35). Formula (4.46) is termed *Volterra displacement equation* and was proposed in 1907. By Weingarten theorem, S_{cut} in (4.46) can be any directional surface bounded by L. Equation (4.46) states clearly that:

(1) The displacement field for a general dislocation is proportional to the Burgers vector.
(2) This displacement field can be determined through the corresponding Green function (which reflects the property of the elastic medium) and the geometry of the dislocation loop.

For isotropic medium, the following *Burgers displacement equation* can be obtained when substituting the relevant Green function (as expressed explicitly in (4.39)) into (4.46)

$$\mathbf{u}(\mathbf{X}) = -\,(\mathbf{b}/4\pi) \iint_{S_{cut}} (\mathbf{R}/R^3)\cdot d\mathbf{S} - (\mathbf{b}/4\pi)\times \oint_L d\mathbf{l}/R$$

$$-\,1/[8\pi(1-\nu)]\,\nabla\left\{\oint_L [\mathbf{b}\times(\mathbf{R}/R)]\cdot d\mathbf{l}\right\} \qquad (4.47)$$

where $\mathbf{R} = \mathbf{X} - \mathbf{X}'$, and $d\mathbf{l}$ denotes the vectorial arc increment along the dislocation loop.

When taken the gradient with respect to $\mathbf{X}$, formula (4.46) becomes

$$u_{m,r} = L_{kpij}\,b_k \iint_{S_{cut}} G_{im,r'j'}\,dS_{p'} \qquad (4.48)$$

where the differentiation property of the Green function has been used

$$G_{ij,r} = -G_{ij,r'} \qquad (4.49)$$

which is due to the translational invariance of the Green function.

Expression (4.48) can be further reduced to a line integral along the dislocation loop L. By means of (4.42), (4.48) may be rewritten as

$$u_{m,r} = L_{kpij}\,b_k \iint_{S_{cut}} \left\{ dS_p'(\partial/\partial X_r') - dS_r'(\partial/\partial X_p') \right\} G_{im,j'}$$

$$= L_{kpij}\,b_k\,\epsilon_{prn} \oint_L G_{im,j'}\,dX_n' \qquad (4.50)$$

for any point $\mathbf{X}$ not on S_{cut}. ϵ_{rpn} in (4.50) is the alternation symbol. The key point in the derivation of (4.50) lies in the addition of an artificial term corresponding to Dirac delta function acting at a point outside S_{cut}, which enables us to transform the surface integral into a line integral by quoting the Stokes theorem. Equation (4.50) is the famous *Mura formula*[9] derived in 1963. It provides, through a simple and clearly defined line integral, the displacement gradients produced by a spatial dislocation loop in an anisotropic elastic medium. It is straightforward to get the dislocation self-stress field from equation (4.50)

$$\sigma_{ij} = L_{ijmr}\,u_{m,r}. \qquad (4.51)$$

Straight dislocations and planar dislocation loops. Attention is now focused on the particular cases of straight dislocations and planar dislocation loops, as shown in Fig. 4.14. The stress and displacement fields at an arbitrary point $\mathbf{X} = (X, Y)$ in the plane containing the dislocation but away from the dislocation line L will be addressed. A planar coordinate system $X' - Y'$ is set up on the dislocation loop plane with outward normal $\mathbf{m}$, so any point on the dislocation loop is marked by the coordinates $\mathbf{X}' = (X', Y')$. For convenience, the origin of the $X' - Y'$ coordinate system is fixed at $\mathbf{X}$. Let us first consider the straight dislocation line L as drawn in the dashed line of Fig. 4.14. It has a normal distance h from the point **X**, and forms an angle of β with respect to the horizontal datum line (without loss of generality, the datum line is taken along the direction of Burgers vector **b**). Referring to the discussions in subsection 4.3.1, we reach the following conclusions from the dislocation fields described in (4.50) and (4.35)

(1) The stress and the displacement gradients are inversely proportional to the distance h.
(2) The stress and the displacement fields at **X** also relied on the orientation β characterizing the angle formed between the dislocation line and the horizontal datum. The angle β reflects informations such as the angle between **t** and **b**, as well as the material anisotropy.

Therefore, the stress and the displacement gradients at **X** caused by a straight dislocation line L^∞ (the superscript ∞ is used to signify the straight dislocation solutions with infinite radius of curvature) can be written as

$$u_{j,r}^\infty = (1/h)U_{jr}(\beta) \qquad \sigma_{jr}^\infty = (1/h)\Sigma_{jr}(\beta) \qquad (4.52)$$

where $U_{jr}(\beta)$ and $\Sigma_{jr}(\beta)$ are termed *in-plane angular distribution functions*. Metallurgists have already worked out the $U_{jr}(\beta)$ and $\Sigma_{jr}(\beta)$ for various crystal lattices, see reference [3] for further details.

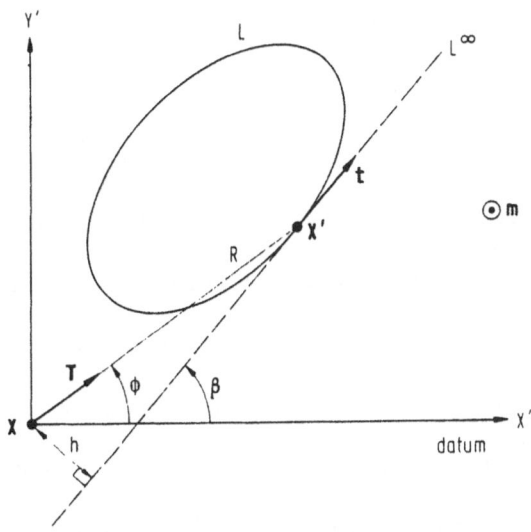

Fig. 4.14. Field descriptions on a straight dislocation (shown in the dashed line) and a planar dislocation loop (shown in the solid line)

As soon as $U_{jr}(\beta)$ and $\Sigma_{jr}(\beta)$ are known, the stress and the displacement gradient at **X** caused by a planar dislocation loop (indicated in Fig. 4.14 by the solid line) can be calculated according to the following formula

$$u_{j,r}(\mathbf{X}) = \oint_L \left\{ U_{jr}(\phi) + U''_{jr}(\phi) \right\} \left[\sin(\phi - \beta)/2R^2 \right] dl$$

$$\sigma_{jr}(\mathbf{X}) = \oint_L \left\{ \Sigma_{jr}(\phi) + \Sigma''_{jr}(\phi) \right\} \left[\sin(\phi - \beta)/2R^2 \right] dl \qquad (4.53)$$

where dl is the arc element in a planar dislocation loop, and R and ϕ form a polar coordinate pair describing the projection of vector $\mathbf{X} - \mathbf{X}'$ on the dislocation loop plane, as elucidated in Fig. 4.14. Formula (4.53) is named *Brown formula* for in-plane field distribution of planar dislocation loops after the innovative contribution of Brown[10]. It provides a simple line integral linking the solution of a general planar dislocation loop with that of a straight dislocation line. The detailed derivation of (4.53) is given in Appendix 4C.

4.3.4 Dislocation Energy

Fig. 4.15 shows an arbitrary solid occupying a domain V. A dislocation loop L and the other defect sources (including dislocations other than L) are contained in V. A prescribed traction field t^A is applied along the exterior boundary of V, S_{ext}. For brevity, it is assumed that the whole exterior boundary S_{ext} is under fixed loading condition by a globally equilibrium traction field t^A. From the superposition principle of linear elasticity, the stress and displacement fields in the domain V can be decomposed into

$$Y = Y^L + Y^I + Y^O + Y^A \qquad (4.54)$$

where Y represents collectively the fields of $\{\sigma_{ij}, \epsilon_{ij}, u_i\}$. Various terms in (4.54) have the meanings of

Y^L: dislocation self-field generated by the dislocation loop L in an infinite medium.

Y^I: image dislocation field, introduced artificially to make S_{ext} a free surface.

Y^O: self-equilibrium field caused by the other defect sources such as voids, inclusions and dislocations other than L.

Y^A: field caused by the applied load, while the domain V is regarded as free of defects.

In the absence of body force, the total potential energy in the domain V is

$$\Pi = (1/2) \int_V \sigma_{ij} \epsilon_{ij} \, dV - \int_{S_{ext}} t_i^A u_i \, dS \qquad (4.55)$$

where the first term refers to the stored elastic strain energy inside V and the second term corresponds to the work done by the applied load. The latter is only relevant to Y^A. On the exterior boundary S_{ext}, one always has

$$(\sigma_{ij}^L + \sigma_{ij}^I)n_j = 0, \quad \sigma_{ij}^O n_j = 0, \quad \sigma_{ij}^A n_j = t_i^A \qquad (4.56)$$

where the first equation is actually used as a definition of the image dislocation field Y^I, the second equation states the self-equilibrium nature of the Y^O field, and

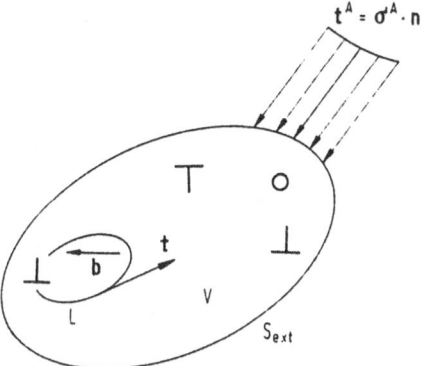

Fig. 4.15. Evaluation of dislocation energy

the last equation signifies the boundary condition of the Y^A field (as well as the whole Y field).

The field decomposition formula (4.54) is then substituted into the expression (4.55) for total potential energy. The corresponding energy decomposition formula is obtained by the utilization of (4.56). Its derivation is somewhat lengthy, and is documented in Appendix 4B.

$$\Pi = \Pi^L + \Pi^{L-I} + \Pi^{L-O} + \Pi^{L-A} + \text{ terms irrelevant to dislocation loop } L$$

(4.57)

where

$$\{\Pi^L, \Pi^{L-I}, \Pi^{L-O}, \Pi^{L-A}\} = b_i \int_S \{\sigma_{ij}^L/2, \sigma_{ij}^I/2, \sigma_{ij}^O, \sigma_{ij}^A\} \, dS_j$$

(4.58)

representing the dislocation line energy, the interaction energy with image dislocations, the interaction energy with other defect sources and the interaction energy with the applied load. The integration surface domain S in (4.58) refers to the surface bounded by the dislocation loop L. Its element can be taken as $dS_j = m_j dS$ for a planar dislocation loop, with $\mathbf{m}$ again referred to the outward unit normal of the slip plane.

Despite the general similarity in various terms in (4.58) composing the total potential energy, cautions should be paid to the approximation which leads to the expression of Π^L. Rigorously speaking, Π^L should be stated as

$$\Pi^L = (b_i/2) \int_{S(r_o)} \sigma_{ij}^L dS_j + (1/2) \int_{\text{Tube}} \sigma_{ij}^L u_j^L \, dS_i$$

(4.59)

instead, where $S(r_o)$ is the remaining part of S from cutting off an annulus of half width r_o around the dislocation loop L. The second term in the above formula is often called as the *core-traction term*. Its integration is carried out on a tubular surface of radius r_o surround the dislocation loop L. Similar to the circumstance of simple dislocations discussed in subsection 4.3.1, this term cannot be evaluated through continuum mechanics approach, and it only contributes about 10 to 15 per cent to the dislocation self energy Π^L. One has to bear in mind that the neglect of the core-traction term, as engaged in (4.58), could cause a certain error. Moreover, the integral in the first term of (4.59) becomes dependent on the selection of r_o. Π^L diverges when r_o approaches zero.

For straight dislocation lines, one can substitute (4.52) into the first equation of (4.58) to obtain the line density of strain energy of straight dislocations (or strain energy per unit dislocation line, as presented in subsection 4.3.1)

$$W(\beta) = (b_i/2)\Sigma_{ij}(\beta)m_j \ln(R/r_o). \tag{4.60}$$

During the derivation of the above formula, the integration domain is taken as the area swept by a straight dislocation segment of unit length with dislocation core area excluded. Thus, the factor $1/h$ in (4.52) is integrated from r_o to R.

For a general dislocation loop, its self energy could be estimated from the result for a straight dislocation.

$$\Pi^L = \oint_L W(\beta) \, dl \tag{4.61}$$

where β is the angle between the tangential vector of dislocation loop with a certain datum line. The critical assumption behind (4.61) is to treat the energy possessed by an arc element dl with local tangential vector $\mathbf{t}$ in the dislocation loop as the energy of a straight dislocation with identical $\mathbf{t}$, $\mathbf{b}$ and dl.

The strain energy line density for a straight dislocation, W, can be written in a general form of

$$W(\beta) = E(\beta) \ln(R/r_o). \tag{4.62}$$

Similar to (4.20), $E(\beta)$ is termed prelogarithmic energy coefficient. By comparison between (4.61) and (4.62), the prelogarithmic energy coefficient must have the following form

$$E(\beta) = \frac{1}{2} b_i \Sigma_{ij}(\beta) m_j. \tag{4.63}$$

Besides, the function E should be in a quadratic form of either stress or strain. The linearity of either stress or strain with respect to the Burgers vector then sets E as a quadratic form of the Burgers vector

$$E(\beta) = K_{ij}(\beta) b_i b_j \tag{4.64}$$

where K_{ij} is often called the *prelogarithmic energy tensor*. A relationship between the prelogarithmic energy tensor and the in-plane angular distribution of stress can be established as

$$b_i \Sigma_{ij} m_j = 2K_{ij} b_i b_j \tag{4.65}$$

by comparing (4.63) and (4.64).

4.3.5 Forces on Dislocations

Forces on dislocations are characterized by non-Newton, thermodynamic energetic forces. To reveal this kind of forces, let us examine the energy variation associated with a virtual movement $\delta\mathbf{r}$ of the dislocation loop L, as delineated in Fig. 4.16. It is assumed that the exterior boundary conditions (including both loads and displacements prescribed along the exterior boundary S_{ext}) are maintained invariant during this virtual movement. Then according to the definition of the energetic force, one has

$$\delta\Pi \equiv - \oint_L \mathbf{f} \cdot \delta r \, d\mathbf{l} \tag{4.66}$$

where $\mathbf{f}$ is the total *Peach-Keohler force* acting on an arc element $d\mathbf{l}$ of the dislocation loop L. Based on the energy decomposition formulae presented in (4.57) and (4.58), parallel decomposition can be written for the energetic force

$$\left\{ \delta\Pi^L, \delta\Pi^{L-I}, \delta\Pi^{L-O}, \delta\Pi^{L-A} \right\} \equiv - \oint_L \left\{ \mathbf{f}^L, \mathbf{f}^I, \mathbf{f}^O, \mathbf{f}^A \right\} \cdot \delta r \, dl \tag{4.67}$$

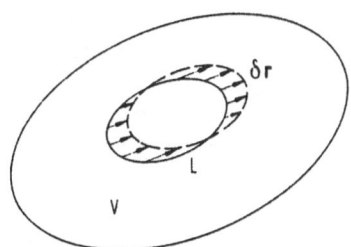

Fig. 4.16. Virtual movement of the dislocation loop

where
$\mathbf{f}^L$: self-interaction force of a dislocation loop produced by the curvature effect, which vanishes for straight dislocation lines;
$\mathbf{f}^I$: image force, or drag force by the free surface;
$\mathbf{f}^O$: interaction force between dislocation loop L and other defect sources;
$\mathbf{f}^A$: driving force on dislocation loop L by the applied load.
We now prove the following *Peach-Koehler formula* for the interaction forces

$$\left\{ \mathbf{f}^I, \mathbf{f}^O, \mathbf{f}^A \right\} = \left(\mathbf{b} \cdot \left\{ \sigma^I, \sigma^O, \sigma^A \right\} \right) \times \mathbf{t}. \tag{4.68}$$

Proof: The proofs for $\mathbf{f}^O$ and $\mathbf{f}^A$ are quite similar. The fields Y^O and Y^A are independent of the location of the dislocation loop L, that implies the variations $\delta\Pi^{L-O}$ and $\delta\Pi^{L-A}$ are only caused by the change in dislocation boundary δS, distinguished by the shaded area in Fig. 4.16. Accordingly, the variations on the last two surface integrals in (4.58) can be presented in the following line-integral forms

$$\{\delta\Pi^{L-O}, \delta\Pi^{L-A}\} = - \oint_L [\mathbf{b}\cdot \{\sigma^O, \sigma^A\} \times \mathbf{t}] \cdot \delta\mathbf{r}\, dl \tag{4.69}$$

The last two expressions in (4.68) can be derived by comparing (4.69) and (4.67).

The proof for the image force expression is more complicated. A virtual movement $\delta\mathbf{r}$ of the dislocation loop not only changes the integration area, but also brings about a re-adjustment of the image dislocation field Y^I so that the first condition in (4.56) is satisfied for a shifted dislocation loop. Then from (4.58)

$$\delta\Pi^{L-I} = - (1/2) \oint_L [(\mathbf{b}\cdot\sigma') \times \mathbf{t}] \cdot \delta\mathbf{r}\, dl$$
$$+ (1/2)\mathbf{b}\cdot \int_S \delta\sigma' \cdot d\mathbf{S}. \tag{4.70}$$

However, it was prove by Gavazza and Barnett[11] in 1975 that

$$\mathbf{b}\cdot \int_S \delta\sigma' \cdot d\mathbf{S} = \mathbf{b}\cdot \int_{\delta S} \sigma' \cdot d\mathbf{S}. \tag{4.71}$$

The proof for (4.71) is too lengthy to be reproduced here. Consequently, the first expression of (4.68) can be derived from the second equation of (4.67), equation (4.70) and (4.71). Q.E.D.

The *dislocation gliding force* (denoted by a scalar f) is the projection of a dislocation force along the slip direction $\mathbf{s} = \mathbf{m}\times\mathbf{t}$, where $\mathbf{m}$ is the outward unit normal of the gliding plane. After straightforward algebra, one gets

$$\begin{bmatrix} f^I \\ f^O \\ f^A \end{bmatrix} \equiv \begin{bmatrix} \mathbf{f}^I \\ \mathbf{f}^O \\ \mathbf{f}^A \end{bmatrix} \cdot (\mathbf{m}\times\mathbf{t}) = \mathbf{b}\cdot \begin{bmatrix} \sigma^I \\ \sigma^O \\ \sigma^O \end{bmatrix} \cdot \mathbf{m} \equiv \begin{bmatrix} \tau^I \\ \tau^O \\ \tau^A \end{bmatrix} b \tag{4.72}$$

where τ^I, τ^O and τ^A are the corresponding resolved shear stress on slip plane and along slip direction, they are also called as *Schmid resolved shear stress*.

By means of the second expression in (4.53) and the equation (4.63), the following *Brown formula*[10] *for the self-gliding force*, f^L, is derived

$$f^L \equiv b_i\sigma_{ij}^L m_j = \oint_L [E(\phi) + E''(\phi)]\,[\sin(\phi - \beta)/R^2]\, dl. \tag{4.73}$$

Utilizing the above Brown formula, Gavazza[12] was able to prove the so-called *Line Tension Formula* frequently used in the dislocation theory

$$f^L = -\mathfrak{x}\,\Gamma \tag{4.74}$$

where $\mathfrak{x}$ is the curvature of dislocation line at the considered point, and Γ denotes the dislocation line tension. The latter can be evaluated through

$$\Gamma = [E(\beta) + d^2E(\beta)/d\beta^2]\, \ln(R/r_o) \tag{4.75}$$

where β is the angle between the local tangential vector $\mathbf{t}$ and the Burgers vector $\mathbf{b}$. Formula (4.74) clearly reflects the influence of curvature to the dislocation self-gliding force, as well as its disappearance for a straight dislocation. The line tension formula (4.74) was first proposed by de Wit and Keohler[13] in 1959 as an approximate formula.

The equilibrium shape of a planar dislocation loop under general loading condition is worth to pay a short discussion. From the principle of minimum potential energy, the first variation of the potential energy should vanish

$$\delta\Pi = 0 \tag{4.76}$$

when the dislocation loop assumes its equilibrium shape. From (4.66), (4.67) and the arbitrary assignment of $\delta\mathbf{r}$, an equilibrium condition for the total dislocation gliding force is derived

$$f \equiv f^L + f^I + f^O + f^A = 0. \tag{4.77}$$

A differential equation controlling the equilibrium shape of L can be obtained by substituting (4.72) and (4.74) into (4.77).

Example: *Interaction between parallel straight dislocations.* As an example to illustrate the above theory, the interaction between two parallel straight dislocations separated by a normal distance h, as shown in Fig. 4.13, will be discussed. The two straight dislocations have the identical tangential vector $\mathbf{t}$, and their corresponding Burgers vectors are denoted as $\mathbf{b}^{(1)}$ and $\mathbf{b}^{(2)}$, respectively. The force $\mathbf{f}^{1-2}$ signifies the action on dislocation $\mathbf{b}^{(1)}$ from the stress field $\sigma^{(2)}$ caused by dislocation $\mathbf{b}^{(2)}$. It can be computed through the Peach-Koehler formula (i.e., taken the second expression of (4.72), and chosen $\sigma^{(2)}$ as $\Sigma^{(2)}/h$ because the force is evaluated at the location of dislocation $\mathbf{b}^{(1)}$) as

$$\mathbf{f}^{1-2} = (1/h)\left\{\mathbf{b}^{(1)}\cdot\Sigma^{(2)}\right\}\times\mathbf{t} \tag{4.78}$$

while the dislocation gliding force exerted on dislocation $\mathbf{b}^{(1)}$ by dislocation $\mathbf{b}^{(2)}$ is

$$f^{1-2} = (1/h)\mathbf{b}^{(1)}\cdot\Sigma^{(2)}\cdot\mathbf{m} = (2/h)\mathbf{b}^{(1)}\cdot\mathbf{K}^{(2)}\cdot\mathbf{b}^{(2)}. \tag{4.79}$$

The last step of derivation is benefited by equation (4.65). It is worthwhile to point out that the formula (4.79) for dislocation interaction force f^{1-2} is sensible only when $\mathbf{t}$, $\mathbf{b}^{(1)}$ and $\mathbf{b}^{(2)}$ are co-planar. For the special case of isotropic solid, the prelogarithmic energy tensor $\mathbf{K}$ can be written down simply as (from (4.20), (4.21) and (4.64))

$$\{K\} = (G/4\pi)\begin{bmatrix} 1/(1-\nu) & & \\ & 1/(1-\nu) & \\ & & 1 \end{bmatrix} \tag{4.80}$$

where the coordinate system in Fig. 4.17 is taken. This coordinate system is formed by right hand, orthonormal base vectors $\{\mathbf{e}_1, \mathbf{e}_2, \mathbf{e}_3\}$, where $\mathbf{e}_2$ and $\mathbf{e}_3$ are aligned with the slip plane normal **m** and the tangential vector **t**, respectively. The inter-action force is expressed by

$$f^{1-2} = (G/2\pi h) \left[\left(b_1^{(1)} b_1^{(2)} + b_2^{(1)} b_2^{(2)} \right) / (1-\nu) + b_3^{(1)} b_3^{(2)} \right] \qquad (4.81)$$

after substituting equation (4.80) into (4.79). According to the proportionality rela-tion between the stress (or the corresponding in-plane angular distribution function Σ) and the Burgers vector **b** (see relation (4.50) and (4.51)), it is straightforward to verify that

$$\mathbf{f}^{1-2} = -\mathbf{f}^{2-1}. \qquad (4.82)$$

Equation (4.82) is similar to Newton's third law stating the action and reaction are equal in magnitude by opposite in direction.

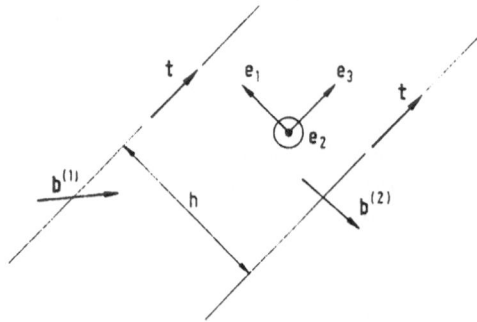

Fig. 4.17. Interaction between two parallel straight dislocations

Two particular cases are worth to be addressed further :
(1) $\mathbf{b}^{(1)} = \mathbf{b}^{(2)}$, this leads to $f^{1-2} > 0$ and implies that *dislocations of the same sign repel each other*;
(2) $\mathbf{b}^{(1)} = -\mathbf{b}^{(2)}$, this leads to $f^{1-2} < 0$ and implies that *dislocations of the opposite signs attract each other.*

4.4 Dislocation Induced Plastic Deformation

4.4.1 Orowan Equation

We turn to discuss the plastic deformation induced by massive dislocation move-ments. Let us consider a volumetric domain V containing N parallel mobile dislo-cations as exhibited in Fig. 4.18. It is assumed that all dislocations are stationary

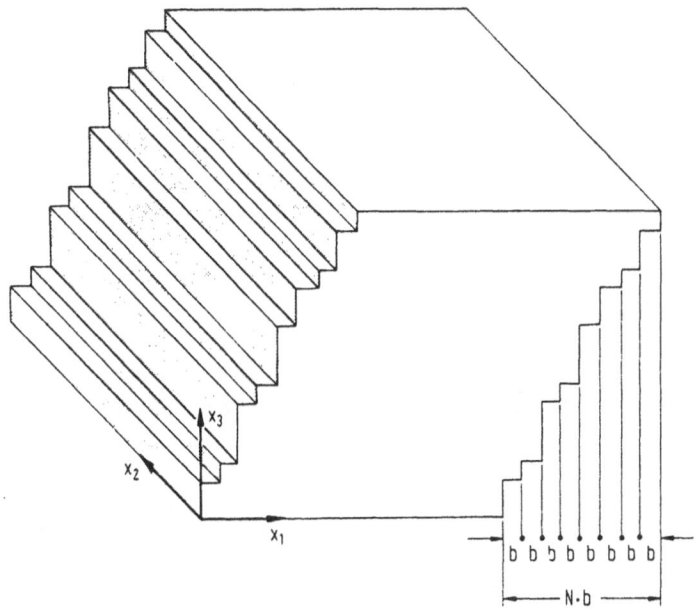

Fig. 4.18. Plastic shear deformation induced by massive uniform dislocation motion

prior to deformation. The domain V starts to deform after a certain threshold applied load is surpassed, and the individual areas swept by each mobile dislocation are denoted as $A_1, \ldots, A_N$. For simplicity, we assume the crystal in V has uniform lattice structure and orientation, and slip operates along the same slip system (which is characterized by the common slip plane normal $\mathbf{m}$ and slip direction $\mathbf{s} = \mathbf{b}/|\mathbf{b}|$, see subsection 4.5.1 in the sequel). Therefore, the total slipped area can be estimated from to the following formula

$$A_{\text{slip}} = \sum_{i=1}^{N} A_i \tag{4.83}$$

which signifies that A_{slip} is the total area swept by all mobile dislocations. Due to the above dislocation motion, an average plastic shear strain is generated in the domain V

$$\gamma = b A_{\text{slip}} / V \tag{4.84}$$

where b is the length of Burgers vector. Taking time derivative of the above expression, one gets the shear strain rate as follows

$$\dot{\gamma} = b \dot{A}_{\text{slip}} / V = b L_{\text{disl}} <v> / V = \rho_d b <v>. \tag{4.85}$$

This equation is named as the generalized *Orowan equation*. In (4.85), L_{disl} is the total length of mobile dislocations, $\rho_d = L_{\text{disl}} / V$ represents the *dislocation line density*, and $<v>$ corresponds the average speed of the dislocation movement. If the motion of each dislocation remains the same (or the areas swept by various dislocations are identical), see Fig. 4.18, then (4.85) is reduced to

$$\dot{\gamma} = \rho_d b v \tag{4.86}$$

which is the famous Orowan (or Taylor-Orowan) equation. This equation provides a link between microscopic measurements (such as dislocation density and speed of dislocation movement) and the corresponding macroscopic plastic deformation rate.

4.4.2 Dislocation Dynamics

Attention is then focused on the dynamic behavior of dislocations driven by gliding stress τ. Limited by the scope of the present treatment, only the relationship between the dislocation gliding stress τ and the dislocation velocity v is addressed here. Several typical dislocation dynamics curves (τ versus v) are plotted in Fig. 4.19, where solid line corresponds to rate independent material response (i.e., the deformation response is not affected by the deformation rate) and the other lines show rate dependent behavior.

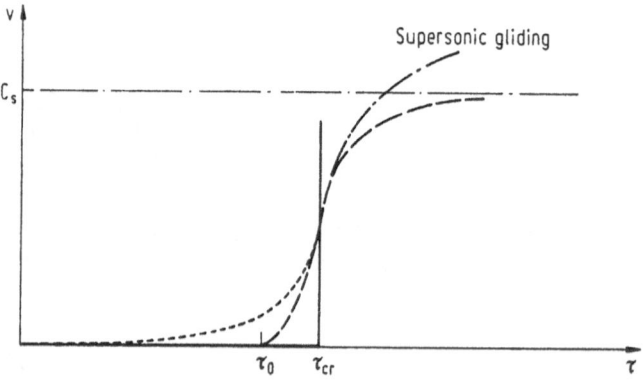

Fig. 4.19. Dislocation dynamics curves. Solid line: rate independent response; dashed line: rate dependent behavior with elastic range and saturated velocity bound; dot line: response predicted by the thermal fluctuation model of dislocation line; dot-dash line: supersonic dislocation movement

For the former rate independent case, it is usually observed that a critical stress τ_{cr} exists, and at which the dislocation can operate in a variety of speed range. The critical stress τ_{cr} is in general a function of the accumulated plastic shear. When τ is less than τ_{cr}, the disability of the dislocation movement deters the occurrence of plastic deformation and supports the existence of an elastic domain frequently observed in macroscopic deformations. The condition $\tau = \tau_{cr}$ for plastic deformation is equivalent to the consistency condition that a stress point should always locate at the current yield surface during loading. Under this type of dislocation dynamics curves, the required plastic deformation can be accomplished instantaneously by rapid dislocation movement.

Typical dislocation dynamics curve in rate dependent solid is shown as the dashed line in Fig. 4.19. It is featured by the unique correspondence between driving stress τ and dislocation speed v. Dislocation glide can be divided into two categories: glide at conventional speed and glide at high speed. The two cases are separated at roughly one tenth of the elastic shear wave speed $(C_s = (G/\rho)^{\frac{1}{2}})$. For dislocation glide at conventional speed, the major resistance for dislocation movement is the Peierl-Nabarro force τ_{P-N}

$$\tau_{P-N} = [2G/(1-\nu)] \exp\{-2\pi d/[b(1-\nu)]\} \tag{4.87}$$

where d is the distance between the adjacent slip planes. The metallurgists modeled this dislocation resistance mechanism by various dislocation dynamics curve descriptions.

(1) *Power-law type* Stein and Low[14] proposed that

$$v = (\hat{\tau}/\hat{\tau}_{cr})^n \tag{4.88}$$

(2) *Exponential type*

$$v = v_o \exp\{-\hat{\tau}_{cr}/\hat{\tau}\} \qquad \text{Gilman[15]} \tag{4.89}$$

$$v = v_o \exp\{-(\hat{\tau}_{cr}/\hat{\tau})^n\} \qquad \text{Kelly and Gillis[16]} \tag{4.90}$$

(3) *Combined power-law and exponential type*

A better correlation to the dislocation dynamics curve data was reported by Frost and Ashby through the following expression[17]

$$v = v_o(\hat{\tau}/G)^2 \exp\left\{-\alpha\left[1 - (\hat{\tau}_{cr}/\hat{\tau})^{3/4}\right]^{4/3}\right\} \tag{4.91}$$

based on micromechanics.

In various expressions from (4.88) to (4.91), v_o, n and α are material constants. For the material with an elastic range, one has

$$\hat{\tau} = \tau - \tau_o, \quad \hat{\tau}_{cr} = \tau_{cr} - \tau_o \tag{4.92}$$

where τ_o denotes the shear stress threshold below which dislocations cease to operate. On the other hand, according to the thermal fluctuation theory of dislocation lines, such as that proposed by Earmme and Weiner[18], certain amount of dislocation lines can always shift forward by thermal fluctuation when subject to a non-zero biased shear stress. Consequently, we have $\tau_o = 0, \hat{\tau} = \tau$ and $\hat{\tau}_{cr} = \tau_{cr}$ when modelled by the thermal fluctuation theory of dislocation line. The dislocation dynamics curve under such models would appear as the dot curve shown in Fig. 4.19. The association of various expressions listed above with the Orowan equation (4.86) would lead to several representative macroscopic visco-plastic constitutive equations in the literatures. For example, the dislocation dynamic model described by (4.88) corresponds to Perzyna's over-stress visco-plastic constitutive equation[19], whereas the curve (4.90) proposed by Kelly and Gillis relates to the popular Bodner-Partom visco-plasticity model[20], etc..

For the dislocation glide at high speed, three possible extra resistances could be confronted beside the Peierls-Nabarro stress. The first is the *thermal elastic resistance*, due to the fluctuation of hydrostatic tension stress in the crystal when dislocations rash by. The induced temperature gradient stimulates the increase of system entropy, so a certain amount of mechanical energy has to be dissipated. The second is the *radiation resistance*. During the high speed glide process, dislocation shifts from one symmetric position to another with rapid successions of its configurations. Elastic wave is emitted from the relevant locations so certain amount of energy is compensated. The last resistance, possibly the most important, comes from the scattering of the crystal lattice to the acoustic wave. The effect produced is termed *phonon drag*. In the phonon drag regime, the dislocation dynamics curve can be described by a linear relationship between v and τ (referring to the high speed regime in Fig. 4.19)

$$v = v_o + \alpha(\tau - \tau_t)/G \tag{4.93}$$

where α and τ_t are material constants. The testing data of Campbell and Ferguson[21] in 1970 for the high rate deformation response of mild steel supported the above expression.

We next consider the possibility of an ultra-high deformation rate regime. Frank[22] and Eshelby[23] arrived at the result that the dislocation velocity in an elastic medium can not exceed the shear wave speed C_s. This statement was supported by a three dimensional continuum analysis. They concluded that the equivalent moving mass for unit dislocation line would become infinitely large when v approaches C_s. Their result predicted a horizontal asymptote of ordinate C_s for the dislocation dynamics curve, as shown in the dashed line of Fig. 4.19. The existence of a limiting velocity for the dislocation movement is still a controversial issue. Some scientists believe that dislocations can move steadily from subsonic to supersonic regime, as conjectured in Fig. 4.19 by the dot-dash line. However, it was pointed out by Earmme and Weiner[18] that a dislocation cannot move in a conventional manner when sound speed is approached. Collision of atoms will inevitably occur and a breakdown velocity is expected.

4.5 Crystallography of Dislocations

In reality, dislocations move in crystals with specific lattice structures, and the crystallography of the lattices surrounded will have effects on the behavior and mobility of dislocations. Due to the vast content of crystallography, the coverage here is rather concentrated in some selected materials relating closely to the subsequent discussions.

4.5.1 Lattice Burgers Vectors

Slip planes. The possibility of non-conservative climb motion of dislocations is usually ruled out because it only occurs at high energy excitation state. Conse-

quently, the *lattice Burgers vectors* can be defined with respect to the glide motion of dislocations. The dislocation gliding planes have the following characteristics:
(1) A dislocation gliding plane must be a closely-packed plane of atoms.
(2) The distance between the adjacent slip planes should be as large as possible to ease the bond-switch adjustment due to dislocation glide.

For cubic lattice, the distance between the adjacent crystallographic planes with Miller index (hkl) can be calculated by analytical geometry as

$$d_{hkl} = a/\sqrt{h^2 + k^2 + l^2} \tag{4.94}$$

where a is the lattice parameter. Checked by the criteria listed above, the major slip planes for FCC, BCC and HCP lattices are

FCC : {111} family with 4 planes,
BCC : {110} family with 6 planes,
HCP : {0001} family with 1 plane.

Slip directions. To minimize the distance for each slip step, dislocations always glide toward the closely-packed directions. The *slip directions* for typical crystal lattices are

FCC : <110> 3 slip directions per slip plane,
BCC : <111> 2 slip directions per slip plane,
HCP : <10$\bar{1}$0> 3 slip directions per slip plane.

Slip systems. Combination of a slip plane and a slip direction forms a *slip system.* The number of slip systems for typical crystal lattices are:

FCC : 12 ,   BCC : 12 ,   HCP : 3 .

Only three slip systems exist in a HCP crystal lattice, causing the difficulty for a HCP-crystal to deform plastically. Beside {110}, it is reported in the literature that BCC crystals can occasionally slip in the crystallographic planes {112} (with 4 planes) and in {123} (with 8 planes). The slip directions remain to be <111> for those additional slip planes. If those secondary possible slip systems are included, BCC lattice would have a total of 48 slip systems.

Burgers vectors for complete dislocations. *Complete dislocations* refers to as the dislocation movements in which a complete shift of the minimum lattice vector is accomplished. For complete dislocations, Burgers vectors are the vectors along the slip direction connecting adjacent lattice sites. For three typical crystal lattices listed above, they are

$$\text{FCC}: \ \mathbf{b} = \frac{1}{2}a<110>,$$

$$\text{BCC}: \ \mathbf{b} = \frac{1}{2}a<111>, \tag{4.95}$$

$$\text{HCP}: \ \mathbf{b} = \ a<10\bar{1}0>.$$

4.5.2 Stacking Faults

Both FCC and HCP lattices are closely-packed lattice structures. Their difference
only lies in their stacking sequences, as depicted in Fig. 4.20. If the three possible
in-plane center positions for a layer of closely-packed atoms are denoted by A, B,
and C, successively. Then from Fig. 4.20, their packing sequences are

FCC : ABCABCABCABC ...
HCP : ABABABABABAB ...

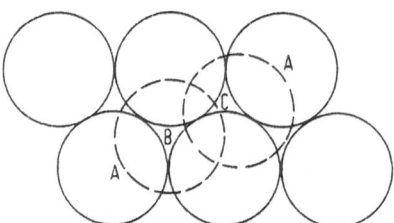

Fig. 4.20. Stacking sequences in FCC and HCP lattices

However, in actual FCC lattices, faults on stacking sequence frequently occur.
One example is the symmetric faulting on the stacking sequence with respect to a
specific layer

▲
...ABCABCBACBA...

which characterizes a twinning distortion with respect to the closely-packed {111}
plane of FCC. The crystal formations on the two sides of ▲ like a twin crystal
lattice.

The other faulting patterns of the stacking sequence of {111} planes are called
stacking faults. Frank[24] pointed out that staking faults can be further divided into
the *intrinsic faults* and the *extrinsic faults*.

Intrinsic faults: faults produced by *subtracting* one layer of atoms from the
normal ABC stacking sequence of FCC lattices. It could be formed by a double
twinning operation at two successive layers marked by ▲, as elucidated in Fig. 4.21.

Extrinsic faults: faults produced by *inserting* one layer of atoms to the normal
ABC stacking sequence of FCC lattices. It could be formed by a double twinning
operation at the two layers marked by ▲ (separated by a central layer), as shown
in Fig. 4.22.

Various quantities representing stacking fault energies per unit closely-packed
plane (or surface tensions), such as γ_T (for twinning), γ_I (for intrinsic fault), γ_E
(for extrinsic fault), can be either experimentally measured or theoretically predicted. The following considerations should be taken into account in a theoretical
formulation of stacking fault energy:

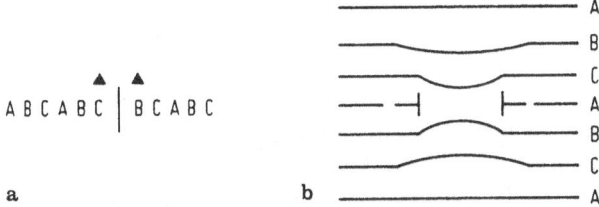

Fig. 4.21. Intrinsic stacking fault. a Stacking sequence; b formation of a prismatic loop

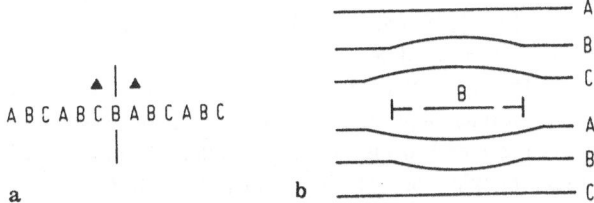

Fig. 4.22. Extrinsic stacking fault. a Stacking sequence; b formation of a prismatic loop

(1) distortion energy for every mis-joined atom-pairs,
(2) specific surface energy of the fault,
(3) entropy of the fault.

Up to now, computations based on the theoretical models for the stacking fault energy are still in the verge of development. The theoretical predictions deviate from the experimental data roughly by a factor of two.

4.5.3 Partial Dislocations and Dislocation Reactions

Frank postulated that a complete dislocation $\mathbf{b}^{(1)}$ can be decomposed into two (complete or partial) dislocations $\mathbf{b}^{(2)}$ and $\mathbf{b}^{(3)}$ whenever energetically favorable. For the general anisotropic case, the above *Frank energy criterion* can be interpreted in terms of the related prelogarithmic energy coefficients. That is, a given dislocation $\mathbf{b}^{(1)}$ will decompose into two different dislocations $\mathbf{b}^{(2)}$ and $\mathbf{b}^{(3)}$ whenever the sum of the prelogarithmic energy coefficients of the latter two dislocations is less than that of $\mathbf{b}^{(1)}$

$$K_{ij}(\beta_1)b_i^{(1)}b_j^{(1)} > K_{ij}(\beta_2)b_i^{(2)}b_j^{(2)}$$
$$+ K_{ij}(\beta_3)b_i^{(3)}b_j^{(3)} \tag{4.96}$$

where β_1, β_2 and β_3 are the angles formed between $\mathbf{b}^{(1)}$, $\mathbf{b}^{(2)}$ and $\mathbf{b}^{(3)}$, respectively, and the tangential vector $\mathbf{t}$. For the special case of isotropy, above criterion is reduced to

$$b^{(1)^2} > b^{(2)^2} + b^{(3)^2}. \tag{4.97}$$

The term *partial dislocation* was used to contrast the concept of a complete dislocation. It refers to a part of the complete dislocation or to partially executed complete dislocation. From the picture of intrinsic stacking fault shown in Fig. 4.21(b), a partial dislocation loop can be formed by removing a disk of atoms from the perfect crystal lattice and its subsequent collapse. The dislocation so-produced is called *Frank partial dislocation* of removing type. The Frank partial dislocation of inserting type can be defined in a parallel manner as shown in Fig. 4.22(b). The dislocation loops in both Fig. 4.21(b) and 4.22(b) are termed *prismatic dislocation loops*. The Burgers vector for a Frank partial dislocation is $(a/3)<111>$. It is thus unable to glide on the slip plane $\{111\}$. Hence, only climb movement is associated with the Frank partials, and they are termed *sessile dislocations*.

Another type of partial dislocations, termed *Shockley partial dislocations*, is shown in Fig. 4.23. Shockley partials play an important role in the dislocation movement in FCC crystals. As presented in Fig. 4.23, a complete dislocation $\mathbf{b}^{(1)}$ can be decomposed into two Shockley partials $\mathbf{b}^{(2)}$ and $\mathbf{b}^{(3)}$ according to the following dislocation reaction

$$\mathbf{b}^{(1)} \longrightarrow \mathbf{b}^{(2)} + \mathbf{b}^{(3)} \tag{4.98}$$

where

$$\begin{aligned}
\mathbf{b}^{(1)} &= (a/2)[\bar{1}01], \\
\mathbf{b}^{(2)} &= (a/6)[\bar{2}11], \\
\mathbf{b}^{(3)} &= (a/6)[\bar{1}\bar{1}2].
\end{aligned} \tag{4.99}$$

To verify the satisfaction of Frank energy criterion, we estimate this dislocation reaction under the isotropic assumption

$$b^{(1)^2} = a^2/2, \qquad b^{(2)^2} + b^{(3)^2} = a^2/3. \tag{4.100}$$

Therefore, the dislocation decomposition (4.98) is energetically favorable.

Repulsive interaction prevails for a Shockley partial pair $\mathbf{b}^{(2)}$ and $\mathbf{b}^{(3)}$. For the purpose of a demonstration, $\mathbf{b}^{(2)}$ and $\mathbf{b}^{(3)}$ are taken as parallel straight dislocations with a common tangential vector $\mathbf{t}$. Their mutual interaction force f^{2-3} can be calculated from equation (4.81) as

$$f^{2-3} = (G/2\pi h) \left\{ \left(\mathbf{b}^{(2)} \times \mathbf{t} \right) \cdot \left(\mathbf{b}^{(3)} \times \mathbf{t} \right) / (1-\nu) + \left(\mathbf{b}^{(2)} \cdot \mathbf{t} \right) \left(\mathbf{b}^{(3)} \cdot \mathbf{t} \right) \right\} \tag{4.101}$$

where h is the distance between two partial dislocation lines. This repulsive force is balanced by the surface tension γ_I caused by an intrinsic stacking fault

$$f^{2-3} \big|_{h=h_e} = \gamma_I. \tag{4.102}$$

That is, the repulsive force between a pair of almost parallel Shockley partials will be canceled out exactly at the equilibrium separation h_e by the surface tension

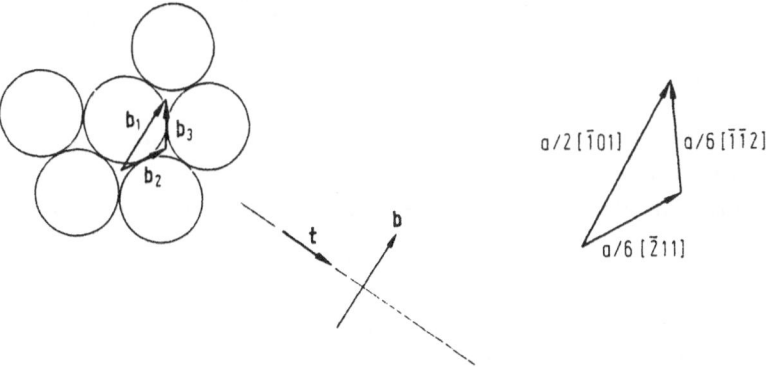

Fig. 4.23. Shockley partial dislocations

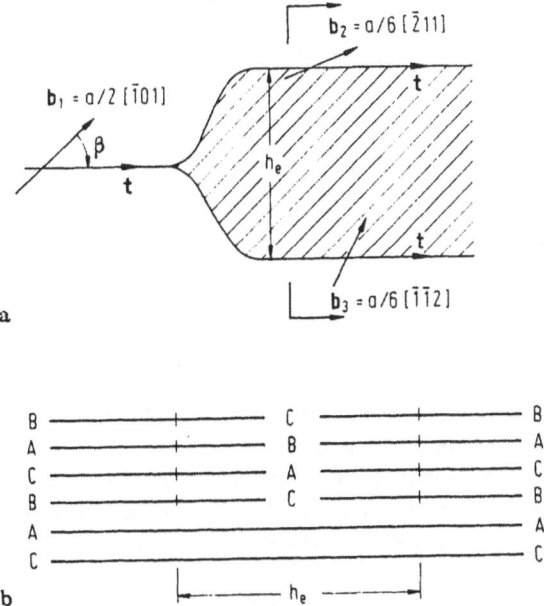

Fig. 4.24. Extended dislocation. **a** A screw dislocation branches into an extended dislocation with a equilibrium separation h_e; **b** stacking sequence of the extended dislocation

induced by the high stacking fault energy. If the angle formed by $\mathbf{t}$ and $\mathbf{b}^{(1)}$ is denoted as β, then the angles formed between $\mathbf{t}$ and $\mathbf{b}^{(2)}$ or $\mathbf{b}^{(3)}$ will be $\beta + \pi/6$ or $\beta - \pi/6$, respectively. The equilibrium distance between the two Shockley partials can be computed by (4.101) as

$$h_e = \left(Ga^2/48\pi\gamma_s\right)\left[(2-\nu)/(1-\nu)\right]\left\{1 - \left[2\nu/(1-\nu)\right]\cos 2\beta\right\}. \quad (4.103)$$

Two shockley partials separated by the above equilibrium distance, as well as a strip of stacking fault between them, are called collectively as an *extended dislocation*. The extended dislocation has to move as a whole, so the dislocation kinetics will be significantly effected by its existence. Taken the example of $b^{(1)}$ being a pure screw dislocation, so the cross-slip is easy to proceed for the complete dislocation $b^{(1)}$. For the extended dislocations $b^{(2)}$ and $b^{(3)}$, however, cross-slip will be blocked. Therefore, the formation of extended dislocations in general limits the mobility of dislocations.

Appendix 4A   Line integral expression of Green function

A derivation will be outlined in this appendix for transforming the three dimensional inverse integral of the Green function

$$G_{km}(\mathbf{X} - \mathbf{X}') = \left(1/8\pi^3\right) \int\!\!\!\int\!\!\!\int_{-\infty}^{\infty} \left(M_{km}^{-1}/K^2\right) \exp\left\{-i\,\mathbf{K}\cdot(\mathbf{X} - \mathbf{X}')\right\} d^3\mathbf{K} \tag{4.34}$$

into a single definite integral. Defining

$$\begin{aligned} R &= |\mathbf{X} - \mathbf{X}'| & \mathbf{T} &= (\mathbf{X} - \mathbf{X}')/R \\ \mu &= \mu\mathbf{z} = R\mathbf{K} & \mu &= KR \end{aligned} \tag{4A.1}$$

we can convert (4.34) into

$$G_{km} = \left(1/8\pi^3\right) \int\!\!\!\int\!\!\!\int_{-\infty}^{\infty} M_{km}^{-1}(\mathbf{z}) \left[\cos(\mu\mathbf{z} \cdot \mathbf{T})/\mu^2 R\right] d^3\mu \tag{4A.2}$$

where $\mathbf{T}$ and $\mathbf{z}$ are unit vectors parallel to $\mathbf{X} - \mathbf{X}'$ and to $\mathbf{K}$, respectively. Without loss of generality, the imaginary part in (4.34) is discarded.

The triple volume integral (4A.2) will be evaluated in a spherical coordinate system (μ, Θ, ϕ) with $\mathbf{T}$ as its axis, where μ the length of vector μ, Θ the angle formed between μ (or $\mathbf{z}$) and $\mathbf{T}$, and ϕ the circumferential angle projected on a plane normal to $\mathbf{T}$ and measured from a fixed datum, as shown in Fig. 4A. Obviously, the volumetric element of triple integration can be written as

$$d^3\mu = \mu^2 \sin\Theta \, d\mu \, d\Theta \, d\phi \tag{4A.3}$$

and we also have

$$M_{km}^{-1}(\mathbf{z}) = M_{km}^{-1}(\Theta, \phi) \qquad \mathbf{z} \cdot \mathbf{T} = \cos\Theta. \tag{4A.4}$$

The substitution of above two formulae into (4A.2) results in

$$G_{km} = \left(1/8\pi^3 R\right) \int_0^{\infty} d\mu \int_0^{2\pi} d\phi \int_0^{\pi} M_{km}^{-1}(\Theta, \phi) \cos(\mu \cos\Theta) \sin\Theta \, d\Theta \tag{4A.5}$$

where the infinite integration with respect to μ can be evaluated as

$$\int_0^\infty \cos(\mu \cos\Theta)\, d\mu = \pi \sum_{n=1,3,5,\ldots}^\infty \delta(\Theta - n\pi/2)/\sin(n\pi/2) = \pi\delta(\Theta - \pi/2)$$

$$(4\text{A}.6)$$

with $\delta(.)$ denoting Dirac delta function. The simplification in the last step of (4A.6) is granted because Θ only takes value in the range from 0 to π. Substituting (4A.6) into (4A.5) and taking advantage the property of Dirac delta function, one immediately obtains the desired expression of the Green function

$$G_{km}(\mathbf{X} - \mathbf{X}') = \left[1/8\pi^2 R\right] \int_0^{2\pi} M_{km}^{-1}(\pi/2, \phi)\, d\phi.$$

$$(4.35)$$

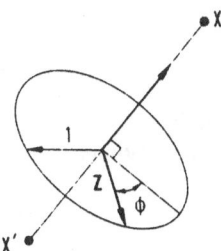

Fig. 4A. Inverse integral under spherical coordinates

Appendix 4B Decomposition of dislocation energy

We start from the general expression of total potential energy

$$\Pi = (1/2)\int_V \sigma_{ij}\,\epsilon_{ij}\, dV - \int_{S_{\text{ext}}} t_i^A u_i\, dS$$

$$(4.55)$$

to derive the decomposition formula of dislocation energy. Field variables in (4.55) are denoted collectively by $Y = \{\sigma_{ij}, \epsilon_{ij}, u_i\}$. They can be decomposed into

$$Y = Y^L + Y^I + Y^O + Y^A.$$

$$(4.54)$$

Substituting (4.54) into (4.55), and taking only the terms relevant to Y^L (including the quadratic and the cross-product terms of Y^L), we arrive

$$\Pi = \Pi^{L+I} + \Pi^{L-O} + \Pi^{L-A} + \text{terms irrelevant to } Y^L$$

$$(4\text{B}.1)$$

where

$$\Pi^{L+I} = (1/2) \int_V \sigma_{ij}^{L+I} \epsilon_{ij}^{L+I} \, dV$$

$$\Pi^{L-O} = (1/2) \int_V (\sigma_{ij}^{L+I} \epsilon_{ij}^O + \sigma_{ij}^O \epsilon_{ij}^{L+I}) \, dV$$

$$\Pi^{L-A} = (1/2) \int_V (\sigma_{ij}^{L+I} \epsilon_{ij}^A + \sigma_{ij}^A \epsilon_{ij}^{L+I}) \, dV - \int_{S_{ext}} t_i^A u_i^{L+I} \, dS \qquad (4B.2)$$

where the simplified notation

$$Y^{L+I} = Y^L + Y^I \qquad (4B.3)$$

is used to denote the combined field produced by the dislocation loop L itself and the corresponding image dislocation field due to the presence of exterior boundary. From the boundary conditions

$$(\sigma_{ij}^L + \sigma_{ij}^I)n_j = 0$$

$$\sigma_{ij}^O n_j = 0 \qquad \sigma_{ij}^A n_j = t_i^A \qquad (4.56)$$

one concludes that both Y^{L+I} and Y^O are self-equilibrium fields.

The first expression in (4B.2) can be cast into the following gradient form

$$\Pi^{L+I} = (1/2) \int_V (\sigma_{ij}^{L+I} u_i^{L+I})_{,j} \, dV \qquad (4B.4)$$

by means of the equilibrium equation. This volumetric integral can then be transformed into a surface integral by Gauss theorem

$$\Pi^{L+I} = (1/2) \oiint_{S_{ext}+S^++S^-+\text{Tube}} \sigma_{ij}^{L+I} u_i^{L+I} n_j \, dS \qquad (4B.5)$$

where "*Tube*" represents a tubular surface with axis L and radius r_o. A piece of smooth surface $S(r_o)$ is attached to the inner rim of this tubular surface, and the surfaces infinitely close to $S(r_o)$ from above and below are labelled by S^+ and S^-, respectively, as described in Fig. 4B. From the first equation of (4.56), the integral along S_{ext} vanishes. Besides, u_i^I is continuous and bounded when across $S(r_o)$ or on the tubular surface. So equation (4B.5) can be simplified to

$$\Pi^{L+I} = (1/2) \left\{ \iint_{S^++S^-} \sigma_{ij}^{L+I} u_i^L \, dS_j + \int_{\text{Tube}} \sigma_{ij}^{L+I} u_i^L \, dS_j \right\}. \qquad (4B.6)$$

This result implies the following energy decomposition formula of $L+I$

$$\Pi^{L+I} = \Pi^L + \Pi^{L-I} \qquad (4B.7)$$

where

$$\Pi^L = (1/2) \left\{ \iint_{S^+ + S^-} \sigma_{ij}^L u_i^L dS_j + \int_{\text{Tube}} \sigma_{ij}^L u_i^L dS_j \right\}$$

$$= (b_i/2) \iint_{S(r_o)} \sigma_{ij}^L dS_j + (1/2) \int_{\text{Tube}} \sigma_{ij}^L u_i^L dS_j \qquad (4B.8)$$

$$\Pi^{L-I} = (1/2) \left\{ \iint_{S^+ + S^-} \sigma_{ij}^I u_i^L dS_j + \int_{\text{Tube}} \sigma_{ij}^I u_i^L dS_j \right\}$$

$$= (b_i/2) \iint_S \sigma_{ij}^I dS_j . \qquad (4B.9)$$

The dislocation condition (4.40), as well as the boundedness of σ_{ij}^I on the tubular surface when r_o tends to zero, are utilized in the derivation of the above two formulae.

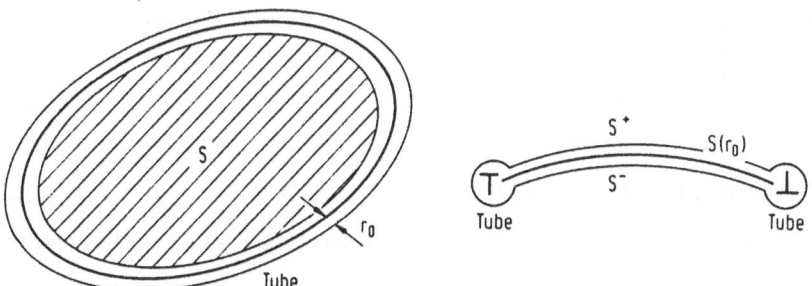

Fig. 4B. Surface integrals used in dislocation energy decomposition

The form of (4.57) is arrived by substituting (4B.7) into (4B.1). The last two expressions of (4.58) can be recovered from the last two expressions in (4B.2) and the reciprocal theorem of linear elasticity

$$\Pi^{L-O} = (1/2) \int_V (\sigma_{ij}^O u_i^{L+I})_{,j} \, dV = b_i \iint_S \sigma_{ij}^O dS_j \qquad (4B.10)$$

$$\Pi^{L-A} = (1/2) \int_V (\sigma_{ij}^A u_i^{L+I})_{,j} \, dV - \int_{S_{\text{ext}}} t_i^A u_i^{L+I} dS$$

$$= b_i \iint_S \sigma_{ij}^A dS_j . \qquad (4B.11)$$

The last two expressions in (4.56), the dislocation condition (4.40), as well as the boundedness of σ_{ij}^O and σ_{ij}^A along the dislocation line L, are used to derive the above expressions.

Appendix 4C Brown formula of dislocation field

For planar dislocation loops, Brown established simple line integral expressions for the stress-strain fields, with their integrands only involving solutions for a straight dislocation. The configuration of a planar dislocation loop is shown in Fig. 4.14 of the text, where $\mathbf{X}'$ denotes a generic point on the dislocation loop L, and $\mathbf{X}$ denotes a point of interest on the plane containing L. We further adopt the notations

$$\mathbf{X}' - \mathbf{X} = R\mathbf{T} \qquad x = X' - X \qquad y = Y' - Y. \tag{4C.1}$$

Equation (4.35) infers that the integrands in (4.48) are the homogeneous functions of R of -3 degree. That is

$$L_{k3ij}\, b_k\, G_{im,r'j'}(R\mathbf{T}) \equiv I_{mr}(R\mathbf{T}) = R^{-3}\Theta_{mr}(\phi) \tag{4C.2}$$

where L_{kpij} and the surface element dS_p' in (4.48) are written as L_{k3ij} and dS', respectively, for a planar dislocation. R and ϕ in (4C.2) are the polar coordinate counterparts of x and y. For $\mathbf{T}(\phi \pm \pi) = -\mathbf{T}(\phi)$ and the symmetry of the Green function (4.36), one concludes that Θ_{mr} is a periodic function of ϕ

$$\Theta_{mr}(\phi + \pi) = \Theta_{mr}(\phi). \tag{4C.3}$$

Furthermore, because I_{mr} is a homogeneous function of R of degree -3, one has

$$(\partial/\partial x)\,\{xI_{mr}\} + (\partial/\partial y)\,\{yI_{mr}\} = -I_{mr} \tag{4C.4}$$

by Euler's theorem of homogeneous functions. Accordingly, (4.48) can be rewritten as

$$u_{m,r}(\mathbf{X}) = -\iint_{S_{cut}} \{(\partial/\partial x)\,[xI_{mr}] + (\partial/\partial y)\,[yI_{mr}]\}\, dx dy \tag{4C.5}$$

which can be converted to a line integral by Green formula

$$u_{m,r}(\mathbf{X}) = \oint_L I_{mr}(y\, dx - x\, dy). \tag{4C.6}$$

From the geometry of the planar dislocation loop (as shown in Fig. 4.14), one has the following relationships

$$\begin{aligned} x &= R\cos\phi \qquad y = R\sin\phi \\ dx &= x'(l)\, dl = \cos\beta\, dl \\ dy &= y'(l)\, dl = \sin\beta\, dl \end{aligned} \tag{4C.7}$$

where l the arc length of the dislocation loop and β the angle between $\mathbf{t}$ and x-axis (its direction coincides with that of the Burgers vector, as specified in the main text). The substitution of (4C.7) into (4C.6) leads to

$$u_{m,r}(\mathbf{X}) = -\oint_L \Theta_{mr}(\phi) \sin(\beta - \phi)(dl/R^2). \qquad (4C.8)$$

Next, we consider the special case of a straight dislocation with an identical Burgers vector, as described by the dashed line of Fig. 4.14. The stress and displacement at a point $\mathbf{X}$ (away from the straight dislocation line by a distance h) are denoted by σ_{mr}^{∞} and u_m^{∞}, respectively. The geometry of this straight dislocation is described by

$$R = h/\sin(\beta - \phi), \qquad l = h\cot(\beta - \phi),$$
$$dl = h\,d\phi/\sin^2(\beta - \phi) \qquad (4C.9)$$

where β remains constant for a straight dislocation. (4C.8) can be simplified by (4C.9) as

$$u_{m,r}^{\infty}(h, \beta) = -(1/h) \int_{\beta-\pi}^{\beta} \Theta_{mr}(\phi) \sin(\beta - \phi)\,d\phi. \qquad (4C.10)$$

The integral in the above expression is irrelevant to h, rendering the solution form for the straight dislocation field as described in (4.52) of the text.

Taking the first and the second derivatives of the above expression with respect to the inclination angle β, one arrives

$$(\partial/\partial\beta)u_{m,r}^{\infty} = -(1/h) \int_{\beta-\pi}^{\beta} \Theta_{mr}(\phi) \cos(\beta - \phi)\,d\phi$$

$$(\partial^2/\partial\beta^2)u_{m,r}^{\infty} = -(1/h)\left\{-\int_{\beta-\pi}^{\beta} \Theta_{mr}(\phi) \sin(\beta - \phi)\,d\phi \right.$$

$$\left. + \Theta_{mr}(\beta - \pi) + \Theta_{mr}(\beta)\right\}$$

$$= -(2/h)\Theta_{mr}(\beta) - u_{m,r}^{\infty}. \qquad (4C.11)$$

The periodic condition (4C.3) of Θ_{mr} is used in the last step of manipulation. Consequently, Θ_{mr} can be expressed by the straight dislocation solution as

$$\Theta_{mr} = -(h/2)\left\{u_{m,r}^{\infty} + (\partial^2/\partial\beta^2)u_{m,r}^{\infty}\right\}. \qquad (4C.12)$$

Denoting

$$u_{j,r}^{\infty} = (1/h)U_{jr}(\beta) \qquad \sigma_{jr}^{\infty} = (1/h)\Sigma_{jr}(\beta) \qquad (4.52)$$

where U_{jr} and Σ_{jr} are the functions solely prescribed by β, one finally arrives at the Brown formulae for stress and displacement fields of a planar dislocation loop

$$u_{j,r}(\mathbf{X}) = \oint_L \left\{U_{jr}(\phi) + U_{jr}''(\phi)\right\} \left[\sin(\phi - \beta)/2R^2\right] dl$$

$$\sigma_{jr}(\mathbf{X}) = \oint_L \left\{\Sigma_{jr}(\phi) + \Sigma_{jr}''(\phi)\right\} \left[\sin(\phi - \beta)/2R^2\right] dl \qquad (4.53)$$

from (4C.8), (4C.12), (4.51) and (4.52).

4.6 References

1. Taylor, G.I., Proc. Roy. Soc. London, A145(1934), p.364.
2. Burgers, G.M., Proc. Kon. Ned. Akad. Wetenschap., 42 (1939), p.293, p.384.
3. Hirth, J.P. and Lothe, J., Theory of Dislocations, McGraw-Hill, 2nd Ed., (1982).
4. Weingarten, G., Atti. Accad. Lincei Rend., 5(1901), p.57.
5. Nabarro, F.R.N., Adv. Phys., 1(1952), p.384.
6. Rice, J.R., J. Mech. Phys. Solids, 19(1971), p.433.
7. Bacon, D.J., Barnett, D.M. and Scattergood, R.O., Progr. in Mater. Sci., 23(1978), p.51.
8. Timoshenko, S.P. and Goodier, J.N., Theory of Elasticity, McGraw-Hill, (1970).
9. Mura, T., Phil. Mag., 8(1963), p.843.
10. Brown, L.M., Phil. Mag., 15(1967), p.363.
11. Gavazza, S.D. and Barnett, D.M., Scripta Metall., 9(1975), p.1263.
12. Gavazza, S.D., Energy Release Rates and Associated Forces on Singular Dislocations, Ph.D. Thesis, Stanford Univ., (1975).
13. de Wit, G. and Koehler, J.S., Phys. Rev., 116(1959), p.113.
14. Stein, D.F. and Low, Jr. J.R., J. Appl. Phys., 31(1960), p.362.
15. Gilman, J.J., Microplasticity, (1968), p.17.
16. Kelly, J.M. and Gillies, P.P., Int. J. Solids & Struct., 10(1974), p.45.
17. Frost, H.J. and Ashby, M.F., Deformation Mechanism Maps, Pergamon Press, (1982).
18. Earmme, Y.Y. and Weiner, J.H., 45(1974), P.603.
19. Perzyna, P., Adv. Appl. Mech., 9(1966), p.243.
20. Bodner, S.R. and Parton, Y.J., Appl. Mech., 42(1975), p.385.
21. Campbell, J.D. and Ferguson, W.D., Phil. Mag., 21(1970), p.63.
22. Frank, F.C., Proc. Phys. Soc., A62(1949), p.131.
23. Eshelby, J.D., Proc. Phys. Soc., A62(1949), p.307.
24. Frank, F.C., Phil. Mag., 42(1951), p.809.

5 Deformation Mechanisms II: Miscellaneous

Beside the gliding motion of dislocations, plasticity can be induced by a variety of other deformation mechanisms, such as *deformation twinning, shear banding, grain boundary sliding and phase transformation*. These deformation mechanisms will be described herein to give a more complete picture of plasticity.

5.1 Deformation Twinning

5.1.1 Twinning versus Slip

Twins may be produced as a result of mechanical deformation or annealing. The first type is known as *mechanical or deformation twin* and the second type is called *annealing twin*. Deformation twinning is another mechanism of plastic deformation. Like slip, deformation twinning enables a crystal to undergo a permanent shape change with negligible volume variation. Unlike slip, twinning involves the shear of a part of a crystal by a fixed magnitude, characteristics of the crystal structure. Twins form at a very high speed (in the order of microseconds) and this produces a sound sometimes known as tincry. As shown in Fig. 5.1, twinning involves movement of planes of atoms in the lattice parallel to a specific (twinning) plane so that the lattice is divided into two symmetrical parts which are differently oriented. The amount of movement of each plane of atoms in the twinned region is proportional to its distance from the twinning plane, so that a mirror image is formed across the twin plane. In slip, the orientations of the crystal above and below the slip plane are the same. The difference in orientation between the twinned and untwined regions makes the twins visible on a polished and etched surface.

Twinning occurs in a definite crystallographic direction and on a specific crystallographic plane for each crystal structure. Some common *twinning systems* are shown in Table 5.1 below:

A detailed analysis of the crystallography of twinning can be found in the work of Christain[1].

Deformation twinning is favoured by low temperature, high strain rate and low stacking-fault energy, as well as in HCP metals at orientations which are unfavorable for basal slip. Twinning occurs in many bcc transition metals such

Table 5.1. Some common twinning systems in metals

Crystal structure	Metals	Twinning plane	Twinning direction
FCC	Cu, Ag, Au	$\{111\}$	$\langle 11\bar{2}\rangle$
BCC	α-Fe, Ta	$\{112\}$	$\langle 111\rangle$
HCP	Zn, Mg, Ti	$\{10\bar{1}2\}$	$\langle\bar{1}011\rangle$

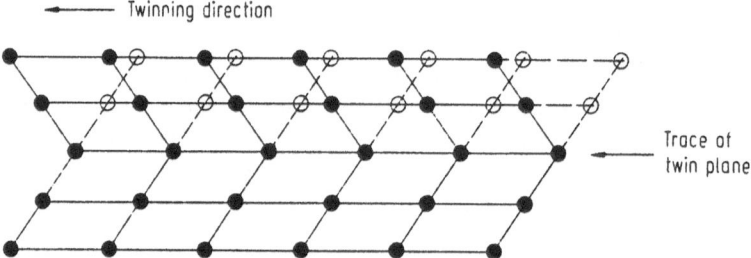

Fig. 5.1. Schematic diagram of twin formation

as Fe, V and Nb. Aluminium has a high stacking fault energy and deformation twins have not been observed. In FCC solid solutions alloys such as Cu-Zn, Cu-Al and Ni-Cu alloy where the stacking fault energy is low, deformation twins forms during rolling at room temperature under intermediate to large rolling strain. An example of profuse deformation twin formation in cold-rolled α-brass is shown in the electron micrograph of Fig. 3.33. In copper, deformation twins have been produced by tensile deformation or by shock loading.

5.1.2 Kinematics of Twinning

In twinning there is an orientation difference across the twin plane. The lattice of the twinned part is a reflection of the non-twinned part with respect to the twinning plane. Referring to Fig. 5.2, the unit vector on twinning plane is $\mathbf{v}$ and $\mathbf{w}$ is a vector on an arbitrary crystal direction before twinning. The components of $\mathbf{w}$ in the twinning plane is w_p and the component normal to the twinning plane is w_n. After twinning, the vector $\mathbf{w}$ becomes vector $\mathbf{w}''$ which equals $-\mathbf{w}'$. This transformation is equivalent to a 180° rotation around the twinning plane normal $\mathbf{v}$. The transition between the two lattices can be described by a vector transformation law changing $\mathbf{w}$ to $\mathbf{w}''$, i.e.,

$$\mathbf{w}'' = \mathbf{R}\,\mathbf{w} \tag{5.1}$$

where the elements of the second rank tensor $\mathbf{R}$ are

$$R_{ij} = 2v_i v_j - \delta_{ij}, \tag{5.2}$$

and δ_{ij} is the Kronecker delta. The values of v_i can be obtained from the indices of the twinning plane referring to the coordinate system of the crystal $(x_1 x_2 x_3)$. For cubic metals, $x_1 = [100], x_2 = [010]$, and $x_3 = [001]$, and the twinning plane is denoted by $(h_1 h_2 h_3)$, then

$$v_i = h_i / (\Sigma h_i^2)^{1/2}. \tag{5.3}$$

From (5.1), the twinning matrices around the $\{111\}$ twinning poles in FCC metals are as follows:

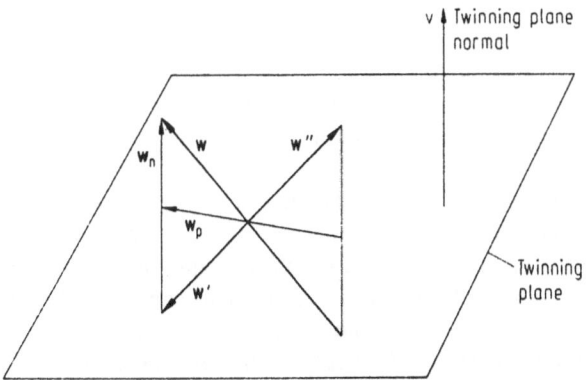

Fig. 5.2. A vector diagram showing the twin transformation. After twinning, an arbitrary crystal direction $\mathbf{w}$ changes to $\mathbf{w''}$ and $\mathbf{w'}$ is the reflection of $\mathbf{w}$ in the twinning plane. $\mathbf{w''} = -\mathbf{w'}$

Pole	Twinning matrix	Pole	Twinning matrix
[111]	-1 2 2 2 -1 2 2 2 -1	[1Ī1]	-1 -2 -2 -2 -1 -2 2 -2 -1
[Ī11]	-1 -2 -2 -2 -1 2 -2 2 -1	[ĪĪ1]	-1 2 -2 2 -1 -2 -2 -2 -1

For hexagonal metals, $x_1 = [2\bar{1}\bar{1}0], x_2 = [01\bar{1}0]$, and $x_3 = [0001]$, and the twinning plane is $(hkil)$, then

$$v_1 = h/m^{1/2}, \quad v_2 = (2k + h)/(3m)^{1/2}, \quad v_3 = 1/[(c/a)m^{1/2}],$$

with

$$m = h^2 + 1/3(2k + h)^2 + l^2/(c/a)^2 \tag{5.4}$$

where c and a are the dimensions of the hexagonal unit cell referred in Fig. 3.9 of Chap. 3.

The orientation relationship of the twinned part of the crystal $\mathbf{T'}$ expressed in directional cosines of the crystal system and the matrix T of the origin is given by

$$\mathbf{T'} = \mathbf{R}\,\mathbf{T}. \tag{5.5}$$

5.1.3 Directionality of Twinning

The amount of macroscopic shear $d\gamma^t$ of the crystal due to twinning on the i-th twinning system is given by

$$d\gamma^t = \gamma_o\,df_i^t \tag{5.6}$$

where γ_o is the shear associated with the i-th twinning system and is a constant for a crystal structure, and df_i^t is the volume fraction of material twinned in the incremental strain step. In cubic metals, the twinning shear is 0.707 and in HCP metals, the twinning shear will depend on the c/a ratio of the lattice[2]. A quantitative treatment which permits prediction of which slip and twinning systems operate to a given shape change for FCC crystals deformed by plane strain compression has been given by Chin et al.[3]. Although the plastic strain contributed by deformation twins is small, twinning exerts effect on the yielding behaviour, and on the development of the crystallographic texture in certain metals and alloys.

If twinning occurs during a tensile test, serrations in the tensile stress-strain curve will result. The resolved shear stress τ^t on a twinning system under a uniaxial stress σ is,

$$\tau^t = m^t\sigma \tag{5.7}$$

where m^t is the Schmid's factor for the twinning system and is analogous to that (m^s) used for slip, i.e.

$$\tau^s = m^s\sigma. \tag{5.8}$$

A twinning system will be activated when

$$\tau^t = \tau_c^t \tag{5.9}$$

where τ_c^t is the critical resolved shear stress for twinning. The value of m^t is different in tension and compression for each crystal orientation and therefore could contribute to the Bauschinger effect on yielding due to the directionality of mechanical twinning.

Whether twinning or slip is favoured in a particular grain orientation depends on the ratio of the critical resolved shear stress for twinning and that for slip. An example of calculation is shown by Backofen[4] for the case of a FCC crystal which is stressed along the [111] and [001] direction in both tension and compression. The Schmid's factors for slip and twinning are listed below in Table 5.2.
It can be seen that twinning will be most favored when compression is along [001] and most unlikely when compression is along [111] of the crystal. Twinning will

Table 5.2. List of the Schmid's factors

Orientation	m^s	m^t	
		tension	compression
[1 1 1]	0.272	0.314	0.157
[0 0 1]	0.408	0.235	0.471

be totally suppressed if

$$\tau^t/\tau_c^t < \tau^s/\tau_c^s \tag{5.10}$$

or

$$\tau_c^t/\tau_c^s > m^t/m^s. \tag{5.11}$$

There shall be no twinning if the critical resolved shear stress ratio for twinning and slip is greater than $0.471/0.408 (= 1.152)$ calculated for the [001] direction. On the other hand, there shall be no slip if

$$\tau_c^t/\tau_c^s < m^t/m^s \tag{5.12}$$

i.e. when the stress ratio is less than $0.157/0.272 \ (= 0.577)$, calculated for the most favorable orientation for slip.

Although the amount of gross deformation produced by twinning is small, twinning is an important deformation mechanism since it may bring potential slip planes into a more favorable position for slip. It has been shown by Wassermann[5] that if mechanical twinning is an available mode of deformation in addition to slip, the crystallographic texture components near $\{112\}<111>$ can be transformed to $\{552\}<115>$ by twinning and hence by further slip to $\{110\}<001>$.

5.2 Shear Banding

5.2.1 Structure of Shear Bands

Shear banding is a highly localized form of plastic deformation frequently observed in a variety of metals and alloys deformed at high strain rates to large plastic strains. They appear as zones of intense shear with or without microstructural modification of the original materials. Flow localization in the form of shear bands is an important part of the ductile fracture mechanism as fracture eventually occurs along these bands. There has been a tendency to regard shear bands as a failure mode. However, under compressive loading, shear banding delays rather than promotes the occurrence of cracking. Shear banding provides a mechanism through which large plastic strain can be accommodated within microstructures without fracture.

According to Backman and Finnegan[6] there are two types of shear bands. These are *deformed shear bands* and *transformed shear bands*. The former often occur in non-ferrous alloys such copper alloys, aluminium alloys, nickel alloys, and the latter are observed frequently in alloy steels, titanium and uranium alloys which are capable of phase transformation at elevated temperature. The temperature attained within shear bands is extremely high and martensite can form from the high temperature phase on rapid cooling to room temperature. In materials of low thermal diffusivity, the rate of heat generation resulting from the localized plastic band exceeds its rate of dissipation into the surrounding materials. Shear bands thus formed are thermoplastic in origin when the effect of thermal softening outweighs the effects of strain and strain rate hardening.

Transformed shear bands are up to hundreds of micrometers wide with distinct microstructures resulting from the phase transformation, while deformed shear bands appear as thin dark-etching bands under the optical microscope (Fig. 3.34) and are very sharp and narrow even observed under an electron microscope. Fig. 5.3 is a schematic drawing of a heavily rolled brass showing the way in which shear bands appear in the main planes of the rolling geometry.

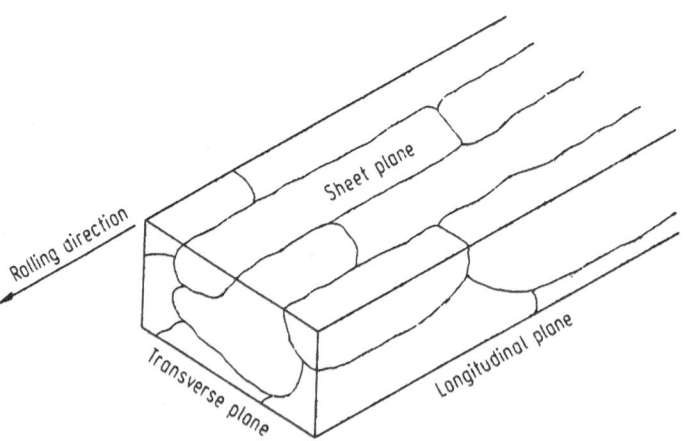

Fig. 5.3. Schematic diagram of shear bands in the rolling geometry

Although there is no phase transformation in the deformed shear band, the fine crystallite within the band reveals a dynamically recovered structure due to the effect of adiabatic heating. The strain which can be accommodated in such band is enormous and has been estimated by Duggan et. al.[7] to be in the order of $\sim$ 3-4. It must be emphasized that shear bands are macroscopic features in the sense that the direction of shearing can be non-crystallographic (i.e. the shear band planes are not necessary crystallographic slip planes). Despite the classification of shear bands being either transformed or deformed, the former follow on from deformed shear bands which represent an earlier stage in the adiabatic strain localization process. In the following subsections, the discussion will focus on the mechanism of formation of deformed shear bands only.

5.2.2 Crystallographic Model of Shear Band Formation

Physically, shear bands form as a result of the gradual exhaustion of the crystallographic slip process in an imperfection free material. Shear-band localization is very sensitive to material characterization. The propagation of shear bands in textured polycrystals requires a constitutive model which accounts for the material anisotropy. An important source of material anisotropy is the preferred orientations of the grain aggregates. For simplicity, the material constitutive equations for an individual grain can be written as

$$\sigma = M\tau_c \tag{5.13}$$

where σ is the equivalent stress or the plastic work per unit volume strain and τ_c is the critical resolved shear stress on the active slip systems. M is the Taylor factor which relates the effective strain $d\epsilon_w$ (subscript w is employed to emphasize its correspondence with respect to the workpiece coordinates) and the total dislocation shear $d\Gamma$

$$M = d\Gamma/d\epsilon_w. \tag{5.14}$$

The Taylor factor can be calculated readily from the maximum work principle of Bishop and Hill[8] and is related to the orientation of the crystals with respect to the principal stress.

In a strain rate sensitive material, the stress also depends on the density of mobile dislocations N and the velocity of dislocation v. Assuming the active slip plane has a hardening rule of the form

$$\tau = k\Gamma^n \tag{5.15}$$

where n is the hardening exponent which is assumed to be the same for all slip planes and k is a material constant, the macroscopic stress σ is written by Dillamore [9] in the form of

$$\sigma = \frac{k}{(bv)^m} M^{(1+n+m)} \epsilon^n \cdot \dot{\epsilon}^m \cdot N^{-m} \tag{5.16}$$

where $b =$ the length of Burgers vector, $v =$ dislocation gliding velocity, $m =$ strain rate sensitivity, $N =$ density of mobile dislocation, $\epsilon =$ strain rate.

Either equation (5.13) or (5.16) gives the material constitutive equation for an individual grain. When more than one grain is considered in the sample volume, M will be averaged over all grain orientations being considered. A large value of M will indicate a large shear strength of the grain.

The geometry of shear band formation is shown in Fig. 5.3. Referring to the sample axis $(X - Y)$, the displacement gradient E is denoted by

$$E_+ = d\gamma \begin{bmatrix} \sin\beta\cos\beta & 0 & \cos^2\beta \\ 0 & 0 & 0 \\ -\sin^2\beta & 0 & -\sin\beta\cos\beta \end{bmatrix} \tag{5.17}$$

for a positive shear band, where $d\tau$ is the incremental shear strain in the band inclined at angle β to the specimen axis, and

$$E_- = d\gamma \begin{bmatrix} \sin\beta\cos\beta & 0 & -\cos^2\beta \\ 0 & 0 & 0 \\ \sin^2\beta & 0 & -\sin\beta\cos\beta \end{bmatrix} \tag{5.18}$$

for a negative shear band.

Sheet surface

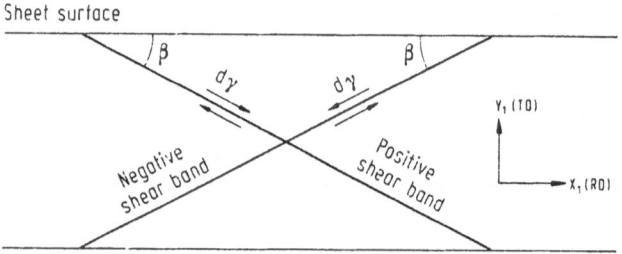

Fig. 5.4. Geometry of shear band formation

The symmetric strain tensor of both the positive and negative shear band referred to the sheet geometry of Fig. 5.4 is given by

$$\epsilon_w = d\gamma/2 \begin{bmatrix} \sin^2\beta & 0 & \cos^2\beta \\ 0 & 0 & 0 \\ \cos^2\beta & & -\sin^2\beta \end{bmatrix} \tag{5.19}$$

The macroscopic imposed strain tensor can be transformed from the workpiece coordinate system $(X - Y)$ to the crystallographic axis of the crystal, i.e.

$$\epsilon_c = P\epsilon_w P^{-1} \tag{5.20}$$

where ϵ_c is the strain tensor referred to the cube axes of the crystals and P is the transformation matrix given by

$$P = \begin{bmatrix} r_1 & u_1 & n_1 \\ r_2 & u_2 & n_2 \\ r_3 & u_3 & n_3 \end{bmatrix} \tag{5.21}$$

and P^{-1} is the transpose of P with

$$\begin{array}{ll} r_1 = u/(u^2 + v^2 + w^2)^{1/2}, & n_1 = h/(h^2 + k^2 + l^2)^{1/2} \\ r_2 = v/(u^2 + v^2 + w^2)^{1/2}, & n_2 = k/(h^2 + k^2 + l^2)^{1/2} \\ r_3 = w/(u^2 + v^2 + w^2)^{1/2}, & n_3 = l/(h^2 + k^2 + l^2)^{1/2} \end{array} \tag{5.22}$$

where $\{hkl\}<uvw>$ are the Miller indices of the grain orientation, and the unit vector

$$\mathbf{u} = \mathbf{n} \times \mathbf{r} \tag{5.23}$$

is perpendicular to both slip plane normal $\mathbf{n}$ and slip direction $\mathbf{r}$.

For an isotropic material the orientation of the shear band should coincide with the direction of maximum shear stress and makes an angle 45° with the principal stress axis. If the shear angle deviates from 45° by an angle α, the shear strain in the band will be increased by a factor of $1/\cos 2\alpha$ in order to produce the same amount of macroscopic strain. Since the resolved shear stress on the inclined plane decrease as $1/\cos 2\alpha$ from the 45° plane, an *effective Taylor factor* is defined by $M/\cos 2\alpha$. The shear is considered by Dillamore et al.[9] to occur at angle such that the plastic work in deforming the material will be minimum, i.e. at which the value of $M/\cos 2\alpha$ is minimized. Very often the variation of $M/\cos 2\alpha$ with shear band angles is associated with a plateau and a range of shear angle is possible based on the principle of minimum work alone. An example showing the variation of the effective Taylor factor with potential shear band angles for the ideal grain orientations $\{112\}<111>$ is shown in Fig. 5.5. The shear strength is found to be a minimum from 19.47° to 54.74°.

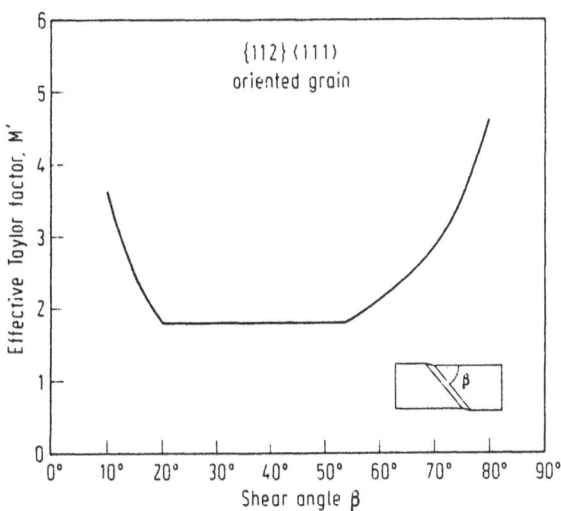

Fig. 5.5. Variation of Taylor factor for a positive shear band in $\{112\}<111>$ oriented grains

Another criterion used for the determination of shear band angle is based on *load instability*[10]. A shear band will form when

$$\frac{1}{\sigma}\frac{d\sigma}{d\epsilon} < 0. \tag{5.24}$$

Differentiating equation (5.13), we have

$$\frac{1}{\sigma}\frac{d\sigma}{d\epsilon} = \frac{1}{M}\frac{dM}{d\epsilon} + \frac{M}{\tau_c}\frac{d\tau_c}{d\gamma} \leq 0. \tag{5.25}$$

For a strain rate sensitive material, instability occurs when

$$\frac{1}{\sigma}\frac{d\sigma}{d\epsilon} = \frac{n}{\epsilon} + \frac{m}{\dot{\epsilon}}\frac{d\dot{\epsilon}}{d\epsilon} + \frac{1+n+m}{M}\frac{dM}{d\epsilon} - \frac{m}{N}\frac{dN}{d\epsilon} \leq 0. \qquad (5.26)$$

The term $S = (1/M)(dM/d\epsilon)$ represents the *texture softening factor* when S is negative or *texture hardening factor* when S is positive, and $(M/\tau_c)(d\tau_c/d\Gamma)$ represents the slip hardening contribution. The term $d\dot{\epsilon}/d\epsilon$ is generally positive while $dN/d\epsilon$ is negative (i.e. the number of mobile dislocation decreases with strain), and the only term that cause instability in equation (5.25) or (5.26) is $1/M(dM/d\epsilon)$. At the early stages of deformation, $(M/\tau_c)(d\tau_c/d\Gamma)$ is positive and a shear band will develop when $(1/M)(dM/d\epsilon)$ is most negative. This texture softening factor is computed numerically for both the positive and negative shear band by determining the rate of lattice rotation $(d\Omega/d\epsilon)$ and the associated change in the Taylor factor for each grain orientation, i.e.

$$\frac{dM}{d\epsilon} = \frac{d\Omega}{d\epsilon}\frac{dM}{d\Omega}. \qquad (5.27)$$

The texture softening factor calculated for single orientations sometimes contains several minima such that no unique solution can be found. An example of this behaviour is shown for the $\{112\}<111>$ oriented grains (Fig. 5.6).

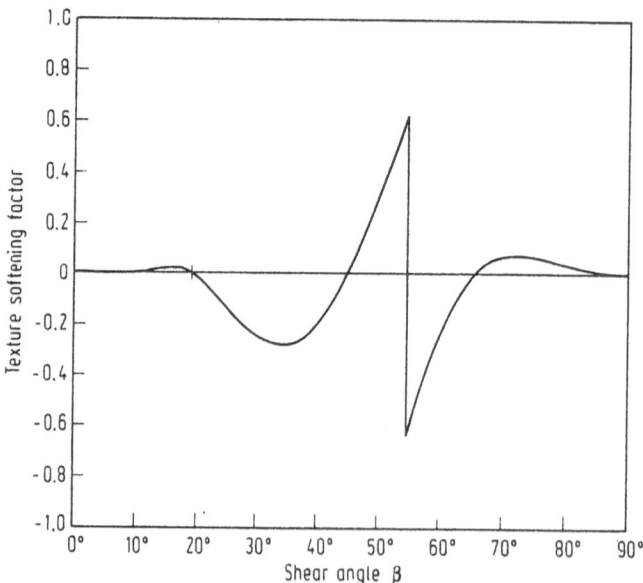

Fig. 5.6. Variation of texture softening factor for a positive shear band in $\{112\}<111>$ oriented grains

It has been shown by Lee and Chan[11] that a minimum texture softening is not a sufficient condition for the shear bands to occur. The most likely shear band propagation angle is given by the one which has the minimum softening factor among the ones which has the same minimum number of slip systems and shear strength. The prediction agrees well with the frequent occurrence of $\sim 19°$ and $\sim 35°$ shear bands found in aluminium alloys and accounts for shear band angles in other alloy system.

The critical strain at which a shear band will form can be solved from equation (5.24). From (5.13) and (5.14), the material constitutive equation can be written as

$$\sigma = Mk(M\epsilon_w)^n. \tag{5.28}$$

Load instability occurs when

$$\frac{n}{\epsilon_w} + (n+1)S \le 0 \tag{5.29}$$

and the critical strain is given by

$$\epsilon_w^* = -n/(n+1)S. \tag{5.30}$$

The critical strain at which grain-scale shear band will appear thus depends on the n-exponent and the texture softening factor for the particular grain orientation. For sample-scale shear bands S will be the average texture softening factor calculated for all the texture components considered in the polycrystalline aggregates. Equation (5.30) holds only when the texture softening factor S is negative which is consistent with the instability condition. A more complex expression for the critical strain can be derived from (5.16) but this would depend on a detailed knowledge of dislocation dynamics.

5.2.3 Symmetry Requirement in Shear Band Formation

Shear bands occur either in single sets, unsymmetrical pairs or symmetrical pairs. Under plane strain compression, the redundant shears caused by the single or the unsymmetrical sets of shear bands cannot be cancelled out and this would cause strain accommodation problem. From (5.18), the ratio (k) of the shear strain to the nominal strain equals $\cos 2\beta / \sin 2\beta$. The redundant shear equals zero only when $k = 0$ (i.e. at $\beta = 45°$). A single set of shear bands at angles smaller than $45°$ and set of unsymmetrical shear bands are often observed in rolled metals. An example is the $\sim 19°$ and $\sim 35°$ shear bands found in the same grain of in Al-4%Cu alloy[14]. Even for a set of symmetrical shear bands, they do not form simultaneously as shown by the usual large shear displacement in either one of them.

Ways of overcoming the problem include broadening of the shear bands[9], rotation of the materials between the shear bands to adapt to the external prescribed strain, or annihilation of the redundant shears by different sense of shear bands in different layers[10]. The large rotation of material within the shear band is unlikely as judged by the stability of the rolling texture found in polycrystalline aluminium[11]. Only very small rotation was detected in Ni-single crystals[12] when shear bands were formed. The broadening of the shear bands for even a moderate strain does not match the observed narrow width of the shear bands. While different senses of shear bands are sometimes observed at different layers in a deformed polycrystalline aggregate, the operation of a single set of shear bands in single crystal cannot be explained in this way.

Shear bands occur when a number of slip systems required to accommodate the imposed shape change coincides with the minima in shear strength and texture softening factor. An example of the slip activities in an $(112)[11\bar{1}]$ oriented grain is given in Fig. 5.7, in which the predicted shear angle occurs at 35.26° for the positive and 19.47° for the negative shear bands as shown in Fig. 5.6. These shear band angles are arrived at by summing the components of the activated slip systems which are found to be a_1, a_2, and c_3, d_3 in Bishop and Hill's nomenclature[8]. The reduction in the number of active slip systems from five to two (i.e. a_1, a_2, or c_3, d_3) for a single set of shear bands is insufficient to accommodate the external shape change. In order to preserve material continuity, the crystal then shears macroscopically on a plane and along a direction that is symmetrical to the resultant of the slip produced by the vectorial addition of the two individual slip systems as shown in Fig. 5.8.

The magnitude and direction of the slips and slip systems are defined by the vectors $\mathbf{p}$ and $\mathbf{r}$ and the resultant of these two vectors is denoted by $\mathbf{s} \cdot \mathbf{s}$ gives the magnitude and direction of the *ghost shear band* which is proposed by Lee and Chan[14] to balance out the redundant shear strain of the observed shear band, i.e

$$\mathbf{p} + \mathbf{r} = \mathbf{s}_{ghost}. \tag{5.31}$$

The observed shear band $\mathbf{s}_{obs}$ is related to the $\mathbf{s}_{ghost}$ by

$$|\mathbf{s}_{obs}| = |\mathbf{s}_{ghost}| \tag{5.32}$$

but rotated $(180 - 2\beta°)$ clockwise about the transverse direction. In a $(112)[111]$ crystal, β is predicted to be 19.47° for a observed negative shear band of which the shear plane and shear direction is (115) and $[\bar{5}\bar{5}2]$ respectively. The shear banding may take place on planes or along directions, depending on the resultant shear deformation produced, and is thus non-crystallographic, i.e the shear planes are not necessarily the (111) slip planes and the shear directions not necessarily the $<110>$ slip directions. Such a macroscopic shearing of the crystal is possible due to the high internal stresses within the work-hardened grain and the rigid external constraints that are imposed upon it, such as those occurring in plane strain rolling or compression. In a tensile test, the external constraint is less severe and shear banding is than less frequent. The relative ease at which shearing can occur also depend on other factors such as the grain size, the stacking fault energy, and the temperature and the rate of deformation.

The non-crystallographic shear banding does not produce any orientation change in the matrix. However the slip that is responsible for the ghost shear band is crystallographic and is distributed in the matrix. When only 2 slip sys-

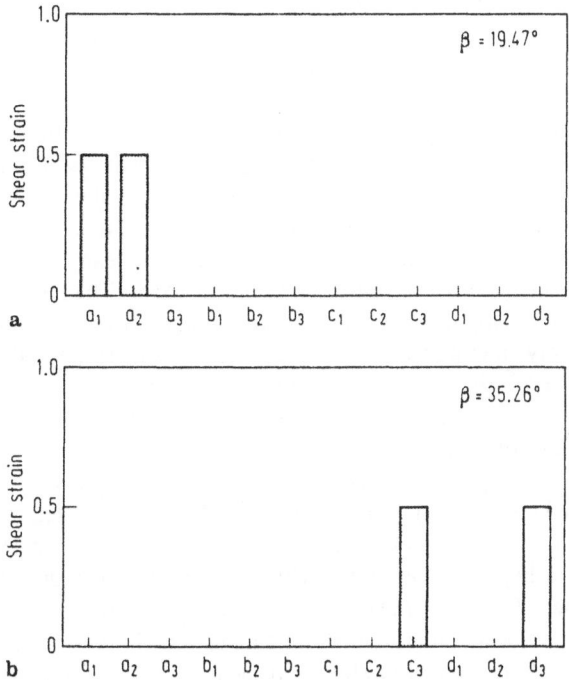

Fig. 5.7. Slip activities in (112)[111] oriented grain. **a** Positive shear band; **b** negative shear band

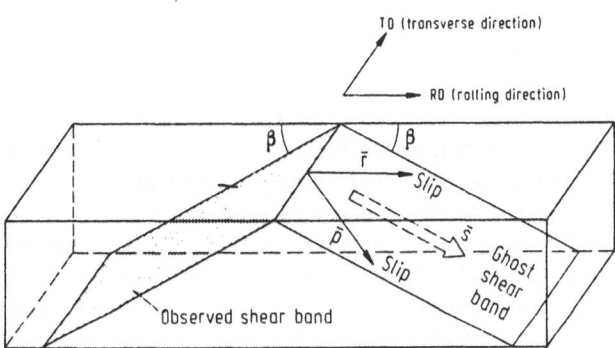

Fig. 5.8. Formation of a shear band which is symmetrical to the resultant of the crystallographic slips in the matrix grain

tems are available in the matrix, there is no crystal rotation as shown in the case of (112)[11$\bar{1}$] oriented grain up to very large rolling strain, based on the Taylor analysis for an imposed gradient given by equation (5.17) or (5.18).

The deformation mechanism operating within the bands themselves is largely unknown and the justification for applying the Taylor theory of homogeneous plastic deformation to the analysis of the localized shear is often a puzzle to researchers. There is no evidence to support that the Taylor mode of deformation occurs within the bands. The dilemma of applying the Taylor theory of homogeneous plastic deformation for the analyzing the highly localized shear bands can be removed if it is understood that the Taylor theory is not applied directly to the non-crystallographic shear bands themselves, but to the crystallographic slip activities that must occur outside the shear bands in order to accommodate for the redundant shear that accompany single set of shear bands or unsymmetrical set of shear bands. The idea of an uniformly distributed shear of opposite sign superimposed on the whole shear-banded structure is not entirely new and has been proposed by Van Houtte[10]. Recent experimental evidence of slip activities outside the band has been shown by Harren et al. [15] and Yeung [16] to exist. The deformation produced by slip activities outside the band and on the reduced number of slip systems will be equivalent to the operation of a single shear band that is symmetrical, but in the opposite sense, to the observed one. It is this "ghost shear band" that is predicted by the Taylor type of analysis. The external constraints that occur under plane strain deformation coupled with the high internal stress within a crystal with reduced hardening capacity will favour *slip induced shearing* (i.e shear banding) as an alternate mechanism of plastic deformation at high strain and high strain rate. When the crystallographic slip is totally prevented or slow to respond to external loading at high strain rate then the only way to relax the constraint of the deformation is by cracking[17].

5.2.4 Shear Banding and Bifurcation

The microscopic hardening event discussed above provides the physical basis of the occurrence of shear band as explained by the bifurcation analysis which aims at finding a state at which the solution of a definite boundary value problem looses its uniqueness. Conditions that govern uniqueness depend sensitively on the hardening rates, the stress state and the number and relative orientation of the active slip systems. For an imperfection free FCC or BCC single crystal subjected to uniaxial tension (Fig. 5.9), Asaro [18] shows that bifurcation into a shear band mode could occur when h/σ fell below a critical value which is given by

$$(h/\sigma) = \frac{\cos 2\Theta - \cos^2 2\Theta / \cos 2\phi}{(1-q)\cos^2 2\Theta / \cos^2 2\phi + (1+q)\sin^2 2\Theta / \sin^2 2\phi} \qquad (5.33)$$

where h is the slip plane hardening rate (i.e. the ratio of the rate of resolved shear stress to the magnitude of shear on the active slip systems), σ is the current tensile stress, Θ is the orientation of the shear band, ϕ is the angle of the double slip system with respect to the tensile stress axis, and $q \equiv h_1/h$. h and h_1 are the elements of the hardening matrix $h_{\alpha\beta}$ such that $h = h_{11} = h_{22}$ and $h_1 = h_{12} = h_{21}$, the off-diagonal elements of which represent latent hardening[19].

For $q = 1$ and $\phi = 30°$, the shear band angle is found to be 37.2°, i.e., a few degrees from the slip system. When Θ lies in the range from 0° to 45°, (h/σ) is always positive. An interesting point is that (h/σ) will become zero when ϕ approaches 45° which is attributed to the existence of a vertex in the stress space of $(\sigma_{22}, \sigma_{12})$. In other words, the onset of shear banding is influenced by the deviation from a smooth yield surface. Such a distortion of the yield surface (sharp vertices or noses) can be easily attained at a grain-scale level. For a more mathematically vigorous treatment of the subject readers should refer to the work of Asaro in Ref.[19].

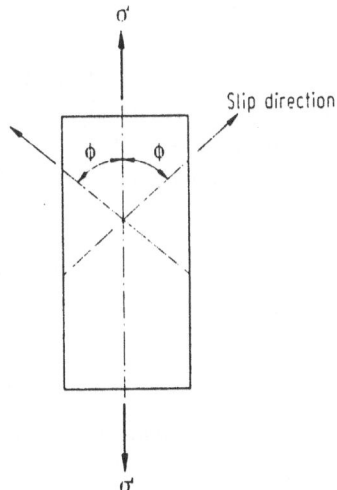

Fig. 5.9. Inclination of slip systems to the tensile stress axis

The importance of slip plane rate on the formation of shear band has been demonstrated by Chang and Asaro[20] in an Al-2.8%Cu single crystal and Harren et al.[15] in polycrystal. For a material with a progressively decreasing rate of strain hardening, uniqueness is no longer certain and bifurcation then occurs. The continuum and the crystallographic model are complementary to each other in the analysis of shear localization. Much research is still to be done in the characterization of shear banding. A link which will bridge the continuum and the crystallographic model of shear band formation is through a proper description of the yield surface constructed from the material variables taking into account the crystallographic textures, lattice rotation and strain hardening of the dislocation slip planes.

5.3 Grain Boundary and Grain Boundary Sliding

5.3.1 Kinematics of Grain Boundary

Metals, as well as some ceramics and crystalized polymers, usually exist in the form of *polycrystal aggregates*, with *grain boundaries* as the interfaces between grains with different orientations. Similar to the definition of surface tension, the energy required to form grain boundary of unit area is defined as

$$\Gamma = G/A \tag{5.34}$$

where G is the Gibbs free energy, and A denotes the surface area.

The grain shapes in a polycrystal should meet the following requirements :

A. Thermodynamic requirements: *the stabilized grain shape should minimize the total grain boundary energy.* Several consequences can be deduced from this requirement.

(1) The equilibrium grain boundaries are planar in shape.
(2) According to the surface tension analogy of grain boundaries, which models grain boundary in a similar way as the interface in the soap-bubble aggregates, only triple joined grain boundaries are stable, as shown in Fig. 5.10.
(3) For the three grain boundaries joined together, the individual angles extended between them should satisfy the following Sine law

$$\Gamma_1/\sin\phi_1 = \Gamma_2/\sin\phi_2 = \Gamma_3/\sin\phi_3 \tag{5.35}$$

as shown in Fig. 5.10. For grain boundaries of identical structure ($\Gamma_1 = \Gamma_2 = \Gamma_3$), the above formula can be simplified to

$$\phi_1 = \phi_2 = \phi_3 = 120°. \tag{5.36}$$

B. Topological requirement: *a group of connected grains should satisfy the Euler's law of topology* as follows

$$C - E + F - B = 1 \tag{5.37}$$

where C, E, F and B represent the numbers of corners, edges, faces and bodies of grains in polycrystal aggregates. Although Euler's law can not be universally applied to all circumstances, (5.37) is valid under certain topological restrictions which do not conflict with the formation law of polycrystal aggregates.

The above requirements can be used to determine the grain geometry of polycrystals. We first discuss the case of an array of two-dimensional grains. The Euler equation is reduced to

$$C - E + F = 1 \tag{5.38}$$

for two dimensional case, where C, E and F are the numbers of corners, edges and faces for planar polycrystal aggregates. By the minimum energy requirement, all corners are joined by three edges, i.e.

$$C/2 = E/3 \tag{5.39}$$

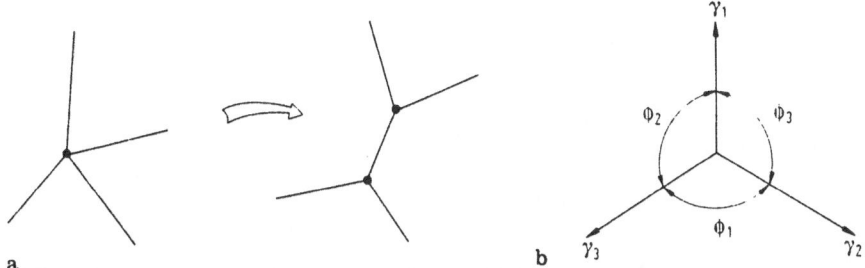

Fig. 5.10. Stability of a grain boundary joint. **a** Unstable joint; **b** stabilized joint

and the number of planar grains relates to the edge number by

$$E = \frac{1}{2}nF \tag{5.40}$$

where n denotes the average number of edges per each planar grains, and each edge is shared by the neighboring two planar grains, as shown in Fig. 5.11(a).

The value of n can be derived by combining the above three expressions as

$$n = 6(1 - 1/F). \tag{5.41}$$

If the grain number is very large, n will approach 6, then the shape of planar polycrystals with uniform grains is composed by an array of hexagons, as shown in Fig. 5.11(a). When the number of grains is finite, however, n would be slightly less than 6 due to the influence of the exterior boundary.

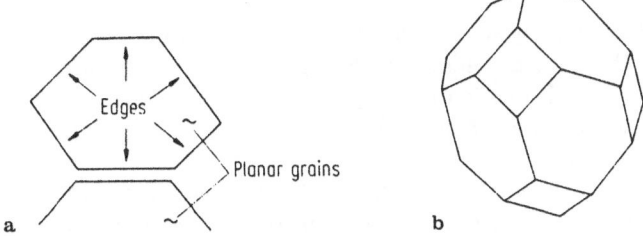

Fig. 5.11. Grain geometry. **a** Hexagonal planar grains; **b** three dimensional Kelvin tetrakaidecahedron

We next discuss the geometry of a three dimensional array of grains, in which (5.37) should be employed instead of (5.38). The minimum energy principle tells us that every corner should be joined by 4 edges, in the shape of a tetrahedron,

$$C/2 = E/4 \tag{5.42}$$

and every edge is shared by 3 faces

$$E = (n/3)F \tag{5.43}$$

where n is the average edge number per each planar polygon. Combining (5.37), (5.42) and (5.43), we have

$$n = 6[1 - (B + 1)/F]. \tag{5.44}$$

The above conditions alone cannot completely determine the shape of a three dimensional grain. If we further require that the polycrystal has to be constructed by grains of identical shape, then only *Kelvin tetrakaidecahedron* shown in Fig. 5.11(b) can satisfy this requirement. Accordingly, there are 14 faces per each grains, and each face is shared by the two neighboring grains. Which yields

$$F = 14B/2 = 7B. \tag{5.45}$$

The substitution of (5.45) into (5.44) would predict an n-value of 5.143 for polycrystals composed of an infinite array of grains. This theoretical prediction agrees well with the experimental data.

5.3.2 Dislocation Model of Grain Boundary

As depicted in Fig. 5.12(a), the grain boundary by the orientation difference of the neighboring (planar) grains can be modeled by scattered edge dislocations. For small mis-orientation angle Θ, termed *low angle grain boundary*, a relation between Θ and the dislocation distribution interval D can be derived from Fig. 5.12 as

$$D = b/2\sin(\Theta/2) \approx b/\Theta \tag{5.46}$$

where b is the length of Burgers vector. The last approximation in (5.46) is valid only for very small Θ. Thus, the mis-orientation angle Θ increases in proportion with the inverse of the dislocation interval D along the grain boundary. This dislocation model breaks down for high angle grain boundaries, due to intensive interaction among dislocation cores.

The grain boundary formed by an edge dislocation array, as shown in Fig. 5.12(a), is called *tilted grain boundary*, in which the rotation axis for mis-orientation Θ is parallel to the dislocation planes, and normal to the picture plane of Fig. 5.12(a). Similarly, one can construct the *twisted grain boundary* as shown in Fig. 5.12(b), where the mis-orientation angle Θ is used to denote the twisting difference between the neighboring grains about the axis perpendicular to the grain boundary. The twisted grain boundary can be described by two arrays of screw dislocations. The grain boundaries formed due to arbitrary mis-orientation between two neighboring grains can be constructed by a dislocation network consisting of both edge and screw dislocations.

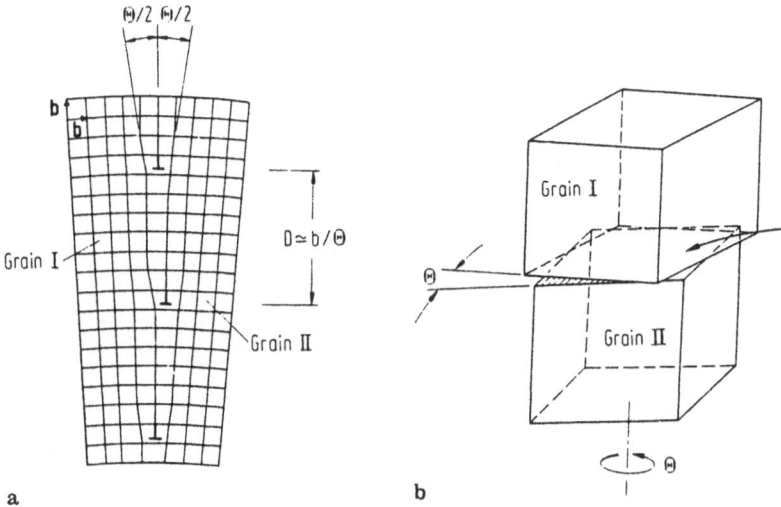

Fig. 5.12. Dislocation models for grain boundaries. a Tilted grain boundary; b twisted grain boundary

5.3.3 Grain Boundary Energy

The grain boundary energy is evaluated in the present subsection by the dislocation model. Let us consider a tilted grain boundary represented by an infinite array of equal-spacing edge dislocations. The separation between neighboring dislocation is D, as shown in Fig. 5.13.

For simplicity, the calculation is proceeded under the assumption of material isotropy. From (4.16) and straight-forward coordinate transformation, the shear stress at (x, y) produced by an edge dislocation at (x', y') is

$$\sigma_{xy} = \{Gb/[2\pi(1 - v)]\}(x - x')[(x - x')^2$$
$$- (y - y')^2]/\{(x - x')^2 + (y - y')^2\}^2 . \tag{5.47}$$

The shear stress at (x, y) produced by an infinite array of edge dislocations as drawn in Fig. 5.13 can be evaluated by superposition. The result is given by the infinite series below

$$\sigma_{xy} = \{Gb/[2\pi(1 - v)]\} \sum_{n=-\infty}^{\infty} x [x^2 - (y - nD)^2]/\{x^2 + (y - nD)^2\}^2 . \tag{5.48}$$

The summation of this infinite series can be performed analytically to give

$$\sigma_{xy} = \{(\pi Gbx)/[D^2(1 - v)]\}\{\cosh(2\pi x/D)\cos(2\pi y/D)$$
$$- 1\}/\{\cosh(2\pi x/D) - \cos(2\pi y/D)\}^2 . \tag{5.49}$$

An estimate of the grain boundary is offered by evaluating the interaction energy of the grain boundary dislocation array. An imaginary edge dislocation with

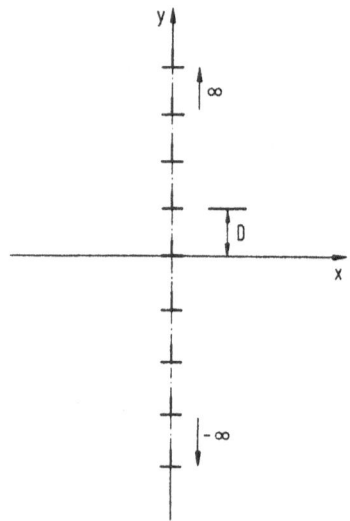

Fig. 5.13. Energy evaluation of a tilted grain boundary

opposite sign is put on the x-axis (with abscissa x) to facilitate this evaluation. Its attraction to the tilted grain boundary can be inferred by the Peach-Koehler formula (4.72)

$$f_x = \sigma_{xy}b = \pi Gb^2 x / \{2D^2(1-\nu)\sinh^2(\pi x/D)\} \tag{5.50}$$

For the grain boundary energy is exactly half of the dislocation interaction energy, one obtains

$$\Pi/2 = (L/2)\int_{r_o}^{\infty} f_x dx = \{Gb^2L/[4\pi(1-\nu)]\}\int_{\eta_o}^{\infty} \eta d\eta/\sinh^2\eta \tag{5.51}$$

where $\eta_o = \pi r_o/D, r_o$ is the radius of dislocation core and L denotes the length of each dislocation. The integral in (5.51) can be evaluated analytically to give the following formula of grain boundary energy

$$\Gamma = \Pi/2LD = \{Gb^2/[4\pi(1-\nu)D]\}\{\eta_o \operatorname{cth}\eta_o - \ln(2\sinh\eta_o)\} \tag{5.52}$$

which offers a general solution for an arbitrary tilted angle $\Theta = 2\sin^{-1}(b\eta_o/2\pi r_o)$. For the case of low angle grain boundary ($\Theta \approx b\eta_o/\pi r_o \ll 1$), the result of Read and Shockley[21] in 1950 is recovered

$$\Gamma \approx \Gamma_o \Theta \ln(e\Theta_o/\Theta) \tag{5.53}$$

where

$$\Gamma_o = Gb/[4\pi(1-\nu)] \qquad \Theta_o = b/(2\pi r_o) \tag{5.54}$$

and e is the base of natural logarithm. From an examination of the low angle grain boundary energy expression (5.53), one finds that the total grain boundary

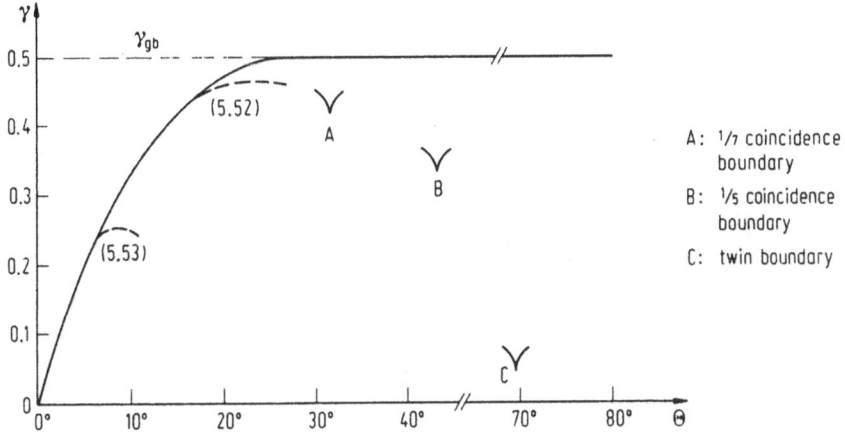

Fig. 5.14. Grain boundary energy versus mis-orientation

energy will decrease whenever two low angle grain boundaries are combined to a grain boundary of relatively large angle, due to the reduction effect caused by the multiplier $-\ln\Theta$ for large Θ. Consequently, the low angle grain boundaries always have the tendency to form boundaries with large mis-orientation.

A plot of specific grain boundary energy (or surface tension) versus Θ is presented in Fig. 5.14, where the solid lines describe the experimental measurements. The approximate estimate based on the low-angle (equation (5.53)) and arbitrary angle (equation (5.52)) expressions are also shown as the dashed lines for comparison reason. A saturated surface tension value γ_{gb} is approached when the mis-orientation angle Θ exceeds 30°. As the value is relatively high, it is instructive to determine the exceptional stable configurations assumed by the grains of a given crystalline solid. As it happens, there are certain special grain boundary structures to which a particular high angle between two adjacent crystals produces a low value of grain boundary energy. These special boundary structures can be divided into two categories: *coincidence boundary* and *coherent twin boundary*. A coincidence boundary, as shown in Fig. 5.15, is incoherent as an ordinary grain boundary; that is, a majority of the atoms of one crystal, in the boundary, do not correspond to the lattice sites of the other crystal. On an average, however, this non-correspondence in a coincidence boundary becomes less as the density of coincidence sites increases. For example, in the mentioned figure, one atom in seven, in a boundary is in a lattice position for both the crystals. We call this boundary a one-seventh coincidence boundary and the atomic sites (black spots in the figure) in question form a coincidence lattice for the two grains. Coincidence lattices occur in all common crystalline structures and have a density of sites varying from 1/3 to 1/9 and less.

A twin boundary is frequently a kind of coincidence boundary, but it is convenient to treat it separately. The energy of a twin boundary γ_{twin} is generally only

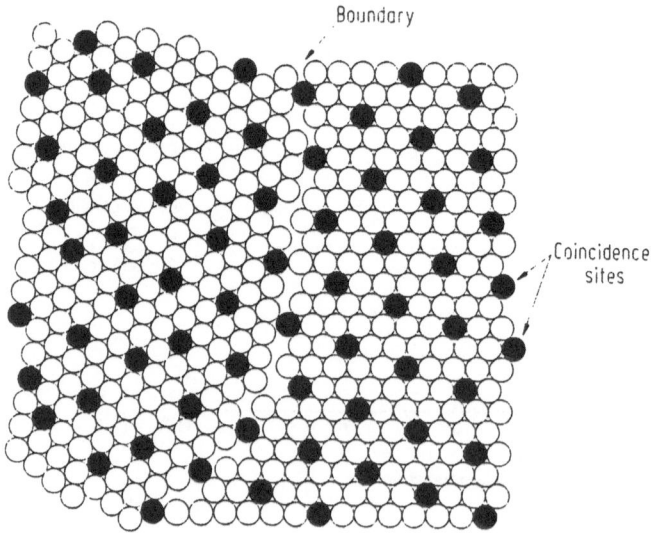

Fig. 5.15. Coincidence lattice made by every seventh atom in the two grains, misoriented 22° by a rotation around the <111> axis

about 10 percent of γ_{gb}, whereas the energy of a coincidence boundary is only slightly less than γ_{gb}. The two most common twin orientations are :
(1) rotation twins, produced by a rotation about the twinning axis;
(2) reflection twins, in which the two lattices maintain a mirror symmetry with respect to the twinning plane.
A description of twinning deformation was provided in Sec. 5.1.

5.3.4 Grain Boundary Sliding

Grain boundary sliding provides another important plastic deformation mechanism. While the progressions of dislocation motion or shear band are directed toward the previously ordered (or defect-free) part of the medium, the grain boundary sliding is rather localized at the disordered lattice structures, such as mismatched grain boundaries, rendering its mathematical and physical descriptions difficult. Grain boundary sliding not only induces the relative shearing along the interlayer which separates the neighboring grains with misorientation, but also produces grain rotation which becomes a dominant issue in the mechanism of *superplasticity*.

Another interesting issue associated with grain boundary sliding is the deformation of sliding polycrystalline aggregates. As will be extensively discussed in section 7.2 that the conventional polycrystal model (as proposed by Taylor[22] in 1938) assumes homogeneous deformation throughout the polycrystalline aggregate, see Fig. 5.16(a), as well as the continuous deformation across the grain

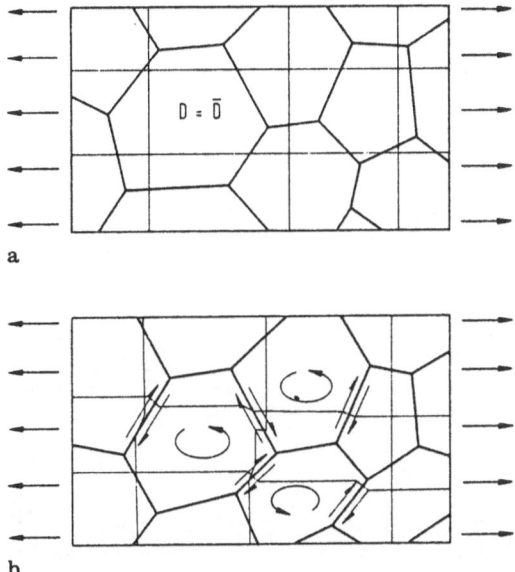

Fig. 5.16. Taylor polycrystal and sliding polycrystal. **a** A Taylor polycrystal with homogeneous deformation and continuous orthogonal grid; **b** a sliding polycrystal with nonuniform deformation within the grain and kinematic jumps across the grain boundaries

boundaries. Consequently, the orthogonal grid engraved on a specimen prior to its deformation will remain to be a uniform orthogonal grid after substantial plastic deformation. The so-called *sliding polycrystal* just states the opposite, namely considerable kinematic jumps occur across the grain boundaries, and in turn perturb the uniform deformation in the grain interior. Accordingly, the previously continuous and orthogonal grid would become zigzag with offset appeared along the grain boundaries, as shown in Fig. 5.16(b). The deformation modes for a sliding polycrystals then consist of nonuniform grain deformation complemented by localized multi-slip near the grain boundary network. The kinematic and constitutional study on sliding polycrystals could promote understandings in many areas including: (1) sliding polycrystals encountered from high temperature creep situation to the plastic processing of superconductive materials; (2) deformation behavior of granular materials like sand, soil and crop grains; (3) mechanics of fibre-reinforced composites with weak interfaces; and (4) failure and storage life prediction of particulate composites like rocket propellants.

To get a vivid picture of grain boundary sliding, one could make a microscopy observation for high purity aluminum specimen (with aluminum content exceeding 99.999%) loaded within the SEM chamber at room temperature, as proceeded by Zhong, Yang and Hwang[23]. The low melting temperature and high ductility of high purity aluminum makes it as an ideal candidate for room temperature observation. Before loading, the specimen surface is carefully mirror polished,

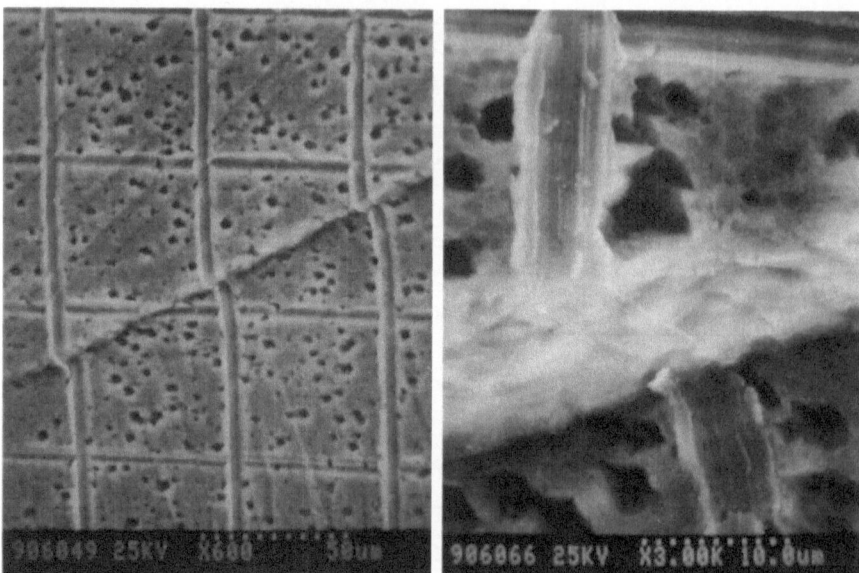

Fig. 5.17. In-plane grain boundary sliding. a Grain boundary sliding revealed by the grid offset; b the intensive shearing along the grain boundary interlayer

etched out the grain boundary and engraved a densely spaced (10, 20 or 50 lines per millimeter) orthogonal grid. A photograph revealing *in-plane grain boundary sliding* is presented in Fig. 5.17(a), where the sliding amount can be measured from the offset of grid along the grain boundary. Many measurements of this kind indicate that the sliding along the grain boundary is nonuniform. Usually, the triple points of the polycrystal are pinned together. Away from a triple point, the sliding rapidly increases from the zero value to its maximum near the middle of the grain boundary segment. For pure aluminum with a grain size of 1mm and subject to remote strain of 10 to 15 percent, the amount of in-plane grain boundary widening is in the order of 10 microns. The intensive shear deformation along the interlayer of two neighboring grains is highlighted in Fig. 5.17(b). As a byproduct of grain boundary sliding, the number of active slip system within the neighboring grains is reduced, due to the addition of the grain boundary sliding mode to accommodate the prescribed remote deformation.

Beside in-plane grain boundary sliding, *out-of-plane sliding* also occurs along the plane inclined to the specimen testing plane. As shown schematically in Fig. 5.18(a), the out-of-plane sliding results in *apparent grain boundary widening*. The scanning electronic microscopies on out-of-plane sliding are shown in Fig. 5.18(b) and (c), for the overall strain of 1% and 35%, respectively. The amount of apparent grain boundary sliding is also in the order of 10 microns, and it increases as the overall deformation intensifies. Only those boundary segments with favorable mis-orientations can observe out-of-plane sliding. As a result of out-of-plane sliding and induced grain rotation, the previously flat specimen surface

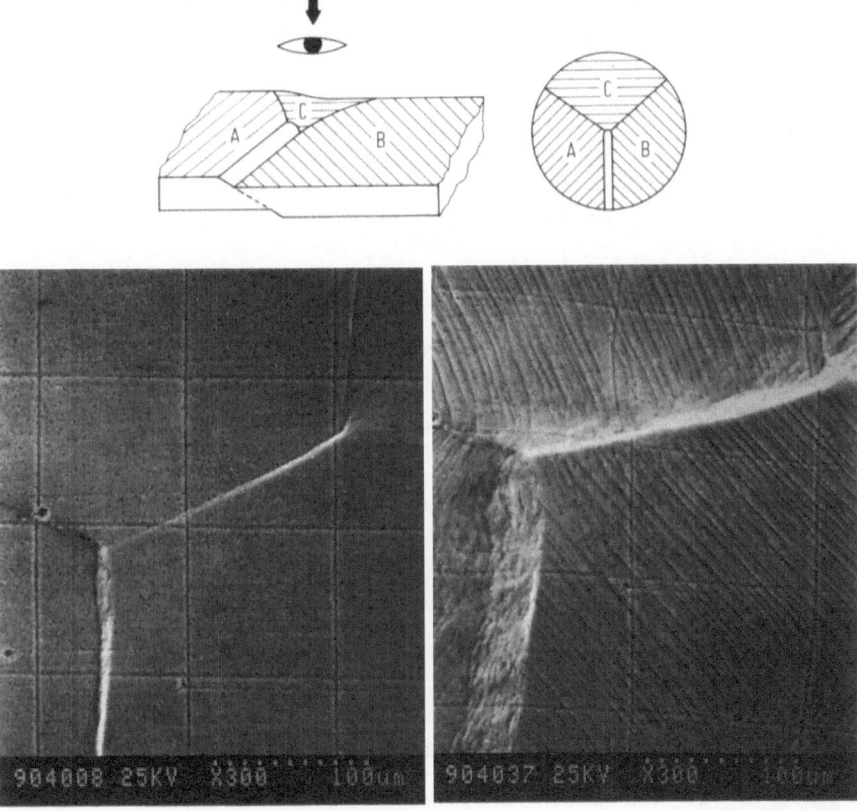

Fig. 5.18. Out-of-plane grain boundary sliding. a Schematics of apparent grain boundary widening; b triple-joined grain boundary under slight plastic deformation. $\epsilon =1\%$; c substantial grain boundary widening under large plastic deformation, $\epsilon = 35\%$

would become zigzag. The measurement on high purity aluminum specimen under 35 percent of overall strain indicates a surface profile variation as high as 180 microns[23].

The mathematical description of grain boundary sliding can be carried out by special dislocation model, such as by *Somigliana dislocation*, see Mura and Furuhashi[24]. A methodology to evaluate the constitutive relations of sliding polycrystals is provided by Yang, Zhong and Hwang[25]. To conclude this section, we would like to provide a basic kinematic formula for the overall deformation with the participation of grain boundary sliding. If a polycrystal of volume V is composed of n grains separated by m grain boundary segments (both n and m are large integers), then the macroscopic velocity gradient $\mathbf{L}$ defined by

$$\mathbf{L} = (1/V) \int_V \mathbf{l}\, dV, \qquad \mathbf{l} = \partial \mathbf{v}/\partial \mathbf{x} \qquad (5.55)$$

can be presented by the following additive decomposition of the grain interior contribution $\mathbf{L}_g$ and the grain boundary contribution $\mathbf{L}_b$,

$$\mathbf{L} = \mathbf{L}_g + \mathbf{L}_b. \tag{5.56}$$

In (5.56)

$$\mathbf{L}_g = (1/V) \sum_{i=1}^{n} \int_{V_{(i)}} \mathbf{l}\, dV$$

$$\mathbf{L}_b = (1/V) \sum_{j=1}^{m} \int_{S_{(j)}} [[\mathbf{v}_{(j)}]]\,\mathbf{n}_{(j)}\, dS \tag{5.57}$$

where $\mathbf{l}$ the microscopic velocity gradient, $V_{(i)}$ the volume of the i-th grain, $S_{(j)}$ the surface area of the j-th grain boundary, $\mathbf{n}_{(j)}$ its outward unit normal and $[[\mathbf{v}_{(j)}]]$ the total velocity jump across the grain boundary layer. We refer to the paper by Zhong, Yang and Hwang[23] for the proof of this formula. The same additive decomposition can be performed to the deformation rate tensor $\mathbf{D}$ and the material spin tensor $\mathbf{W}$ as follows :

$$\mathbf{D} = \mathbf{D}_g + \mathbf{D}_b, \qquad \mathbf{W} = \mathbf{W}_g + \mathbf{W}_b, \tag{5.58}$$

because they are the symmetric and skew-symmetric parts of the velocity gradient $\mathbf{L}$.

5.4 Transformation Induced Plasticity

5.4.1 Phase Transformation

Phase transformation refers to the intrinsic micro-structural change under the excitation of mechanical or thermodynamic driving force. The transformed region is separated from the non-transformed region (termed *parent phase*) by a surface or an interlayer which undergoes abrupt micro-structural change. This *phase transformation surface* is in general non-material, and its shifting velocity can be either very slow or up to the order of sound speed according to the characteristics of phase transformation. The *diffusive transformations* in which the migration of matter (or vacancy) necessarily occurs would result in rather slow shifting velocity unless at very high temperature, for the atomistic diffusion cannot proceed rapidly in the conventional temperature range. On the other hand, the *non-diffusive transformations* which take place without the help of matter migration can produce fast propagation of transformation boundary at room or elevate temperature range. The latter case is frequently referred as *martensitic transformation*, and will be focused upon in this section.

Typical example of non-diffusive transformation is the phase transformation occurred during the quenching of steel. Driving by the thermodynamic force provided by rapid quenching, the high temperature micro-structure of austenite (parent phase) in steel changes to martensite through a non-diffusive topological alternation on the crystal lattice. Relatively modern examples of martensitic transformation include the toughened ceramics containing ZrO_2, and the pseudo-elastic and shape memory effects displayed by certain structural or functional alloys. The *driving force for the phase transformation* could be mechanical (such as in the case of stress-induced martensitic transformation of ZrO_2) or thermal (a typical example is the austenite-martensite transformation proceeded during quenching of steels) or thermal-mechanical (like the case of shape memory alloy). In conclusion, martensitic transformation is driven by *energetic force* which forms a work conjugate with the shifting velocity of the transformation boundary and can be phrased under a thermodynamics framework. Two processes of phase transformation should be distinguished, namely the *forward transformation* during which the parent phase is transformed into the martinsite of a fixed orientation, and the *reverse transformation* in which the transformed martinsite is subsequently transformed back to the parent phase. Symbolically, we denote the former as $P \longrightarrow M$ and the latter $M \longrightarrow P$. In terms of the plasticity features, the forward transformation has less history dependence whereas the reverse transformation depends critically on the history of the previous transformation. In fact, only those material particles which have been transformed into martensites can undergo reverse transformation. This historic dependence makes the constitutive modelling for reverse transformation more difficult. *Transformation induced plasticity* (TRIP) opens an exciting new field of mesoplasticity. We refer to the work by Bondaryev and Weyman[26] for the foundation of constitutive modelling, and the works by Budiansky, Hutchinson and Lambropoulos[27], and by Evans[28] for the toughening description, and the recent work by Sun and Hwang[29] for its micromechanics formulation.

5.4.2 Mechanics of Martensitic Transformation

Both temperature and stress as macroscopic variables affect the martensitic transformations due to their influence on the thermodynamics and kinematics of the transformation. Moreover, the thermodynamic and kinetic effects are strongly dependent on the direction of stress with respect to lattice orientation. The thermoelastic transformation is realized if the martensite forms and grows continuously as the temperature is lowered (or the applied stress is increased), and shrinks and vanishes continuously as the temperature is raised (or the stress is decreased). The transformation proceeds essentially in equilibrium between the chemical driving energy and the resistive energy whose dominating component is the stored elastic energy. The material undergoing thermoelastic martensitic transformation may display two major kinds of mechanical behavior, i.e., *pseudoelasticity* (PE) and *shape memory effect* (SME) at high and low temperatures where austenite and martensite, respectively, are stable or metastable.

In terms of the three dimensional characteristics of transformation strain, it can be decomposed into the dilatational part (transformation induced dilatancy, denoted by a scalar ϵ^{pv}) and the deviatoric part (transformation induced shear, denoted by a deviatoric tensor ϵ^{pd}). Some earlier works on this subject were focused on the dilation effect, because it was believed that a majority part of the microscopic shear deformation can be accommodated by the twinning structure of martensitic assembly. But recent experimental investigations seem to indicate that the shear component of transformation has equal importance in the transformation plasticity and the toughness enhancement.

A number of stress-induced phase transformation criteria have been introduced. The early criteria (e.g., see McMeeking and Evans[30], Budiansky, Hutchinson and Lambropoulos[27]) emphasized the importance of lattice dilation ϵ^{pv} induced by hydrostatic stress $\Sigma_m = \Sigma_{ii}/3$,

$$\Sigma_m = \Sigma_c \longrightarrow E_{ii}^p = f\epsilon^{pv} \tag{5.59}$$

where second rank tensor Σ and $\mathbf{E}^p$ stand for macroscopic stress and transformation strain, respectively. The later investigations result in a spectrum of transformation criteria, and their mechanical interpretations in the constitutive formulation is parallel to the yield surface in conventional plasticity theory. Summarized below are several representative criteria:

(1) Maximum shear induced transformation, see Evans and Cannon[31]

$$\tau_{max} = \tau_c \longrightarrow E_{21}^p = f\epsilon^{pv}. \tag{5.60}$$

(2) Maximum tension induced transformation, Lambropoulos[32]

$$\Sigma_1 = \sigma_c \longrightarrow E_{11}^p = f\epsilon^{pv}. \tag{5.61}$$

(3) Criterion based on "liquid drop" hypothesis, also see Lambropoulos[32]

$$m(\sigma_e^M/p_o)^2 + (\sigma_m^M/p_o)^2 = P(a) \tag{5.62}$$

where

$$m = (5/3)(1 - \nu^2)/[(1 - 2\nu)(7 - 5\nu)]$$

$$p_o = E\epsilon^{pv}/(3(1 - 2\nu)) \tag{5.63}$$

and σ_e^M and σ_m^M are the J_2 equivalent and mean stresses of the matrix stress tensor σ^M, respectively. $P(a)$ on the right hand side of (5.63) is a material function of the transformed particle size a.

(4) Stress sensitive phenomenological criterion, Chen and Reyes-Morel[33]

$$\Sigma_e/\Sigma_e^* + \Sigma_m/\Sigma_m^* = 1 \tag{5.64}$$

where Σ_e^* and Σ_m^* are experimentally measurable material constant, and Σ_e represents the macroscopic J_2 equivalent stress.

Fig. 5.19 delineates the kinematics accompanying with a pseudoelastic transformation. The pseudoelastic behavior is a close analogue to the thermoelastic transformation, the transformation from austenite to martensite (Fig. 5.19(a), (b) and (c)) is driven by the increment of stress. As shown in the loading part of the stress-strain curve in Fig. 5.19, the parent phase appears to be linear elastic as long as the stress is below the critical value required to excite $P \to M$ transformation.

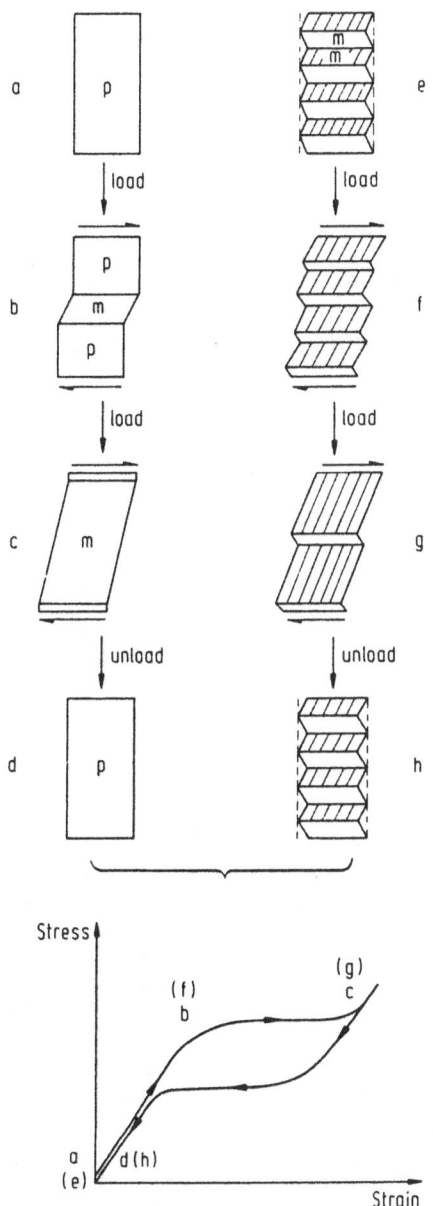

Fig. 5.19. Schematic illustration of the mechanisms of PE by transformation (graphs a, b, c, d) and reorientation (graphs e, f, g, h)

Additional strain induced by phase transformation occurs when the applied stress exceeds this critical value. By the increment of applied load, the martensitic phase

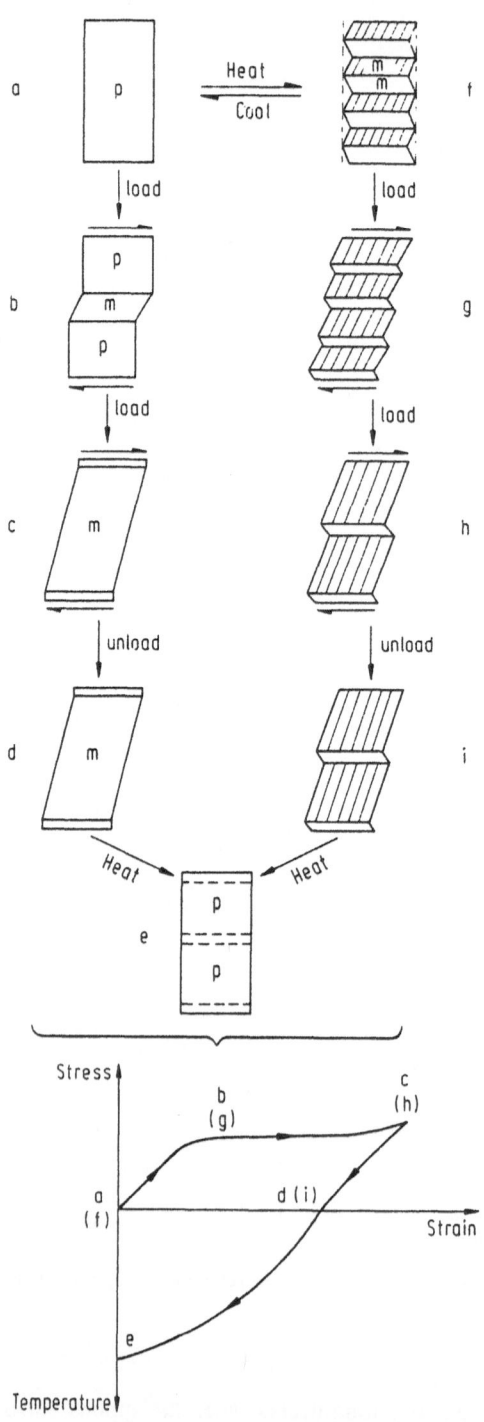

Fig. 5.20. Schematics illustrations of the mechanisms of SME by transformation (graphs a, b, c, d, e) and by reorientation (graphs f, g, h, i, e), respectively

grows which consequently lowers the stress-strain curve. At a certain stress level, the forward transformation is completed and further application of load would result in a linear response relevant to the elastic property of the martensite. A uniaxial unloading process would cause a reversal of the above-mentioned mechanism. Due to the thermodynamic dissipation in phase transformation, however, the reverse $M \rightarrow P$ transform starts at a lower stress level than the finishing stress of the forward transformation $P \rightarrow M$. Accordingly, a hysteresis loop is formed in the loading-unloading curve of Fig. 5.19. It is assumed in PE that the stress level to finish the reverse transformation $M . \rightarrow P$ is positive, so that the final part of the unloading curve coincides to the initial loading curve and both are related to the elastic behavior of the parent phase. Beside the hysteresis loop, the other portion of the loading-unloading curve would just like the curve for an elastic body, so comes the name of pseudoelastic behavior. Another mechanism of PE is featured by a reorientation process, as shown in Fig. 5.19 through graph (e) to graph (h), involving the gradual reorientation of twinning martensite to the unidirectional martensite during loading and its reverse transformation back to the twinning martensite upon unloading.

The shape memory effect arises if the macroscopic deformation of the above processes remains after unloading and then disappeared by heating to cause reverse transformation. As illustrated in Fig. 5.20, the previous three steps in SME (for either transformation or reorientation processes) are similar to PE, except the fourth step related to inverse transformation does not occur after the complete removal of the applied load. However, the reverse transformation can be accomplished by heating. That is to say, the material sample would resume its previous shape (before load application) by put it in an appropriate temperature environment, so comes the name of shape memory effect. It must be recognized that the pseudoelastic behavior and shape memory effects are interrelated, i.e., if the hysteresis in the case of PE is so large that the reverse transformation or reorientation is incomplete when the applied stress is removed, the residual martensite can be reverted by heating, and which exactly identifies the process of SME.

5.4.3 Micromechanics Modelling

In establishing the micromechanics constitutive law under transformation induced plasticity, a constitutive element (a very small representative material sample) as shown in Fig. 5.21 is usually taken as the subject of study. The constitutive element consists of a great number of untransformed grains of parent phase (as the matrix) and the transformed grain as disperse second phase martensitic inclusions (with volume fraction f) embedded coherently in the elastic matrix. Temperature is uniformly distributed everywhere in the element and the external macroscopic stress (Σ) or strain (E) is applied on the boundary. Microscopic observation reveals that under a critical thermal-mechanical condition the martensite/parent interface propagates so fast that only two states exist for a grain: the untransformed state and transformed state. The intermediate state of transformation (i.e., $0 < f < 1$)

is only meaningful in the sense of macroscopic average over the constitutive element. Representative local transformation criteria are listed from equations (5.59) to (5.64).

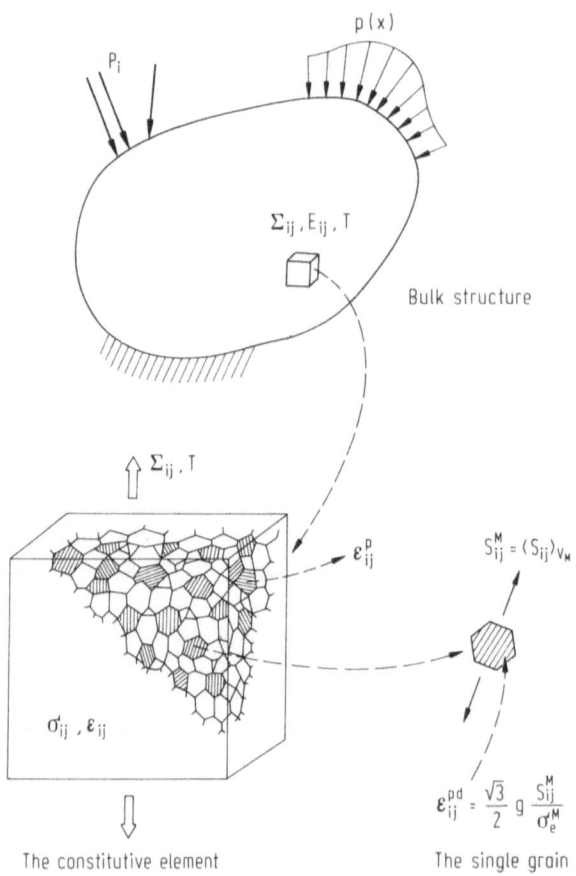

Fig. 5.21. Micro-structure and micromechanics model of the constitutive element

In the process of stress induced transformation, the volumetric part ϵ^{pv} of transformation strain is the constant lattice dilation and independent of stress state. The deviatoric part ϵ^{pd}, however, is stress state dependent because of the directionality in the crystallographic orientation and its shear nature. It is well known that there are 24 possible variants of martensite. Among them, only the one which is energetically most favorable can be preferentially formed and grow up to the grain boundary in a polycrystal. It is assumed that at any instant of transformation the deviatoric strain ϵ^{pd} for a particular grain is always along the direction of the average deviatoric matrix stress S^M evaluated at the point where this grain is located at that instant. Accordingly, Sun and Hwang[29] proposed

$$\epsilon^{pd} = (\sqrt{3}/2)\,g\,(S^M/\sigma_e^M)$$

(5.65)

where $\sigma_e^M = (3S^M : S^M/2)^{\frac{1}{2}}$ stands for equivalent matrix stress and $g = \sqrt{3}(2\epsilon^{pd} : \epsilon^{pd}/3)^{\frac{1}{2}}$ represents the intensity of deviatoric transformation strain. Through an averaging technique involving the application of Mori-Tanaka theorem[34], Sun, Hwang and Yu [35] were able to derive the following incremental macroscopic constitutive relations for combined dilatant and shear transformation plasticity

$$\dot{E} = \dot{E}^e + \dot{E}^p$$
$$= M : \dot{\Sigma} + \dot{f}\left\{ \left(\epsilon^{pd}/3\right)\mathbf{1} + (\sqrt{3}/2)g\left(S^M/\sigma_e^M\right)\right\}$$

(5.66)

where M denotes the elastic compliance tensor and the transformation rate (in terms of $\dot{f}$) can be determined by the consistency condition[35]. The generalization of this method to reverse transformation, as well as to SME, can be found in reference [29].

5.4.4 Transformation Toughening

Transformation induced plasticity plays an important role in the recent progresses on toughened structural ceramics, such as the tetragonal-monoclinic martensitic transformation in the partially stabilized zirconia (PSZ) containing structural ceramics and the tetragonal zirconia polycrystal (TZP). The transform plastic strain induced by the high stress in the crack tip zone of the above material systems would provide shielding to the crack tip, in terms of the reduction of the near tip stress intensity factor

$$\Delta K = K_{appl} - K_{tip},$$

(5.67)

see McMeeking and Evans[30] and Budiansky, Hutchinson and Lambropoulos[27], where the latter is abbreviated as the BHL solution. However, both of the above-mentioned solutions are only pertinent to the case of pure dilatant transformation. It was found in the BHL solution that ΔK would be zero for a stationary crack, and the crack tip shielding by dilatant transformation plasticity is possible only when crack starts to propagate. As a matter of fact, it is the dilatancy-induced closure along the propagating crack wake that furnishes the desired shielding effect.

Attention is now focused on the explicit formulae to estimate the shielding effect on transformation toughening. For a propagating crack, the following energy flux integral can be shown as path independent[27]

$$I = \int_{\Gamma} (Un_1 - \mathbf{n} \cdot \Sigma \cdot \mathbf{u}_{,1})\,ds$$

(5.68)

where Γ is the integration contours with ds being an arc increment and $\mathbf{n}$ being its outward unit normal. The quantity U in the integrand refers to the strain energy

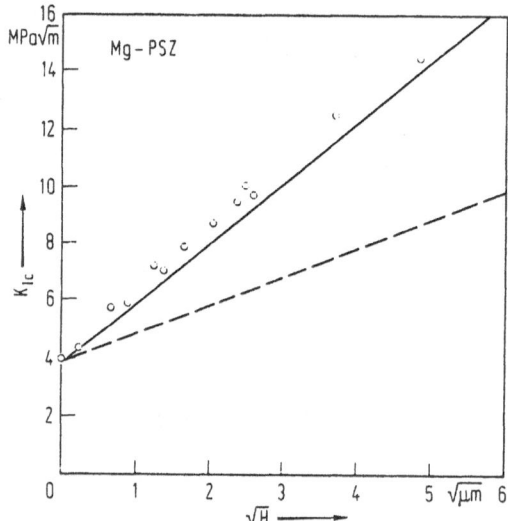

Fig. 5.22. Comparison between the toughening predictions and the experimental data. (After Sun, Q.P. et. al., J.Mech.Phys.Solids, 39, 1991, p.507.)

density defined by

$$U = \int_0^{\mathbf{E}} \Sigma \cdot d\mathbf{E}. \tag{5.69}$$

For a crack propagating in the x_1 direction, the integration contour Γ can be deformed as two circles centered at crack tip but one with infinitesimal and the other with infinite radius and linked by two straight lines running along the upper and lower crack faces. After straightforward fracture mechanics calculation, one can establish the following relation between K_{appl} (which dominates the remote field) and K_{tip} (which prevails near the crack tip) :

$$K_{appl}^2 - K_{tip}^2 = \left[2E/\left(1 - \nu^2\right)\right] \int_0^H U\left(x_2\right) dx_2 \tag{5.70}$$

where for propagating cracks, the strain energy density U in the integral of (5.70) is evaluated across the transformed zone of height H at the crack wake and appears as a function of the vertical coordinate x_2. It was shown that[27] the toughness increment can be expressed in the following form :

$$\Delta K = \beta E \epsilon^{pv} f \sqrt{H}/(1 - \nu) \tag{5.71}$$

for crack growing in steady state. For purely dilatant transformation governed by equation (5.59), the scalar β in (5.71) can be evaluated as 0.214[27]. However, this β value underestimate the actual toughness increment, as shown in Fig. 5.22 where the data points represented by circles are the testing results for Mg-PSZ[33] and the dashed line gives the prediction by BHL solution[27]. Substantial discrepancy exists between the theory and the experiment. Other phase transformation criteria have also been engaged to improve the theoretical prediction. For example, it is documented in the literature that

$\beta = 0.38$ maximum shear induced transformation[31]

$\beta = 0.50$ maximum tension induced transformation[32]

$\beta = 0.48$ phenomenological criterion[33].

Nevertheless, the physical interpretation under which the above criteria are proposed seems still ambiguous. Based on the micromechanics constitutive law (5.66) which takes into the consideration of both dilatant and shear transformation strain, Hwang, Sun and Yu[35] were able to derive

$$\beta = 0.46. \tag{5.72}$$

As shown in the solid curve of Fig. 5.22, this prediction agrees well with the experimental data measured by Chen and Reyes-Morel[33].

5.5 References

1. Christian, J.W., Theory of transformations in Metals and Alloys, Pergamon Press (1965).
2. Schmid, E. andBoas, W., Plasticity of Crystals, London(1950).
3. Chin, G.Y., Hosford, W.F. and Mendorf, D.R., Proc. Roy. Soc., A309(1969), p.433.
4. Backofen, W.A., Deformation Processing, Addision Wesley (1972), p.78.
5. Wassermann, G., Z. Metallke, 54(1963), p.61.
6. Backman, M.E. and Finnegan, S.A., Metallurgical Effects at High Strain Rates (edited by Rohde et al.) Plenum Press (1973), p.531.
7. Duggan , B.J., Hatherly,W.B., Hutchinson, W.B., and Wakerfield, Metal Science, 8(1978), p.343.
8. Bishop, J.F.W. and Hill, R., Phil.Mag. 42(1951), p.414.
9. Dillarmore, I.L. Roberts, J.G., and Bush ,A.C., Metal Science, 13(1979), p.73.
10. Van Houtte, P., Gil Sevillano, J and Aernoudt, Z.Metallk., 70(1979), p.503.
11. Lee, W.B. and Chan K.C., Acta.Metall. Mater. 39(1991), p.411
12. Brown,K., J. Inst of Metals, 100(1972), p.341.
13. Feller-Kniepmeier, M. and Wanderka, N., Proc. 8th Int. Conf. on Textures of Materials, New Mexico (1988), p.517.
14. Lee, W.B. and Chan, K.C., Scripta Metall. Mater., 24(1990), p.997.
15. Harren, S.V., Dave, H.E. and Asaro, R.J., Acta Metall., 36(1988), p.2435.
16. Yeung, W.Y., PhD Thesis, University of Hong Kong (1985).
17. Lee, W.B. and Chan. K.C., Int.J.Fracture, 51(1992), p.207.
18. Asaro, R.J., Acta Metall., 27(1979), p.445.
19. Asaro, A.J., Advances in Applied Mechanics, 2(1983), p.1.
20. Chang, Y.W. and Asaro, R.J., Acta Metall., 9(1961), p.513.
21. Read, W.T. and Shockley, W., Phys. Rev., 78(1950), p.275.
22. Taylor, G.I., J. Inst. Metals, 62(1938), p.307.
23. Zhong, Z., Yang, W. and Hwang, K.C., Acta Mech. Sinica, 7(1991), p.360.
24. Yang, W., Zhong, Z. and Hwang, K.C., Meso-deformation and constitutive relations of sliding polycrystals, IUTAM Sym. on CRFDPM, July 18-21, 1991, Beijing, China, Springer-Verlag, p.136.
25. Mura, T. and Furuhashi, R., J. Appl. Mech., 51(1984), p.308.
26. Bondaryev, E.N. and Wayman, C.M., Metall. Trans., 19A (1988), p.2407.
27. Budiansky, B., Hutchinson, J.W. and Lambropoulos, J.C., Int. J. Solids & Structs. 19(1983), p.337.

28. Evans, A.G., J. Am. Ceram. Soc., 69(1990), p.181.
29. Sun, Q.P. and Hwang, K.C., Micromechanics constitutive description of thermoelastic martinsitic transformation, to appear in Adv. Appl. Mechanics, 31(1993).
30. McMeeking, R. and Evans, A.G., J. Am. Ceram. Soc., 65 (1982), p.242.
31. Evans, A.G. and Cannon, R.M., Acta Metall., 34(1986), p.761.
32. Lambropoulos, J.C., Int. J. Solids Structs., 22(1986), p.1083.
33. Chen, I.W. and Reyes-Morel, P.E., J. Am. Ceram. Soc., 69(1986), p.181.
34. Mori, T. and Tanaka, K., Acta Metall., 21(1973), p.571.
35. Sun, Q.P., Hwang, K.C. and Yu, S.W., J. Mech. Phys. Solids, 39(1991), p.507.

6 Strengthening Mechanisms

6.1 Strain Hardening

Strain hardening is an *intrinsic strengthening mechanism* caused by internal multiplication and interaction of dislocations. This feature forms a clear distinction to the other *extrinsic strengthening mechanisms* examined in the sections to follow. An extensive review in this topic was given by Nabarro, Basinski and Holt[1].

6.1.1 Taylor's Theory

In as early as 1934, a pioneering investigation on the modelling of strain hardening was accomplished by G.I. Taylor[2] as the first successful application of the newly-born dislocation theory. A parabolic relationship between stress and strain, in agreement with the strain hardening data for aluminium at that time, was predicted by the Taylor theory. The essence of *Taylor's strain hardening theory* lies in the consideration of interactions between large amount of straight parallel dislocations in crystal. The dislocation motion under applied stress is hindered by the stress fields caused by the other dislocations. Accordingly, the driving stress for each dislocation to operate would increase as dislocation density increases, and flow stress is elevated with the progress of plastic deformation. The above key idea behind the Taylor's theory still dominates in the current study of strain hardening.

We now put Taylor's strain hardening model (as shown in Fig. 6.1) into quantitative prospective. It is assumed that l is the mean gliding distance (also termed mean slip path) of dislocation lines, and ρ denotes the dislocation density at prescribed plastic deformation. Integrating the Taylor-Orowan equation (4.86) from a static non-deformed state, one obtains the shear strain along the primary slip system as

$$\gamma = m\rho bl \qquad (6.1)$$

where m is a factor related to the orientation of primary slip system, usually termed *Schmid factor*, and b signifies the length of Burgers vector.

In the original approach of G. I. Taylor, only parallel straight edge dislocations are considered. Furthermore, he assumed a uniform distribution of those parallel

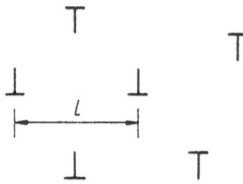

Fig. 6.1. Taylor's strain hardening model. Hardening is caused by self-interaction of parallel straight dislocations with an average spacing of L

dislocations, with the implication that the average distance L between dislocations will be equal to $\rho^{-\frac{1}{2}}$. Taylor also defined the stress for a material to resist plastic deformation as the interaction stress between two parallel edge dislocations separated by a distance L

$$\tau = kGb/L \tag{6.2}$$

where k is a dimensionless constant. The derivation of (6.2) comes from (4.81) and the second equation of (4.72). Combining (6.1) and (6.2) and also the result $L = \rho^{-\frac{1}{2}}$, one arrives

$$\tau = kG\sqrt{b\gamma/ml} \propto \sqrt{\gamma} \tag{6.3}$$

which describes the parabolic relation between shear stress and shear strain in aluminium.

Despite its successful accomplishment, Taylor's theory is nevertheless suffered by the following inadequacies:
(1) The dislocation configurations actually observed from plastically deformed crystals seldom exhibit the assumed pattern of uniformly distributed, straight parallel dislocations.
(2) Screw dislocations and cross-slips are not engaged in Taylor's theory.
(3) The linear hardening behavior observed in some HCP and FCC single crystals cannot be explained by Taylor's model.
(4) The re-operation of the immobile dislocations (such as dislocation pile-ups) under higher driving stress is ignored.

Mott[3] proposed an improved model based on *super-dislocation* idealisation in 1951 to overcome the last inadequacy of Taylor's model. He took into account the *dislocation pileups* which are likely to form when dislocations run into obstacles (such as the *Lomer-Cottrell lock* as described in the next subsection). He used a super-dislocation to replace the dislocation pileup, and carried out the Taylor's analysis for the interaction between super-dislocations.

6.1.2 Lomer-Cottrell Lock

The analysis of G.I. Taylor in 1934 emphasized the hardening due to elastic interaction between non-intersecting straight dislocations in the primary slip systems.

When dislocations from different slip systems interact, however, a dislocation structure distinct from the preceding description can emerge. A dislocation reaction mechanism between dislocations in the *primary slip system* and *conjugate slip system* was proposed by Lomer[4] (1951) and Cottrell[5] (1952), and it is illustrated in Fig. 6.2. We first assume that small amount of immobile (sit-in) dislocations are initially contained in the conjugate slip system $(1\bar{1}1)\frac{1}{2}a[\bar{1}\bar{1}0]$. When the dislocations along the primary slip system $(11\bar{1})\frac{1}{2}a[101]$ sweep by their gliding plane, some of them might be intercepted by the dislocations in conjugate slip system along the common line [011] of the two slip planes. Along this common line, two dislocations from the different slip systems combine to a *Lomer dislocation* with Burgers vector $\mathbf{b}^L = \frac{1}{2}a[0\bar{1}1]$ according to the reaction

$$\frac{1}{2}a[101] + \frac{1}{2}a[\bar{1}\bar{1}0] = \frac{1}{2}a[0\bar{1}1]. \tag{6.4}$$

This dislocation reaction can proceed because the elastic strain energies for the three dislocations in (6.4) are all identical, then the formation of a Lomer dislocation actually reduces the total dislocation energy.

It is observed that the tangential vector for a Lomer dislocation line, $\mathbf{t}^L = [011]/\sqrt{2}$, is normal to the corresponding Burgers vector $\mathbf{b}^L$. That indicates Lomer dislocation is of edge-type. It could only glide in the plane (100) containing both $\mathbf{b}^L$ and $\mathbf{t}^L$. However, gliding on crystallographic planes of $\{100\}$ type is extremely difficult in FCC crystals, leaving Lomer dislocations immobile. Moreover, their existence would confine subsequent dislocation movement on the primary slip plane. As shown in Fig. 6.2(c), Lomer dislocation would exert a repulsive force $Ga^2/[8\pi(1-\nu)d]$ on whatever dislocations coming from the primary or conjugate slip system, where d represents the inter-dislocation distance. Consequently, the primary and conjugate slip systems will be harden roughly to the same extent. A *hardening matrix* $\{h^{\alpha\beta}\}$ is introduced to measure strain hardening developed in various slip systems, where a generic element, $h^{\alpha\beta}$, denotes the hardening effect on dislocations along slip system α by dislocation glide along slip system β, and α or β can take value in all slip systems. When $\alpha = \beta$, the corresponding hardening is called *self hardening*, and represents the hardening effect by dislocation glided in the same slip system. When α is not equal to β, the corresponding hardening is called *latent hardening*, and describes the hardening due to dislocation motion in the other slip systems. For the simple case of two active slip systems, i.e., the primary slip system (with superscript 1) and the conjugate slip system (with superscript 2), the hardening behavior of a Lomer-Cottrell lock can be expressed by

$$\{h^{\alpha\beta}\} = \begin{bmatrix} h & h \\ h & h \end{bmatrix} \tag{6.5}$$

which is in good agreement with the experiment observation by Taylor and Elam[6,7] for latent hardening.

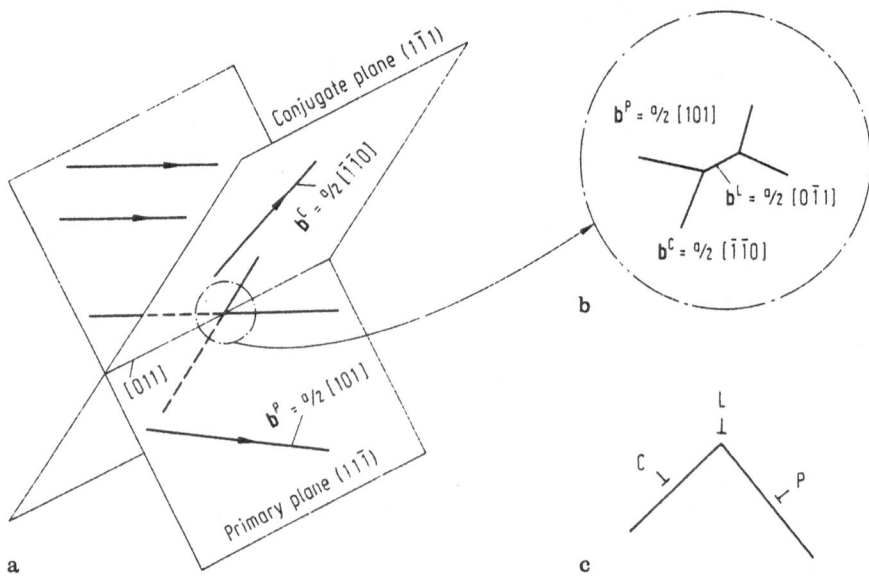

Fig. 6.2. Lomer-Cottrell lock. **a** Intersection of dislocations in primary slip system by sit-in dislocations in conjugate slip system; **b** formation of Lomer dislocation; **c** interactions between immobile Lomer dislocation and the dislocations in primary or conjugate slip system

6.1.3 Forest Dislocations

Another type of dislocation reactions leading to both self and latent strain hardening is demonstrated in Fig. 6.3. This reaction involves the dislocation $\frac{1}{2}a[10\bar{1}]$ along primary slip system $(11\bar{1})\frac{1}{2}a[101]$ and the *forest dislocation* $\frac{1}{2}a[\bar{1}10]$ along another slip system $(111)\frac{1}{2}a[\bar{1}10]$. When these two dislocations intertwine, the reaction

$$\frac{1}{2}a[101] + \frac{1}{2}a[\bar{1}10] = \frac{1}{2}a[011] \tag{6.6}$$

takes place to form a dislocation segment along [110] direction where the two slip planes meet. The Burgers vector for this dislocation segment is $\frac{1}{2}a[011]$, so it can glide on the primary slip plane $(11\bar{1})$. However, the dislocation nodes shown in Fig. 6.3(b) are in a stable low energy state, so they will pin on the movement of both primary and forest dislocations. If the adjacent primary slip plane is further considered, a spatial dislocation network with a configuration shown in Fig. 6.3(c) could be manifested. It would behave differently from the parallel dislocation configuration assumed early in Taylor's theory.

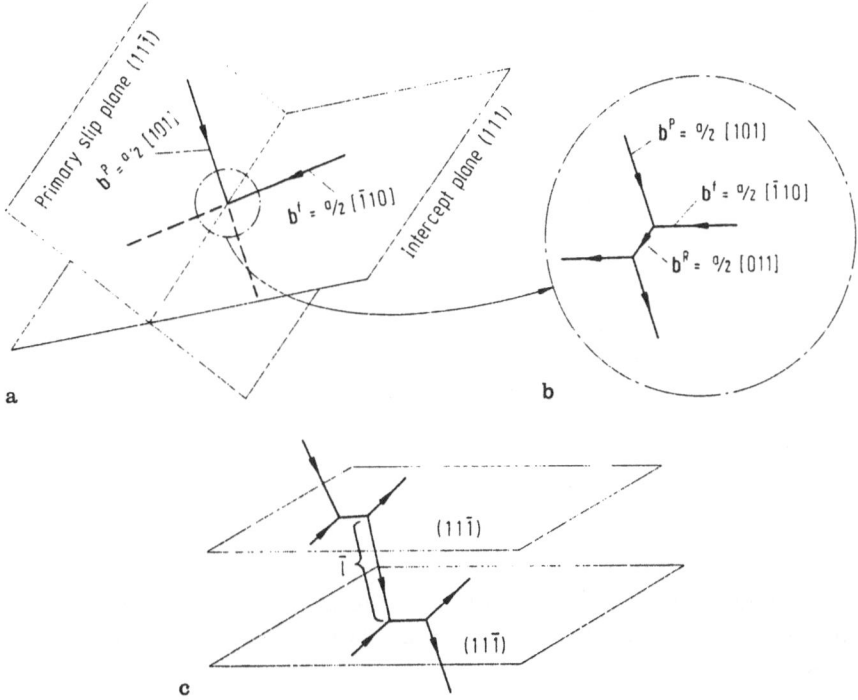

Fig. 6.3. Forest dislocation. a Interaction between active dislocations and forest dislocations; b formation of stable nodes and a dislocation segment; c dislocation network

6.1.4 Frank-Read Source

The above mechanisms make the dislocation source model suggested by Frank and Read[8] in 1951 (abbreviated as *Frank-Read source* in the following) possible. As stated in the previous subsection, the nodes in a dislocation network can be regarded as immobile pinning points, and the dislocation segments between nodes are capable of gliding in the form of *bow-out* when subject to applied stress, as depicted in Fig. 6.4. Facilitated by the calculations on planar dislocation loop in Chapter 4, with the further assumption of material isotropy, one is able to derive (see also the derivation of equation (20-1) in reference [9] for details)

$$\tau = [Gb/4\pi r(1 - \nu)] \left\{ [1 - \nu(3 - 4\cos^2 \beta)/2] \ln(L/r_o) - 1 + \nu/2 \right\} \quad (6.7)$$

where τ the resolved shear stress for dislocation glide, β the angle span between Burgers vector **b** and tangential vector **t**, L the interval between the neighboring pinning nodes, and r the radius of curvature of the dislocation curve. For L usually far exceeds the dislocation core dimension r_o, the above formula can be reduced to

$$\tau \approx [Gb/4\pi r(1 - \nu)] \left\{ 1 - \nu(3 - 4\cos^2 \beta)/2 \right\} \ln(L/b). \quad (6.7)'$$

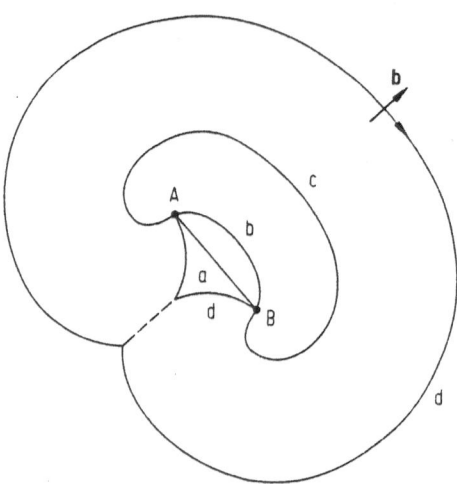

Fig. 6.4. Frank-Read source

The minimum radius of the dislocation arc-segment shown in Fig. 6.4 corresponds to the semi-circular case of $r = L/2$, which yields the following critical value of τ

$$\tau_{cr} \approx [Gb/2\pi L(1 - \nu)] \left\{ 1 - \nu(3 - 4\cos^2 \beta)/2 \right\} \ln(L/b). \qquad (6.8)$$

For the typical case of $L = 1000b$ and $\nu = 0.33$, the above expression gives a τ_{cr} value of $0.5Gb/L$ for a pure edge dislocation ($\beta = 90°$) and a τ_{cr} value of $1.5Gb/L$ for a pure screw dislocation ($\beta = 0°$). When τ exceeds τ_{cr}, the dislocation bow-out becomes unstable, and a dislocation loop is then generated with the original dislocation segment intact. Repetition of the same process by a Frank-Read source can generate sufficient amount of dislocation loops. When the emanating process of dislocation loops is blocked by certain obstacles, the strain hardening phenomenon will occur.

6.1.5 Stress-Strain Curve of a Single Crystal

A representative stress-strain curve for FCC single crystals is shown in Fig. 6.5. The appearance of this curve substantially deviates from the early observation of Taylor[2] for aluminium single crystals. Three distinctive stages, designated by stage I, II and III, are featured in this curve, with angles Θ_I, Θ_{II} and Θ_{III} denoting the curve slopes $(d\tau/d\gamma)$ at different stages.

The first stage in Fig. 6.5 is called the stage of *easy glide*. It immediately follows the elastic deformation, and represents a slightly linear hardening response starting from the initial yield stress τ_o. The deformation in this stage is characterized by the movement of long straight dislocations distributed uniformly in the crystal.

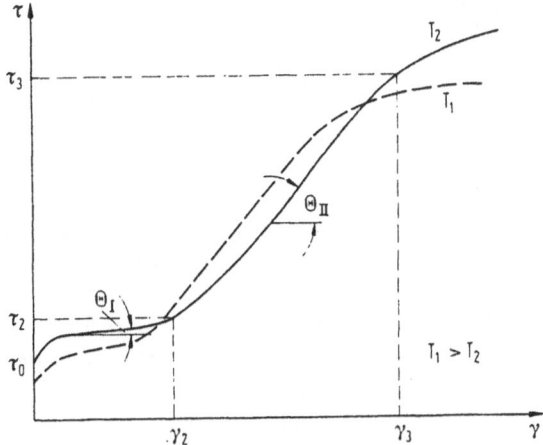

Fig. 6.5. Stress-strain curve for a FCC single crystal, featured by three stages of deformation

The straight line segments of dislocation have a typical length of 1000μm and the inter-dislocation distances range from 10 to 100 μm. This stage is usually absent in polycrystals. The termination of the first stage is marked by the appearance of secondary slip.

The second stage is called linear hardening stage. It is featured by :
(1) High linear strengthening ratio, with $\Theta_{II} \approx 10\Theta_I \approx G/300$.
(2) Co-existence of dislocations running along primary and conjugate slip systems.
(3) Non-uniform dislocation cell structure.
(4) As implied by the experimental evidences, the mean slip distance λ of dislocations in the second stage is inversely proportional to the additional shear amount in this stage

$$\lambda = D/(\gamma - \gamma_2) \tag{6.9}$$

where D is a constant ($\approx 4 \times 10^{-4}$ cm) and $\gamma - \gamma_2$ is the amount of shear strain measured from the beginning of the second hardening stage.
(5) The flow stress is proportional to the square root of the dislocation density

$$\tau = \tau_o + \alpha Gb\rho^{\frac{1}{2}} \tag{6.10}$$

where τ_o is the flow stress with dislocation interaction suppressed and α is a constant ranging between 0.3 to 0.6. The above formula is supported by substantial experiment data collected by Mitchell[10].

The third stage of strain hardening starts from the point (γ_3, τ_3) in Fig. 6.5, from which the strain hardening ratio gradually declines. The stress-strain curve in this stage can be modeled fairly well by a parabolic curve

$$\tau = \Theta_{III}(\gamma - \gamma')^{\frac{1}{2}} \tag{6.11}$$

where γ' is a constant. This hardening stage is featured by the massive occurrence of cross-slips at high stress level.

The flow stress of metal is also affected by temperature and deformation rate. Metallurgists usually regard the flow stress τ as a combination of a temperature independent part τ_G and a temperature dependent part τ^*

$$\tau = \tau_G + \tau^*. \tag{6.12}$$

The ratio τ^*/τ_G usually has an order of unity in the first stage of deformation. As plastic deformation increases, this ratio declines gradually and reaches a constant of 0.1 (independent of temperature) on the second stage. The property of ratio τ^*/τ_G remains constant in stage II is frequently referred as the *Cottrell-Stokes law*.

The rate sensitivity of flow stress is governed by the dislocation dynamics curve. When dislocation motion accelerates from normal speed to high speed, the resistance mechanism shifts from energetic barriers to phonon drag. Consequently, the flow stress will increases dramatically when the strain rate $\dot{\gamma}$ exceeds 10^5 per second.

6.1.6 Seeger Theory

Seeger and his co-workers advanced a comprehensive model to explain the stress-strain behavior of FCC crystals. A discussion on the *Seeger theory* will be carried out in this subsection to explain the hardening behavior from the first stage to the second stage.

First stage of strain hardening. It is mentioned above that the first stage of strain hardening is characterized by the long range interaction between dislocations which are relatively far apart. It is assumed that N dislocation sources exist in a unit volume, and the value of N preserves at the first stage of deformation. Under the excitation from an external stress τ, n dislocation loops are generated from each dislocation source on the primary slip plane. These dislocation loops are piled up in front of certain obstacles (such as grain boundaries) of an approximate circular shape. If the *mean slip path* λ of a typical dislocation loop is much greater than the inter-distance d between two neighboring dislocation pileups, as shown in Fig. 6.6, one can regard these n dislocation loops as a super-dislocation with Burgers vector $n\mathbf{b}$.

The applied stress field σ_{ij} necessary to operate the mentioned super-dislocation is

$$\sigma_{ij} = \sigma_{ij}^o - \sigma_{ij}^b \tag{6.13}$$

where σ_{ij}^o represents the critical stress on which a dislocation starts to run through a dislocation-free crystal, while σ_{ij}^b denotes the back stress applied by the super-dislocation to the Frank-Read source S in Fig. 6.6. For simplicity, the shape of the super-dislocation loop is assumed to be circular with a radius λ. This assumption is only consistent to the case of isotropic medium with a negligible Poisson's

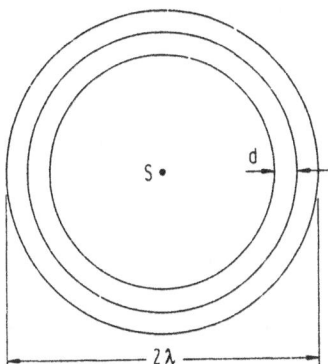

Fig. 6.6. Pileup calculation for a super-dislocation loop

ratio, otherwise the equilibrium shape of the super-dislocation loop will slightly deviate from the assumed circular shape. From the general formula (4.53) for planar dislocation loop, this back stress can be expressed as

$$\sigma_{ij}^b = (1/2) \oint_L \{\Sigma_{ij}(\phi) + \Sigma_{ij}''(\phi)\} \left[\sin(\phi - \beta)/R^2\right] dl. \tag{6.14}$$

For the particular case of the back stress exerted at the Frank-Read source (or equivalently at the origin of a circular dislocation loop), we have

$$\phi - \beta = -\pi/2, \quad R = \lambda, \quad dl = \lambda d\phi. \tag{6.15}$$

Hence,

$$\sigma_{ij}^b = -(1/2\lambda) \int_0^{2\pi} \{\Sigma_{ij}(\phi) + \Sigma_{ij}''(\phi)\} d\phi. \tag{6.16}$$

From (6.16) and the knowledge of section 4.3, one is able to compute σ_{ij}^b for a general anisotropic solid. If the assumption of isotropy is adopted, the Σ_{ij} tensor is simplified to

$$\Sigma_{ij}(\phi) = nG/\left[2\pi(1-\nu)\right] \begin{bmatrix} 0 & 0 & b(1 - \nu\cos^2\phi) \\ & 0 & -\nu b \sin\phi\cos\phi \\ symmtry & & 0 \end{bmatrix}. \tag{6.17}$$

Without loss of generality, we take a coordinate system such that the Burgers vector and the slip plane normal coincide with the 1-direction and the 3-direction, respectively. When substituting (6.17) into (6.16), one finds the only non-trivial component of σ_{ij}^b is

$$\sigma_{13}^b = -\left[nGb(2-\nu)\right]/\left[4\lambda(1-\nu)\right]. \tag{6.18}$$

This is exactly the resolved shear stress along the primary slip system. A combination of (6.13) and (6.18) would lead to the following formula of flow stress

$$\tau = \tau^o - \tau^b = \tau^o + \left[nGb(2-\nu)\right]/\left[4\lambda(1-\nu)\right]. \tag{6.19}$$

The plastic deformation produced by the above mechanism can be evaluated from (4.84)

$$\gamma = bA_{\text{slip}}/V \tag{6.20}$$

where V is equal to unity because a unit volume is referred to in the present model. Furthermore, the above dislocation loop configuration indicates

$$A_{\text{slip}} = \pi\lambda^2 Nn \tag{6.21}$$

when combining with (6.20), one arrives

$$\gamma = \pi\lambda^2 nbN. \tag{6.22}$$

The number of dislocation sources can be computed according to the following formula

$$N = 1/(\pi\lambda^2 s) \tag{6.23}$$

where s is the spacing between adjacent dislocation glide planes. Combining (6.19), (6.22) and (6.23), we arrive

$$\gamma = nb/s, \tag{6.24}$$
$$\tau = \tau_o + (2-\nu)/[4(1-\nu)](s/\lambda)G\gamma \tag{6.25}$$

which predicts a linear relationship between stress and strain. The slope in the easy glide stage is given by

$$\Theta_I = d\tau/d\gamma = (2-\nu)/[4(1-\nu)](s/\lambda)G. \tag{6.26}$$

Apparently, Θ_I is inversely proportional to the mean slip path λ of the dislocation loop.

The second stage of strain hardening. The intensification of strain hardening in the second stage is due to a rapid increase of dislocation sources. The dramatic increase of Frank-Read sources is attributed to the dislocation operation along secondary slip systems, as well as to the formations of Lomer-Cottrell locks. At the same time, the number of dislocation loops generated by each dislocation sources remains steady. Under this dislocation manifestation picture, the plastic strain increment can be derived from (6.22) as

$$d\gamma = \pi\lambda^2 nbdN. \tag{6.27}$$

Quoting the relationship (6.9) which characterizes the experimental findings in the second stage, one is able to integrate the above formula as

$$(\gamma - \gamma_2)^3 = 3\pi D^2 bnN. \tag{6.28}$$

At the second stage, the pileups of n dislocation loop can still be regarded as super-dislocations of Burgers vector $n\mathbf{b}$, with operation stress for dislocation source given by equation (6.10). The density of super-dislocations in (6.10) is given by

$$\rho = 2\pi\lambda N = 2(\gamma - \gamma_2)^2/(3Dnb). \tag{6.29}$$

The substitution of (6.29) into (6.10) (please notice the b in (6.10) should be replaced by nb for super-dislocation) will lead to

$$\tau - \tau_2 = \alpha G \sqrt{2nb/3D}(\gamma - \gamma_2) \tag{6.30}$$

which is in the correct form of linear strain hardening as described phenomenologically from the experimental data. Moreover, the slope on the stress-strain curve is predicted by

$$\Theta_{II} = \alpha G \sqrt{2nb/3D} \tag{6.31}$$

in the second stage of strain hardening. From the typical experimental data of FCC metals, $n \approx 25$, and $D \approx 5 \times 10^{-4}$ cm, it is estimated that

$$\Theta_{II}/G \approx 1/300 \tag{6.32}$$

which correlates the testing results reported in the previous subsection.

Modelling of strain hardening in stage III requires consideration of cross-slip phenomenon. An elaborate model pertinent to this strain stage was advanced by Kuhlmann-Wilsdorf, we leave the readers to explore this aspect by consulting references [11] and [12].

6.2 Solution Hardening

6.2.1 Point Defects

Different types of point defects exist in a crystalline solid, they include :
(1) *vacancies*, referred to the vacant sites in a crystal lattice;
(2) *interstitials*, the impurity atoms of smaller size occupying non-lattice sites;
(3) *self-interstitials*, matrix atoms occupying non-lattice sites;
(4) *substitutional atoms*, some matrix atoms at the lattice sites are replaced by other different atoms.

These point defects, termed simple point defects, are demonstrated in Fig. 6.7. As distinguished from more complicated point defects such as bivacancies and dumbbells, the above-mentioned simple defects can be represented by a *volumetric dilatation center* or a *volumetric contraction center* from the mechanics point of view.

The equilibrium density of point defects, c_o, can be assessed from a thermodynamics analysis as

$$c_o = \exp\{-U_1/kT + S_{\text{vib}}/k\} \tag{6.33}$$

where T the absolute temperature, k the Boltzmann constant, U_1 and S_{vib} the internal energy and the vibration entropy possessed by a single point defect. In the presence of internal or external stress field, the variation of defect density is

$$c = c_o \exp\{-p(V_s - V_m)/kT\} \tag{6.34}$$

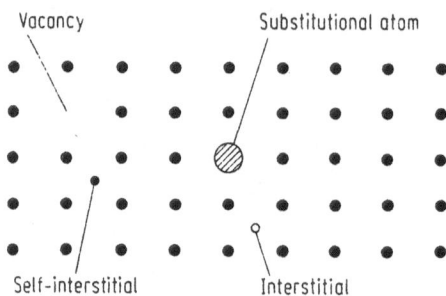

Fig. 6.7. Simple point defects

where p is the hydrostatic pressure (with compression taken as positive) and $V_s - V_m$ is the misfit volume between a solute atom and a matrix particle. The above formula will play an important role in the later analysis on Cottrell atmosphere.

6.2.2 Displacement Field Induced by a Force Dipole

As shown in Fig. 6.8, a point defect, represented either by a volumetric dilatation or by a volumetric contraction center, can be characterized by three force dipoles mutually perpendicular to each other. Therefore, a superposition of the Green function for concentrated force (as discussed extensively in subsection 4.3.2) can be carried out to give

$$\mathbf{u}(\mathbf{X}) = \sum_{q=1}^{3} \{ \mathbf{G}(\mathbf{X} - \mathbf{X}' - a\mathbf{e}_q) - \mathbf{G}(\mathbf{X} - \mathbf{X}' + a\mathbf{e}_q) \} \cdot (F\mathbf{e}_q) \qquad (6.35)$$

where $\mathbf{e}_q$ ($q = 1$, 2 and 3) represent orthonomal base vectors in the three dimensional physical space. Let us consider the case in which the force amplitude tends to infinity while the dipole interval a vanishes, but the dipole strength

$$P = \lim_{a \to 0} 2Fa \qquad (6.36)$$

keeps constant during this limiting process. Then (6.35) is simply reduced to

$$u_i(\mathbf{X}) = -P\, G_{ij,j}(\mathbf{X} - \mathbf{X}') \qquad (6.37)$$

which anticipates the observation from Fig. 6.8 that a positive P corresponds to a volumetric dilatation center whereas a negative P corresponds to a volumetric contraction center. Accordingly, equation (6.37) describes the displacement fields at $\mathbf{X}$ due to a point defect at $\mathbf{X}'$.

Eshelby[3] derived a relation linking the total volume dilatation δV and the dipole strength P

$$\delta V = \int_V u_{i,i}\, dV = P\, C_{iijj} \qquad (6.38)$$

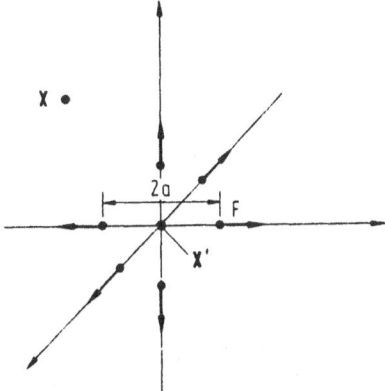

Fig. 6.8. Point defect and corresponding force dipoles

where $\mathbf{C}$ represents the fourth-rank compliance tensor of elasticity, and it is the inverse of $\mathbf{L}$ introduced in (4.22).

For isotropic material, an explicit expression of the dipole displacement field can be derived by utilizing formula (4.39) for the Green function

$$\mathbf{u}(\mathbf{X}) = \{P/[4\pi(\lambda + 2G)]\}\,(\mathbf{R}/R^3) \qquad \mathbf{R} = \mathbf{X} - \mathbf{X}' \qquad (6.39)$$

where λ and G are Lamè constants. A relationship between the actual volume change δv of the volumetric dilatation source and the dipole strength is found as follows

$$\delta v = 4\pi r^2 u_r\big|_{r=0} = P/(\lambda + 2G). \qquad (6.40)$$

Combining (6.38) and (6.40), one obtains a relationship between the actual source dilation δv and the total volumetric dilation δV as

$$\delta V = 4\pi R^2 u_r\big|_{r=R} = 3\delta v(\lambda + 2G)/(3\lambda + 2G) = 3P/(3\lambda + 2G). \qquad (6.41)$$

6.2.3 Interactions Between Point Defects and Dislocations

Attention is now focused on the interaction energy between a specific point defect and the other defect fields. As mentioned above, a point defect can be treated by three mutually perpendicular force dipoles, with the corresponding stress and displacement fields labelled by a superscript P. With this field contribution excluded, the remaining stress and displacement fields (which are caused by the external force and/or by other defects) are labelled by a superscript F. Similar to the derivation engaged in subsection 4.3.4, the *interaction energy of the point defect* is

$$\Pi^{P-F} = -\int_V \mathbf{f}^P \cdot \mathbf{u}^F \, dV \qquad (6.42)$$

where the body force $\mathbf{f}^P$ induced by the point defect can be written as

$$f_i^P = \frac{1}{2}P \lim_{a \to 0}(1/a)\{\delta(\mathbf{X} - \mathbf{X}' - a\mathbf{e}_i) - \delta(\mathbf{X} - \mathbf{X}' + a\mathbf{e}_i)\}. \tag{6.43}$$

Upon substitution of (6.43) into (6.42), one finds

$$\Pi^{P-F} = -\frac{1}{2}P \lim_{a \to 0}(1/a) \sum_{i=1}^{3}\{u_i^F(\mathbf{X}' + a\mathbf{e}_i) - u_i^F(\mathbf{X}' - a\mathbf{e}_i)\}$$
$$= -Pu_{i,i}^F(\mathbf{X}') \tag{6.44}$$

which demonstrates a striking feature of the interaction energy Π^{P-F}, that is, it is only relevant to the dipole strength (so to the total volume change by (6.38)) of the point defect and the external volumetric strain at the point defect (with the self field of point defect excluded). Through a parallel derivation, one can also write down the interaction energy between a point defect A at $\mathbf{X}$ and a point defect B at $\mathbf{X}'$ as

$$\Pi^{A-B} = P^A P^B G_{ij,ji}(\mathbf{X} - \mathbf{X}'). \tag{6.45}$$

For isotropic material, the expression of interaction energy (6.44) can be further reduced to

$$\Pi^{P-F} = p(\mathbf{X}')\delta V \tag{6.46}$$

where $p(\mathbf{X}')$ is the hydrostatic pressure at $\mathbf{X}'$ caused by stress fields other than the self-field of the point defect.

We next consider a virtual variation $\delta\mathbf{r}$ on the location $\mathbf{X}'$ of the point defect. The corresponding variation in the interaction energy can be written in a form

$$\delta\Pi^{P-F} = -\mathbf{f}\cdot\delta\mathbf{r} \tag{6.47}$$

where $\mathbf{f}$ denotes the driving force on the point defect, similar to the Peach-Koehler force on a dislocation. A comparison on the above two formulae leads to an expression of $\mathbf{f}$

$$\mathbf{f} = -\delta V \nabla p(\mathbf{X}'). \tag{6.48}$$

Example: *Interaction between point defects and an edge dislocation.* Let us consider the interaction between a point defect at (x, y) and a straight edge dislocation situated at the origin of the chosen coordinate system. For simplicity, only two dimensional analysis is engaged and the material is assumed to be isotropic. Consequently, equation (4.16) can be used for the hydrostatic pressure p

$$p = (Gb_e/3\pi)[(1 + \nu)/(1 - \nu)](\sin\Theta/r) \tag{6.49}$$

which is substituted back to equations (6.46) and (6.48) to give the interaction energy and the force on the point defect

$$\Pi^{P-F} = Ay/(x^2 + y^2) \qquad A = (Gb_e/3\pi)[(1+\nu)/(1-\nu)]\delta V$$
$$f_x = -(\partial/\partial x)\Pi^{P-F} = (A/r^2)\sin 2\Theta$$
$$f_y = -(\partial/\partial y)\Pi^{P-F} = -(A/r^2)\cos 2\Theta. \tag{6.50}$$

From equation (6.50) one finds the force vector $\mathbf{f}$ is aligned by rotating an angle $\pi/2 - 2\Theta$ clockwise with respect to the position vector of the point defect. If $\hat{\mathbf{f}}$ denotes a unit vector forming an angle of $2\Theta - \pi/2$ with respect to the positive x-axis, then the force on the point defect can be put into a more compact form

$$\mathbf{f} = (A/r^2)\hat{\mathbf{f}} \tag{6.51}$$

where A/r^2 describes the amplitude of the force. This force drives the interstitial atoms ($\delta V > 0$) above the slip plane of an edge dislocation downward, and drives the vacancies ($\delta V < 0$) beneath the slip plane upward.

6.2.4 Cottrell Atmosphere

The non-uniform hydrostatic pressure field near an edge dislocation will lead to the non-uniformity on point defect distribution. If the pressure field (6.49) is substituted back into (6.34), the equilibrium density of point defect distribution around an edge dislocation becomes

$$c = c_o \exp(-p\delta V/kT) = c_o \exp[(-A\sin\Theta)/(rkT)]. \tag{6.52}$$

Accordingly, the interstitials ($A > 0$) would concentrate beneath the slip plane, whereas the vacancies ($A < 0$) would concentrate above the slip plane, as shown schematically in Fig. 6.9. This biased distribution of point defects around an edge dislocation is named *Cottrell atmosphere*.

6.2.5 Solution Hardening and Yield Point Drop

The addition of solute atoms to a solvent matrix would result in solution hardening, due to the elastic distortion caused by the size difference between the solute and the solvent. For substitutional solid state solution, the total volume change due to the replacement by a solute atom is

$$\delta V = V_s - V_m \tag{6.53}$$

provided the elastic compressibility is ignored. In (6.53), V_s and V_m represent the volumes occupied by each solute and matrix solvent atom, respectively. These

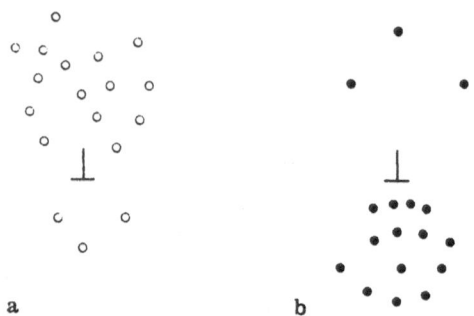

Fig. 6.9. Cottrell atmosphere. **a** Vacancies; **b** interstitials

point defects would form a Cottrell atmosphere around each edge dislocation. This atmosphere produces a pinning force f_D, identical in magnitude but opposite in sign to the resultant force f acting on all point defects, on the edge dislocation

$$f_D = -f. \tag{6.54}$$

This pinning force can be calculated by the methodology proposed in subsection 6.2.3, it greatly confines the mobility of dislocations. Therefore, the material yield strength can be enhanced considerably via the addition of the solute atoms, so comes to the notion of *solution hardening*.

At normal temperature environment, dislocations could break clean from the pinning Cottrell atmosphere when sufficient external stress is applied. The movement of a Cottrell atmosphere is controlled by the rate process. The diffusion of solute atoms are usually too slow to catch up with the non-diffusive dislocation glide. When the dislocation is escaped from the interaction of Cottrell atmosphere, the pinning force on the dislocation diminishes. The dislocation will subsequently glide in a relatively "clean" solvent matrix, leading to the *yield point drop* shown in Fig. 6.10, as frequently observed from uniaxial tensile test curves of mild steel. A low resistance regime for dislocation glide would immediately follow the yield point drop, and result in an "easy slip" deformation stage in the stress-strain curve before efficient strain hardening accumulates. The macroscopic slip bands, also termed *Lüders bands*, are formed in this regime.

6.2.6 Creep Controlled by Point-Defect Diffusion

At relatively high temperature, the linking force between the gliding dislocation and the Cottrell atmosphere would be able to drag the point defects moving along with the dislocation, due to the relative ease of diffusion at high temperature regime. Accordingly, the whole deformation procedure would be controlled by the diffusion process of point defects, the latter consequently apply a drag force on the gliding dislocation. A complete mathematical solution for the combined movement of dis-

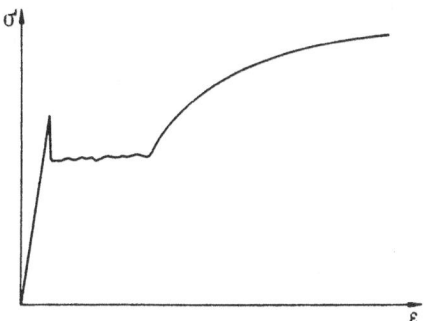

Fig. 6.10. Yield point drop and subsequent easy-slip stage

location and accompany Cottrell atmosphere is quite complicated, and will not be attempted here. In 1949, Cottrell and Jaswon[14] first attacked this problem. Their computation on the dislocation force, however, suffered from some inconsistencies.

The exact solution for this problem was given by Fuentes-Samaniego[15] in his Ph.D. thesis thirty years after the pioneer work of Cottrell and Jaswon[14]. The drag per unit dislocation line on the dislocation gliding, f/L, maintains the following approximate relationship with the velocity v of dislocation movement

$$f/L \approx \frac{1}{2}\pi[vc_oA^2)/(DkT)]\ln[(DkT)/(vA)] \tag{6.55}$$

where D the diffusion constant, k the Boltzmann constant, T the absolute temperature. A in (6.55) has already been defined in (6.50). (6.55) consists of an important formula for the description on the rate dependent process of high temperature creep.

6.3 Precipitation and Dispersion Hardening

6.3.1 Second Phase Particles

In the design of various alloys, second phase particles are frequently utilized to obtain hardening effect. A typical example consists of the *solution-aging treatment* of aluminium alloy, in which substantial hardening is achieved by the precipitation of the second phase particles. The major processing steps for this treatment are :
(1) formation of over-saturated solid solution by rapid cooling from high temperature,
(2) precipitation of metastable, transition structures,
(3) aging in room or elevated temperature, in which the desired fine aging phases are formed.

For aluminium alloy, the structure of the second phase particles gradually changes from the metastable GP-1 zone, to GP-2 zone, then to Θ' phase, and finally to the equilibrium Θ phase ($CuAl_2$) during the aging process. The typical hardness versus aging time curves for Al-Cu alloy are plotted in Fig. 6.11. The most pronounced age-hardening effect occurs when the precipitates size falls in the range of GP-2 zone.

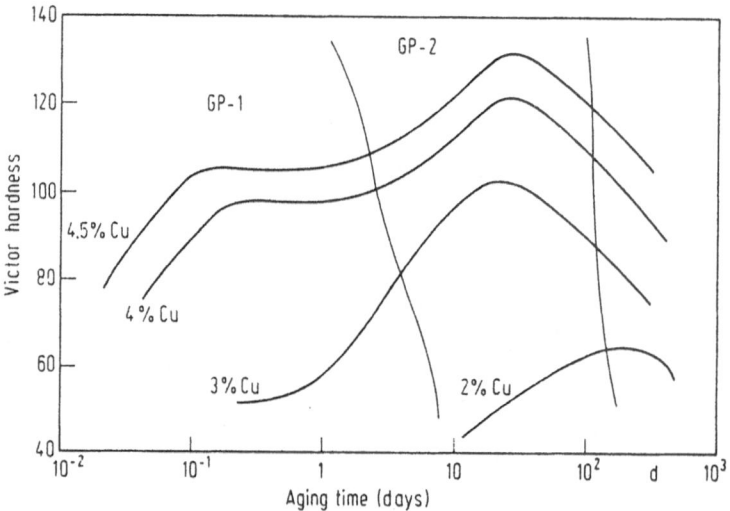

Fig. 6.11. Hardness versus aging-time relationship of Al-Cu alloy during aging at 130°C

Dispersion hardening constitutes another mechanism of strengthening by the second phase particles. The dispersion particles on a ductile matrix can be introduced by blending, sintering or internal oxidation techniques. As a distinction from the precipitates, the dispersion phases will not dissolve into the matrix at high temperature, rendering its capability to resist deformation (especially creep) at high temperature.

The influence of these fine second phase particles to the matrix hardening will be investigated in the subsequent subsections.

6.3.2 Long Range Stress

The *long range stress* is referred to the resistance to a dislocation line that moves to bypass the second phase particles of volume concentration c. During the calculation of long range stress, it is assumed that the dislocation lines situate in between the average spacing l of the second phase particles. The concept of long range stress is sensible only for dilute distribution of the second phase particles ($c \ll 1$). In that case, the particles can be approximately regarded as tiny spheres of radius r_o. An average *transformation strain*, $\epsilon_m^t = \epsilon_{kk}^t/3$, is associated with the second phase particles in a non-constrained state. Due to the constraint of the matrix, however, the actual misfit strain for the second phase particles is $\epsilon_m^* = \epsilon_{kk}^*/3 < \epsilon_m'$. From equation (6.39), the displacement field in the matrix (presumably isotropic) is

$$u_r = \epsilon_m^* r_0 (r_0/r)^2, \quad u_\Theta = u_\phi = 0. \tag{6.56}$$

The corresponding matrix stress field under spherical coordinates can be easily computed from the above displacement field as

$$\sigma_{rr} = -4G\epsilon_m^*(r_0/r)^3$$
$$\sigma_{\Theta\Theta} = \sigma_{\phi\phi} = 2G\epsilon_m^*(r_0/r)^3. \tag{6.57}$$

It is interesting to notice that the radial stress is compressional, while the circumferential and hoop stress components are in tension and have values exactly one-half of the former. On the other hand, the second phase particles are subject to a uniform, hydrostatic pressure stress state

$$\sigma_{rr} = \sigma_{\Theta\Theta} = \sigma_{\phi\phi} = -4G\epsilon_m^* \tag{6.58}$$

which indicates the absence of shear stress in the interior of the second phase particles. In the material matrix, if Cartesian coordinates (x, y, z) are adopted with z-direction in alignment with the dislocation line, then the stress field (6.57) would give rise to the following resistance to dislocation glide

$$\sigma_{xy} = 4G\epsilon_m^*(r_0/r)^3 \sin\Theta \cos\Theta. \tag{6.59}$$

This stress field decays rapidly as r increases.

We next consider the volumetric strain of the second phase particles

$$\epsilon_m^* = \epsilon_m' + \epsilon_m^l \tag{6.60}$$

where ϵ_m' is the transformation strain without any constraining stress, and ϵ_m^l corresponds to the volumetric contraction under the hydrostatic stress of (6.58). After straightforward algebra, one gets

$$\epsilon_m^* = [(1+\nu)/3(1-\nu)]\epsilon_m'. \tag{6.61}$$

If N second phase spherical particles of radius r_0 are present in the unit volume, their volume density can be calculated as

$$c = (4/3)\pi r_0^3 N \tag{6.62}$$

and the average inter-particle spacing l is

$$l = N^{-1/3}. \tag{6.63}$$

Substituting $r = \frac{1}{2}l$ into (6.59), we arrive an expression of the average long range resistance τ_{LR} on a dislocation line outflanking the stress fields induced by various distributed second phase particles

$$\tau_{LR} = G\epsilon_m^*(2r_0/l)^3. \tag{6.64}$$

Combination of the previous three formulae would lead to the famous *Mott-Nabarro formula*[16] of the long range stress

$$\tau_{LR} = (6G/\pi)\epsilon_m^* c \tag{6.65}$$

which clearly exhibits the proportionality between τ_{LR} and the particle concentration c. The long range stress predicted in (6.65) is independent of the size of the second phase particles. More detailed studies on the long range stress has been attempted since the work of Mott and Nabarro, and some of the studies adopted a statistical approach. All of those developments prior to 1977 were surveyed by Nabarro[17]. Under various limiting cases of the above investigations, τ_{LR} could be proportional to $c^{1/2}, c^{2/3}$ or c.

The forgoing discussion on the concentration density effect is also applicable to the previous section on solution hardening.

6.3.3 Orowan Stress

The above discussion on long range stress is only pertinent to the case where the dislocation line outflanks the stress fields induced by various inclusions. As the concentration of the second phase particles increases, the short range interaction between dislocation and the second phase particles on the dislocation glide plane become inevitable. A model concerning this short range interaction was first advanced by Orowan[18] in 1948, featured by the *dislocation bow-out* configuration as presented in Fig. 6.12. Later on, the modern transmission electronic microscopy confirmed the actual existence of this mechanism. If we still use l for the inter-particle spacing, the critical stress for dislocation bow-out can be assessed from equation (6.8) as

$$\tau_{\text{Orowan}} = \{Gb/[2\pi l(1-\nu)]\}\{1-\nu(3-4\cos^2\beta)/2\}\ln(l/b). \qquad (6.66)$$

For a pure edge dislocation before its bow-out, one sets $\beta = \pi/2$ in the above formula to arrive

$$\tau_{\text{Orowan}} = [Gb/(4\pi l)]\{(2-3\nu)/(1-\nu)\}\ln(l/b). \qquad (6.67)$$

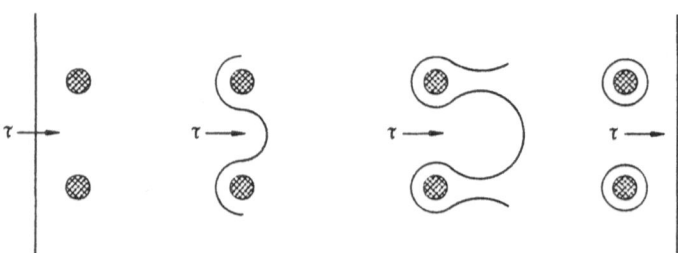

Fig. 6.12. A schematic presentation of dislocation bow-out

It is tacitly implied in the dislocation bow-out mechanism of Orowan that the second phase particles always possess sufficient strength to pin the dislocation line.

While that might be true for large particles, the possibility of particle shearing apart, however, cannot be overlooked for particles of less size. If the interfacial misfit energy between matrix and the second phase particles is denoted by Γ_{m-p} per unit surface, then the surface energy required by shearing-off each particle (creating two surfaces above and below) is $2\pi r_o^2 \Gamma_{m-p}$. During the same process, the work done by the applied stress is approximately $l(2r_o)b\tau_{\text{shear}}$. In conclusion,

$$\tau_{\text{shear}} \approx \pi \Gamma_{m-p} r_o / lb. \tag{6.68}$$

From the comparison between (6.67) and (6.68), one realizes that the occurrence of particle shear-off depends on the particle size r_o and the misfit energy Γ_{m-p}. The latter is relevant to the manner of misfit, such as the extent of lattice coherency. When all the factors discussed above are considered, the flow stress for a precipitation or dispersion hardening solid becomes

$$\tau = \tau_M + \tau_{LR} + \text{Min} \left\{ \tau_{\text{Orowan}}, \tau_{\text{shear}} \right\}. \tag{6.69}$$

In (6.69), τ_M is the matrix flow stress with its calculation elaborated in section 6.1. The flow stress increments $\tau_{LR}, \tau_{\text{Orowan}}$ and τ_{shear} caused by the second phase particles can be evaluated by (6.65), (6.66) and (6.68), respectively. The maximum value of τ usually occurs around the neighborhood where τ_{Orowan} and τ_{shear} are about identical. The particle size under this particular circumstance is termed *critical particle size*. Its existence is often used to explain the peak of the flow stress τ curve at GP-2 zone, as shown in Fig. 6.11.

6.3.4 Ashby Theory

Beside blocking dislocation motion, the second phase particles also excite dislocation loop generation during the plastic deformation, which in turn influences the strain hardening behavior. Ashby[19,20] made a detailed study toward this aspect, he decomposed the total dislocation density into the following two parts

$$\rho = \rho_S + \rho_G \tag{6.70}$$

where ρ_S is the *statistically equilibrium dislocation* density corresponding to the conventional dislocation generation in a uniform plastic deformation and ρ_G is the *geometrically necessary dislocation* density which renders necessary to accommodate local non-uniform deformation. For deformation of a single phase matrix, the latter can often be neglected, i.e., $\rho_S \gg \rho_G$. Thus, the simple strain hardening theory as described in section 6.1 is appropriate. In the presence of large amount of non-deformable, hard second phase particles, however, severe non-uniform deformation in the neighborhoods of those particles would produce high density of geometrically necessary dislocations. Occasionally, ρ_G so-produced even far surpasses ρ_S. This situation was demonstrated by Ashby through a simple example of a cubic hard inclusion.

As shown in Fig. 6.13, a material element subject to remote uniform shear is considered. The material element contains a hard, second phase particle of cubic shape. The resulted plastic deformation can be regarded as a superposition of a uniform matrix shear deformation (visualized as if the hard particle were removed from the center, see Fig. 6.13(b)), and a non-uniform deformation caused by re-insert the cube particle back. The latter deformation can be described either by the square loops of shear dislocations as presented in Fig. 6.13(c); or by the *prismatic dislocation loops* as drawn in Fig. 6.13(d), where linearly varied prismatic loops are inserted to the left hand side of the cube and similar vacant prismatic loops are inserted on the right hand side. For the former shear type loops, the number of geometrically necessary dislocation loops are

$$n = 2r\gamma/b \tag{6.71}$$

where $2r$ corresponds the edge length of the second phase particle. For the latter prismatic loops, a substance of volume

$$\Delta V = \frac{\gamma}{2}V_p = 4r^3\gamma \tag{6.72}$$

should be deposited on the left hand side of the particle shown in Fig. 6.13(d), where V_p is the volume of the second particle. By the same argument, the same amount of vacancy should be engaged on the right hand side of the particle shown in Fig. 6.13(d). Therefore, the total number of prismatic dislocation loops to smear out the non-uniform deformation for each cubic particle is

$$n = 2\Delta V/\bar{A}b = 4r\gamma/b \tag{6.73}$$

where $\bar{A} = 2r^2$ is the average area bounded by the dislocation loops.

Identical total length of dislocation line, L_{total}, for a single second phase particle would be inferred from the expressions of either shear dislocation loops ($n = 2r\gamma/b$, and loop length $= 8r$) or that of prismatic dislocation loops ($n = 4r\gamma/b$, and the average loop length $= 4r$),

$$L_{total} = 16r^2\gamma/b \tag{6.74}$$

where the number of second phase particles per unit volume is

$$N = c/8r^3 \tag{6.75}$$

with c being the density of the second phase particles. The geometrically necessary dislocation density can be derived from preceding two expressions as

$$\rho_G = 2c\gamma/br. \tag{6.76}$$

For the strain hardening material containing massive amount of second phase particles, ρ_G is regarded to form the majority of total dislocation density ρ. By this understanding, a strain hardening correlation can be obtained by combining (6.10) and (6.76)

$$\tau = \tau_o + 2\alpha G[(cb\gamma)/(2r)]^{\frac{1}{2}} \tag{6.77}$$

where $2r$ and c are the average size and the volumetric density of the second phase particles, respectively.

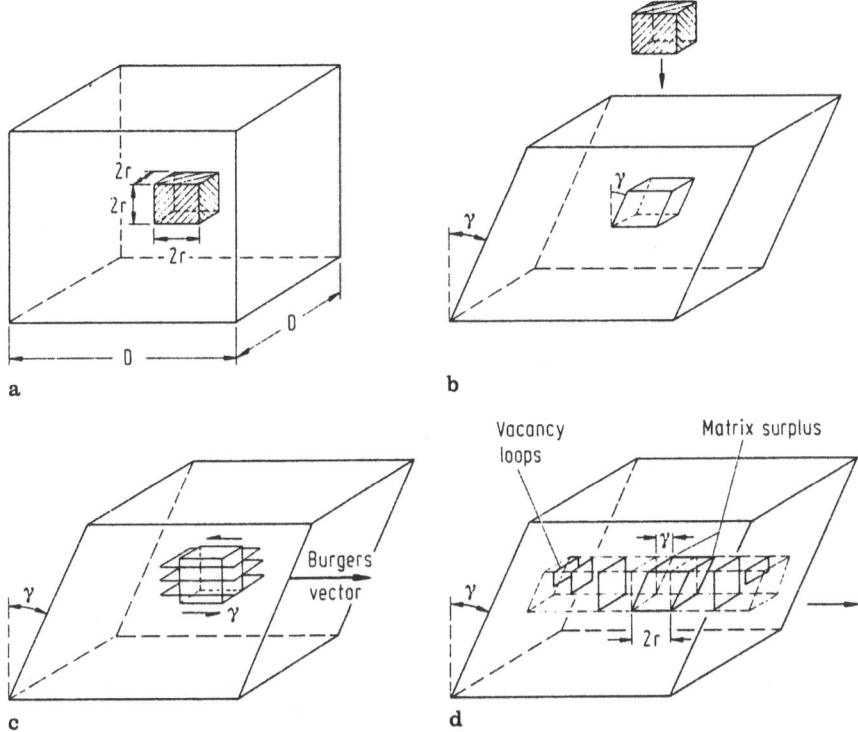

Fig. 6.13. Dislocation accumulation near a hard cubic inclusion. **a** A crystal cell prior to deformation; **b** deformed crystal cell; **c** accumulation of shear loops; **d** accumulation of prismatic loops

6.4 Grain Boundary Hardening

6.4.1 Dislocation Pileups at Grain Boundary

The misorientation between the neighboring grains could promote the formation of *dislocation pileup* stopped at the grain boundary. The strain hardening effect due to this pileup can be estimated by the one-dimensional model shown in Fig. 6.14. In the figure, the origin of Frank-Read source is marked by a capital letter S, and it locates at the central position of the grain. Under applied shear stress τ, two arrays of edge dislocations with opposite signs are emanated from S to the left and to the right. These edge dislocation arrays are blocked and piled up at the two opposite grain boundaries by unfavorable oriented neighboring grains, with the length of double pileup extended to the entire grain dimension d as shown in Fig. 6.14. For simplicity, the medium is assumed to be isotropic except the misorientation barriers along the grain boundary.

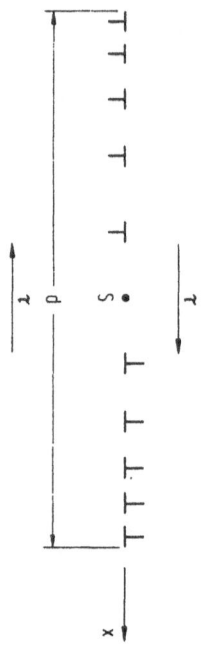

Fig. 6.14. A double pileup formed by dislocations emanating from a Frank-Read source S and being blocked at grain boundary

6.4.2 Hall-Petch Relation

The dislocation distribution portrayed in Fig. 6.14 along x-direction can be modeled by a one-dimensional continuum dislocation model with the following definition on its distribution function $D(x)$

$$D(x) = \pm(1/b)(db/dx) \tag{6.78}$$

where b represents the magnitude of the Burgers vector for the crystal lattice under consideration, and $db = bD(x)dx$ denotes the increment of Burgers vector for continuously distributed dislocations between x and $x + dx$. Quoting from (4.16), one finds a shear stress of the value

$$d\tau = \{Gb/[2\pi(1 - \nu)]\}\{D(x')\,dx'/(x' - x)\} \tag{6.79}$$

at the coordinates $(x, 0)$ by the increment $bD(x')dx'$ of Burgers vector from the dislocation line element centered at $(x', 0)$. An integration on the above formula results in

$$\tau = \{Gb/[2\pi(1 - \nu)]\} \int_{-d/2}^{d/2} D(x')\,dx'/(x' - x) \tag{6.80}$$

which is required to be at constant value in the range $-\frac{1}{2}d \leq x \leq \frac{1}{2}d$. If the following variable transformation is introduced

$$\eta = 2x/d \tag{6.81}$$

and the dislocation distribution function is re-defined as

$$f(\eta) = D(\eta d/2) \tag{6.82}$$

then a Cauchy type integral equation in canonical form is obtained

$$(1/\pi) \int_{-1}^{1} f(\eta')\, d\eta'/(\eta' - \eta) = 2(1 - \nu)\tau/(Gb). \tag{6.83}$$

The solution for this equation can be evaluated analytically as

$$f(\eta) = \{2(1 - \nu)\tau/(Gb)\}\{\eta/\sqrt{1 - \eta^2}\}. \tag{6.84}$$

When substituting back to equation (6.82), one obtains the distribution of the dislocation pileup configuration in Fig. 6.14 as

$$D(x) = \{2(1 - \nu)\tau/(Gb)\}(x/\sqrt{d^2/4 - x^2}). \tag{6.85}$$

The above double pileup can be treated as an imaginary shear crack running from $-\frac{1}{2}d$ to $\frac{1}{2}d$. The dislocation increment $bD(x)dx$ (from x to $x + dx$) slips a distance of x and induces a shear increment of $bD(x)x\, dx$. The total shear amount caused by the double dislocation pileup in Fig. 6.14 is then

$$\gamma = \int_{-d/2}^{d/2} bx\, D(x)\, dx = \{\pi(1 - \nu)\tau d^2\}/(4G). \tag{6.86}$$

Here we point out the final result on the induced shear strain does not contain the magnitude of Burgers vector b, providing a justification to the treatment of double dislocation pileup in terms of a shear crack.

The formation of above shear crack can be visualized by gradually releasing the holding stress $-\tau$. This hypothetical process renders the following expression on the energy release by forming shear crack of unit thickness

$$\Pi = \frac{1}{2}\tau\gamma = \pi(1 - \nu)\tau^2 d^2/(8G). \tag{6.87}$$

Accordingly, the driving force at the tip of postulated shear crack (which also represents the driving force on the forefront dislocation line blocked by the grain boundary) is

$$f = \partial\Pi/\partial d = \pi(1 - \nu)\tau^2 d/(4G). \tag{6.88}$$

It is assumed that slip at adjacent grain will be triggered when f in the above formula reaches a critical value f_{gb} relevant to the grain boundary structure encountered. This criterion will deliver the following expression of the driving stress

$$\sigma = Kd^{-\frac{1}{2}} \tag{6.89}$$

where

$$K = m\{(4Gf_{gb})/[\pi(1-\nu)]\}^{\frac{1}{2}} \tag{6.90}$$

and m signifies the Schmid factor which projects the applied stress tensor to the specific slip system.

The actual testing data of σ of polycrystals can indeed be correlated with respect to $d^{-\frac{1}{2}}$ into a straight line. But this straight line does not approach zero at the limit of an infinite d. As a compromise, (6.89) is commonly modified as

$$\sigma = \sigma_o + Kd^{-\frac{1}{2}} \tag{6.91}$$

and which is the famous *Hall-Petch relation*[22,23]. The first term in (6.91) characterizes the flow stress of a single crystal, whereas the second term indicates the hardening effect due to grain boundaries or grain size. The origin of σ_o could come from either the lattice resistance against dislocation glide or from the internal back stress. Further exploration is required to clarify this issue.

6.5 References

1. Nabarro, F.R.N., Basinski, Z.S. and Holt, D.B., Adv. in Phys., 13(1964), p.193.
2. Taylor, G.I., Proc. Roy. Soc. London, A145(1934), p.362.
3. Mott, N.F., Proc. Roy. Soc. London, B64(1951), p.279.
4. Lomer, W.M., Phil. Mag., 42(1951), p.1327.
5. Cottrell, A.H., Phil. Mag., 43(1952), p.645.
6. Taylor, G.I. and Elam, C.F., Proc. Roy. Soc. London, A102(1923), p.643.
7. Taylor, G.I. and Elam, C.F., Proc. Roy. Soc. London, A108(1925), p.28.
8. Frank, F.C. and Read, W.T., Sympo. on Plastic Deformation of Crystalline Solids, Carnegie Inst. of Tech., Pittsburgh, (1950), p.44.
9. Hirth, J.P. and Lothe, J., Theory of Dislocations, 2nd Ed., McGraw-Hill, (1982).
10. Mitchell, T.E., Progr. Appl. Material Research, 6(1964), p.117.
11. Kulmann-Wilsdorf, D., in Work Hardening, Hirth J.P. and Weertman J. Eds., (1968), p.97.
12. Kulmann-Wilsdorf, D., in Work Hardening in Tension and Fatigue, Thompson A.W. Ed., TMS-AIME, (1977), p.1.
13. Eshelby, J.D., Solid State Phys., 3(1956), p.79.
14. Cottrell, A.H. and Jaswon, M.A., Proc. Roy. Soc. London, A199(1949), p.104.
15. Fuentes-Samaniego, Ph.D. Thesis, Stanford Univ., (1979).
16. Mott, N.F. and Nabarro, F.R.N., Proc. Phys. Soc. London, 52(1940), p.86.
17. Nabarro, F.R.N., Phil. Mag., 35(1977), p.613.
18. Orowan, E., in Internal Stress in Metals and Alloys, Institute of Metal, London, (1948), p.451.
19. Ashby, M.F., Phil. Mag., 14(1966), P.1157.
20. Ashby, M.F., Phil. Mag., 21(1967), P.399.
21. Read, W.T. and Shockley, W., Phys. Rev., 78(1950), p.275.
22. Hall, E.O., Proc. Roy. Soc. London, B64(1951), p.474.
23. Petch, N.J., J. Iron Steel Inst., 174(1953), p.25.

7 Plasticity for Crystalline and Geological Materials

7.1 Single Crystals

7.1.1 Single Crystal Deformation

A majority of plastic deformations take the form of discrete crystalline slip, governed by the glide motion of dislocations along corresponding slip systems as elucidated in length in Chapter 4. A systematic study on plastic slip deformation should start from the simple case of single crystals, where the complications such as grain orientations and grain boundaries could be eliminated. Studies on slip deformation of single crystals was originated from the historic works of Taylor and Elam[1,2], in which plastic slip along various orientations was first observed and examined. The modern formulation on slip-induced plasticity was advanced by Hill[3], and by Hill and Rice[4] in which the constitutive law of crystalline plasticity were elaborated by the mathematical elegance. Comprehensive reviews on this research subject were provided by Asaro[5,6,7] where his contributions to crystalline plasticity were also included.

Decomposition of elastic and plastic deformations. A clear mathematic description on the kinematics of single crystals is now in completion. Distinction on elastic and plastic deformations should be carefully examined at the beginning. The famous multiplication formula of Lee[8], proposed originally as a continuum plasticity postulate, (see equation (2.29)),

$$\mathbf{F} = \mathbf{F}^* \cdot \mathbf{F}^p \tag{7.1}$$

bears new physical interpretations in single crystal plasticity. Equation (7.1) decomposes the local deformation gradient tensor $\mathbf{F}$ into a plastic deformation, $\mathbf{F}^p$, on the right and a combination of elastic deformation and rigid rotation, $\mathbf{F}^*$, on the left of the factorization. From the viewpoint of continuum mechanics, the above decomposition formula is by no means unique. The choice of an intermediate configuration where this decomposition is carried out still remains a controversial issue. A crystallographic delineation, however, can be provided to support the multiplicative decomposition formula (7.1). Let us consider the elastic-plastic deformation of a single crystal lattice illustrated by the three configurations in Fig. 7.1. The

total uniform deformation which carries the single crystal from the reference configuration to the current configuration is denoted by $\mathbf{F}$. Then a quasi-static *lattice
unloading* is engaged in which the average stress measure in the single crystal
is gradually released to zero, corresponding an inverse elastic deformation $\mathbf{F}^{*-1}$
while the undergone material particle flow throughout the lattice is maintained.
When unloading is completed, the residual (or permanent) deformation is given
in accordance to a composition law as $\mathbf{F}^{*-1} \cdot \mathbf{F}$, which should be set equal to the
plastic deformation $\mathbf{F}^p$ inherited in the total deformation $\mathbf{F}$ and unchanged during
the elastic lattice relaxation. This recognition gives us the decomposition formula
(7.1). It is also clear that $\mathbf{F}^p$ corresponds to the material flow caused by plastic
slip motion of material particles above and beneath the slip planes while the lattice
configuration preserved; and $\mathbf{F}^*$ represents collectively the lattice distortion and
rotation.

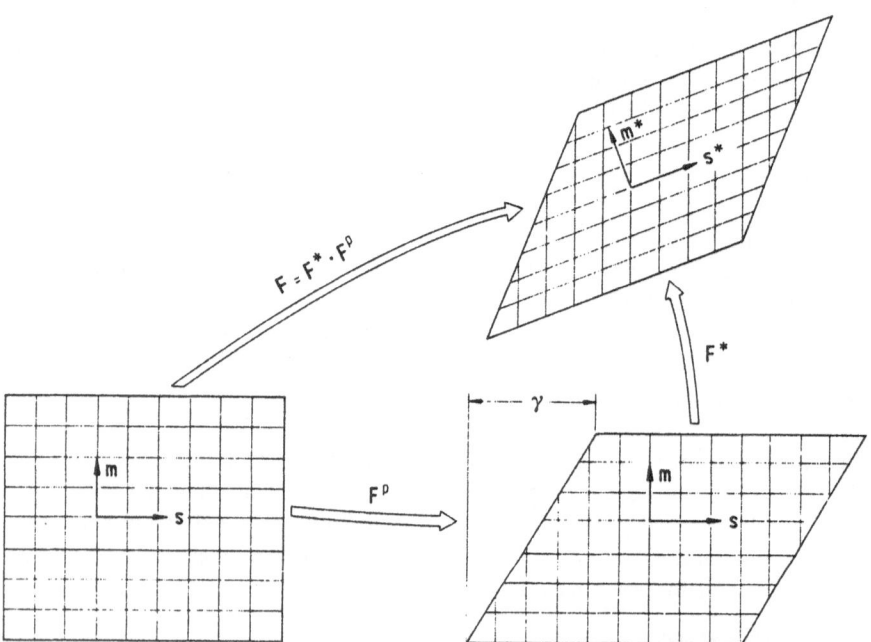

Fig. 7.1. Elastic-plastic deformation of a ductile single crystal. The deformation is carried out from the
reference configuration to the current configuration. An intermediate elastic unloading configuration is
also show to delineate the decomposition formula (7.1)

Description of plastic deformation. A majority of plastic deformation in metals
is resulted from dislocation movement. The plastic deformation generated by an
arbitrary mobile dislocation line with Burgers vector $\mathbf{b}(= b\mathbf{s})$ on a slip plane with
normal $\mathbf{m}$ was quantitatively described in equation (4.12). For distinction, the plastic deformation (4.12) associated with a single dislocation loop is labelled by $\mathbf{F}_c^p$,

with subscript "*c*" emphasizing the attachment to a single crystalline dislocation. For actual plastic deformation in a single crystal of volume V, numerous dislocations are participated in the slip deformation of slip system $(\mathbf{m}, \mathbf{s})$. Therefore, it is sensible to examine the corresponding volume averaging quantities for their global effects. This averaging procedure is mathematically described by a volume averaging operation through the abbreviation introduced below for

$$< \cdot > \equiv (1/V) \int_V (\cdot)\, dV. \tag{7.2}$$

Accordingly, the volume average plastic deformation for a single crystal in the slip system $(\mathbf{m}, \mathbf{s})$ can be evaluated from (4.12) as

$$\mathbf{F}^p \equiv <\mathbf{F}_c^p> = 1 + \gamma \mathbf{s}\,\mathbf{m} \tag{7.3}$$

where γ represents the permanent shear amount along the slip system $(\mathbf{m}, \mathbf{s})$. The value of γ can be calculated through Orowan equation, as discussed in subsection 4.4.1.

The previous derivation only involves one particular slip system. Generally speaking, $\mathbf{F}^p$ would rather be manifested by a combination of slip shear amounts γ^α coming from various slip systems $(\mathbf{m}^\alpha, \mathbf{s}^\alpha)$, where superscript "$\alpha$" serves to label different slip systems. Thus

$$\mathbf{F}^p = 1 + \sum_{\alpha=1}^{n} \gamma^\alpha \mathbf{s}^\alpha \mathbf{m}^\alpha \tag{7.4}$$

where n denotes the number of *active slip systems*. For FCC or BCC single crystals which undergo plastic deformation easily, one has the n-value of 12. The current slip systems (embedded in a lattice subject to elastic distortion and rigid rotation of $\mathbf{F}^*$) are denoted by $(\mathbf{m}^{*\alpha}, \mathbf{s}^{*\alpha})$. Among various choices of $(\mathbf{m}^{*\alpha}, \mathbf{s}^{*\alpha})$, the following definition convected to lattice distortion $\mathbf{F}^*$ is considered as most convenient by Asaro and Rice[9]

$$\mathbf{s}^{*\alpha} = \mathbf{F}^* \cdot \mathbf{s}^\alpha, \quad \mathbf{m}^{*\alpha} = \mathbf{m}^\alpha \cdot \mathbf{F}^{*-1}. \tag{7.5}$$

It is easy to verify that $\mathbf{m}^{*\alpha}$ and $\mathbf{s}^{*\alpha}$ are still perpendicular to each other, but they are slightly different from unit vectors due to elastic lattice distortion.

Additive decomposition of deformation rates. The multiplicative decomposition formula (7.1) for the deformation gradient $\mathbf{F}$ would inevitably lead to an *additive decomposition formula* of the velocity gradient $\mathbf{l}(= \partial \mathbf{v}/\partial \mathbf{x})$

$$\mathbf{l} = \dot{\mathbf{F}} \cdot \mathbf{F}^{-1} = \mathbf{l}^* + \mathbf{l}^p \tag{7.6}$$

where the two terms on the right hand side can be expressed explicitly as

$$\mathbf{l}^* = \dot{\mathbf{F}}^* \cdot \mathbf{F}^{*-1}, \quad \mathbf{l}^p = \sum_{\alpha=1}^{n} \dot{\gamma}^\alpha \mathbf{s}^{*\alpha} \mathbf{m}^{*\alpha} \tag{7.7}$$

where $\mathbf{l}^*$ denotes the distortion rate of crystalline lattice, and $\mathbf{l}^p$ has the physical significance of plastic deformation rate along *current slip systems* in contrast to the ambiguous meaning from its macroplasticity counterpart. The second expression in (7.7) is sometime called *Taylor kinematic equation of crystalline slip*.

The symmetric and skew-symmetric parts of l are again denoted by $\mathbf{D}$ and $\mathbf{W}$, representing the deformation rate and the spin tensors, respectively. A superscript "$*$" or "p" attached to them would in turn relate to the corresponding elastic or plastic part. The plastic parts of $\mathbf{D}$ and $\mathbf{W}$ can be readily written as

$$\mathbf{D}^p = \sum_{\alpha=1}^{n} \dot{\gamma}^\alpha \mathbf{P}^\alpha, \quad \mathbf{W}^p = \sum_{\alpha=1}^{n} \dot{\gamma}^\alpha \mathbf{W}^\alpha \tag{7.8}$$

where the second rank symmetric tensor $\mathbf{P}^\alpha$ and skew-symmetric tensor $\mathbf{W}^\alpha$ only depend on the current slip system α of the single crystal

$$\mathbf{P}^\alpha = (s^{*\alpha} m^{*\alpha} + m^{*\alpha} s^{*\alpha})/2$$
$$\mathbf{W}^\alpha = (s^{*\alpha} m^{*\alpha} - m^{*\alpha} s^{*\alpha})/2 \tag{7.9}$$

and $\dot{\gamma}^\alpha (\alpha = 1, \dots, n)$ describe the shear rates carrying out along various slip systems. They link to dislocation velocities in the corresponding slip systems by Orowan equation, and their determination relies on dislocation dynamics curves.

7.1.2 Elastic Distortion of Lattices

During the deformation of a single crystal, the rate of input work in a unit volume of *reference configuration*, $\dot{W}$, can also be decomposed into a plastic work rate and an elastic work rate

$$\dot{W} = \tau : \mathbf{D} = \tau : \mathbf{D}^* + \tau : \mathbf{D}^p \tag{7.10}$$

where (microscopic) Kirchhoff stress τ forms work conjugate to $\mathbf{D}$ for unit volume in the reference configuration. The first term in (7.10) signifies the rate of lattice distortion energy. By converting to the stress and strain rate measures pertinent to reference configuration, we can phrase the above equation in the following alternative forms

$$\tau : \mathbf{D}^* = \tau : (\dot{\mathbf{F}}^* \cdot \mathbf{F}^{*-1}) = (\mathbf{F}^{*-1} \cdot \tau \cdot \mathbf{F}^{*-1T}) : \dot{\mathbf{E}}^*$$
$$= \mathbf{T} : \dot{\mathbf{E}}^* = \dot{\Phi} \tag{7.11}$$

where $\mathbf{T}$ the second Piola-Kirchhoff stress tensor, and $\mathbf{E}^*$ the *lattice Green strain tensor*. According to the definition of Hill and Havner[10], the latter is given by

$$\mathbf{E}^* = (\mathbf{F}^{*T} \cdot \mathbf{F}^* - 1)/2. \tag{7.12}$$

In equation (7.11), Φ characterizes elastic lattice distortion energy, and hence is regarded a state function of the lattice strain tensor $\mathbf{E}^*$. We mention by passing that Φ can be identified (through thermodynamic argument) as the (elastic) free energy used in Chapter 2. It is henceforth considered as a known function determinable from lattice elasticity. Based on this understanding, the rate of Φ can be written as

$$\dot{\Phi} = (\partial\Phi/\partial\mathbf{E}^*) : \dot{\mathbf{E}}^*. \tag{7.13}$$

When comparing to (7.11), the Kirchhoff stress tensor τ can be expressed in terms of the free energy function Φ by

$$\tau = \mathbf{F}^* \cdot (\partial \Phi / \partial \mathbf{E}^*) \cdot \mathbf{F}^{*T}. \tag{7.14}$$

Equation (7.14) enables us to derive a rate-form constitutive equation for the elastic distortion of crystal lattice, accomplished by taking the *objective rate of (7.14) corotational to lattice spin* as follows

$$\check{\tau}^* \equiv \dot{\tau} - \mathbf{W}^* \cdot \tau + \tau \cdot \mathbf{W}^* = \mathbf{L} : \mathbf{D}^* + \mathbf{D}^* \cdot \tau + \tau \cdot \mathbf{D}^* \tag{7.15}$$

where $\mathbf{L}$ is the forth rank lattice elasticity tensor defined by

$$L_{ijkl} = F_{ip}^* F_{jq}^* (\partial^2 \Phi / \partial E_{pq}^* \partial E_{mn}^*) F_{km}^* F_{ln}^* \tag{7.16}$$

which obviously processes the Voigt symmetry.

The following comments are stated for the elastic constitutive relation (7.15)

(1) Kirchhoff stress $\tau = (\det \mathbf{F}) \sigma$ is adopted as the basic stress measure in the constitutive relation. This formulation is compatible to the framework used earlier in section 2.4 for macroplasticity. In the spirit of Hencky, τ represents a "true tensor" and characterizes "the stress per unit mass". Under the implied plastic incompressibility (the dilatational plasticity theory will be postponed until section 7.4), the difference between the Kirchhoff stress and Cauchy stress is actually negligible. The advantage of using τ consists of the consequential *symmetric numerical formulation* of the problem[11], so that many established results of classical plasticity, including the existence of potentials and related variational principles, can be utilized to facilitate the numerical calculations of practical problems.

(2) The construction of the corotational rate in (7.15) obviously satisfies the objectivity requirement. Besides, the *rate corotational to the lattice spin* is an appealing choice for crystalline plasticity problems. This engagement implies that the rate type constitutive response is traced with respect to the lattice frame. To make it precise, the proposal of (7.15) recognizes that elastic deformation comes as a response due to the distortion of crystalline lattice rather than a response caused by the entire material.

(3) The last two terms in (7.15) can be neglected whenever the stress level is negligible to the elastic modulus, an assumption appropriate for almost all cases of metal deformation. Hereafter, a simplified version

$$\check{\tau}^* \approx \mathbf{L} : \mathbf{D}^* \tag{7.17}$$

will be used instead of (7.15).

(4) The above definition on the corotational rate, though theoretically favorable, is rather inconvenient in actual implementation because lattice spin $\mathbf{W}^*$ has to be traced at all times. Alternatively, the formulation could be transformed to the Jaumann rate of Kirchhoff stress corotational to the material spin $\mathbf{W}$

$$\check{\tau} \equiv \dot{\tau} - \mathbf{W} \cdot \tau + \tau \cdot \mathbf{W}. \tag{7.18}$$

Comparing the first equality of (7.15) and the equation (7.18), one finds the differential spin stemmed from lattice spin and material spin as

$$\beta^\alpha = \mathbf{W}^\alpha \cdot \tau - \tau \cdot \mathbf{W}^\alpha. \tag{7.19}$$

It is easy to observe that β^α will necessarily vanish to give the equivalence of $\mathring{\tau}^*$ and $\mathring{\tau}$ when lattice spin coincides material spin ($\mathbf{W}^\alpha = \mathbf{0}$). The combination of (7.17) and (7.19) gives rise to the elastic constitutive equation in terms of Jaumann rate

$$\mathring{\tau} = \mathbf{L} : \mathbf{D} - \sum_{\alpha=1}^{n} \dot{\gamma}^\alpha \mathbf{R}^\alpha. \tag{7.20}$$

In (7.20) the first term corresponds to a projected elastic response as if plastic deformation (including both straining and spinning) were absent, whereas the second term offers a correction to reduce this "projected elastic stress rate" by subtracting the non-linear corrections from plastic straining and slip-induced spinning, as delineated from the expression of $\mathbf{R}^\alpha$

$$\mathbf{R}^\alpha = \mathbf{L} : \mathbf{P}^\alpha + \beta^\alpha. \tag{7.21}$$

The quantities yet to be determined in (7.20) are the plastic shear rates $\dot{\gamma}^\alpha$ in various slip systems. They have to be forecasted by the plastic constitutive relations.

7.1.3 Schmid Yield Criterion

The local stress field, designated by τ_c, in a single crystal containing dislocations can be decomposed into

$$\tau_c = \tau_c^A + \tau_c^S \tag{7.22}$$

where τ_c^A is caused by the applied load and τ_c^S represents the *self-equilibrium* stress field produced by the internal defects. Referred to the discussion offered in subsection 4.3.4, the field of τ_c^S denotes collectively the stresses associated with Y^L, Y^I and Y^O. The volume averages of the stresses in (7.22) within the single crystal possess the following properties

$$<\tau_c> = <\tau_c^A> = \tau, \quad <\tau_c^S> = 0 \tag{7.23}$$

which addresses the disappearance of the volume average on short range self-equilibrium stress field.

For a given slip system α, the local crystalline gliding force, f_c^α, on a dislocation L is

$$f_c^\alpha = \tau_c^{A\alpha}(\tau) b^\alpha + f_c^{s\alpha}. \tag{7.24}$$

The first term in (7.24) represents the gliding stress driven by the applied stress τ, it should be a linear homogeneous expression of τ as inferred from equation (4.72). The second term $f_c^{s\alpha}$ in the above expression reflects the effect of strain hardening.

It is solely determined by the dislocation distribution pattern and irrelevant to the applied stress.

Driven by the local gliding force f_c^α, the dislocation line L would move with a normal velocity of v^α along the slip direction. The rate of plastic work under the dislocation model can be derived easily as

$$W_c^p = (1/V) \int_{L_{disl}} \sum_{\alpha=1}^{n} f_c^\alpha v^\alpha \, dl \qquad (7.25)$$

in which the line integral must be performed along all mobile dislocation lines in all slip systems in the single crystal.

For any given slip system α in the single crystal, a huge number of dislocations exist and have a common Burgers vector $\mathbf{b}^\alpha = b^\alpha \mathbf{s}^{*\alpha}$ (no sum on α). Consequently, the average gliding force acting on those dislocations is

$$f^\alpha = <f_c^\alpha> = <\tau_c^{A\alpha}>b^\alpha + <f_c^{s\alpha}> = \tau^\alpha b^\alpha \qquad (7.26)$$

where

$$\tau^\alpha = \tau : \mathbf{P}^\alpha = \mathbf{m}^{*\alpha} \cdot \tau \cdot \mathbf{s}^{*\alpha} \qquad (7.27)$$

is the *Schmid resolved shear stress* and no summation will be carried out hereafter for the superscript α. Although the statistic average of the self-equilibrium gliding force $f_c^{s\alpha}$ is zero, it has an effect to the plastic deformation behavior, in the sense that any change in the dislocation distribution pattern (say by the movement of dislocations) would perturb the previous self equilibrium distribution, with the consequence of strain hardening.

An equivalent definition of τ^α will be given below to provide the alternative physical significance of Schmid resolved shear stress. A macroscopic expression on the plastic work per unit mass can be derived by taking the volume average of (7.25) and then utilizing equations (7.23), (7.26), (7.27) and Orowan equation (4.86) successively

$$W^p = \tau : \mathbf{D}^p = \sum_{\alpha=1}^{n} \tau^\alpha \dot{\gamma}^\alpha. \qquad (7.28)$$

It states that Schmid resolved shear stress bears the physical significance of the plastic work conjugate of shear rate $\dot{\gamma}^\alpha$ in the corresponding slip system from the single crystal averaging viewpoint.

Schmid yield criterion refers to the cases where the variations of shear rates $\{\dot{\gamma}^1, \ldots, \dot{\gamma}^n\}$ only depend on the corresponding Schmid resolved shear stresses $\{\tau^1, \ldots, \tau^n\}$. The slip system in which τ^α reaches a critical value of τ_{cr}^α is usually termed the *critical slip system*, in contrast to the non-critical slip system called otherwise. Accordingly

$$\begin{aligned} \tau^\alpha < \tau_{cr}^\alpha &\longrightarrow \dot{\gamma}^\alpha = 0 \\ \tau^\alpha = \tau_{cr}^\alpha &\longrightarrow \dot{\gamma}^\alpha \geq 0 \end{aligned} \qquad (7.29)$$

which provides the yield surface for single crystals in a stress space defined by $\mathbf{m}^{*\alpha} \cdot \tau \cdot \mathbf{s}^{*\alpha} = \tau_{cr}^\alpha, \alpha = 1, \ldots, n$. Apparently, the inferred yield surface has the shape of a convex polyhedron bounded by n hyperplanes. It is non-smooth and has edges and corners. As will be shown later in section 7.3, a normality structure with respect to the mentioned yield surface is implied by Schmid yield criterion.

7.1.4 Hardening Laws for Rate Independent and Rate Dependent Crystals

For a rate independent material, the shear stress rates should be composed by a linear combination of the shear strain rate. The slip deformation within the critical slip system obeys the flow rule of rate independent type, and satisfies the following consistency conditions

$$\dot{\tau}^\alpha \begin{matrix} = \\ < \end{matrix} \dot{\tau}^\alpha_{cr} = \sum_{\beta=1}^n h^{\alpha\beta}\dot{\gamma}^\beta \longrightarrow \dot{\gamma}^\alpha \begin{matrix} > \\ = \end{matrix} 0 \qquad (7.30)$$

for any α, where the equation above or below corresponds to the active or detained slip system. One notices that a critical slip system is not necessarily an active slip system, similar to the case of macroplasticity where a stress state on the yield surface will not necessarily cause plastic flow. The occurrence of plastic flow not only requires that the slip system in a critical state, but also demands that the consistency condition being satisfied. The hardening matrix $h^{\alpha\beta}$ was described in subsection 6.1.2. We emphasize here again that diagonal terms in $h^{\alpha\beta}, h^{\alpha\alpha}$ (no sum for repeated index), are self-hardening coefficients, whereas the off-diagonal terms, $h^{\alpha\beta}$ (α is not equal to β), represent latent hardening coefficients. The determination of latent hardening via measurement on overshooting angle will be discussed in subsection 7.1.6 as an application of the present theory. We also recall that the coefficients of self and latent hardening are identical for the case of a Lomer-Cottrell lock, as elucidated in section 6.1.2. For a majority of metallic systems, the effect of latent hardening could even surpass that of self hardening, rendering the matrix $h^{\alpha\beta}$ non-positive definite.

For rate dependent materials in the regime of dislocation glide with normal speed (referred to the discussion in subsection 4.4.2), the shear strain rates are usually characterized by a power law relation

$$\dot{\gamma}^\alpha = \dot{\gamma}^\alpha_o |\tau^\alpha/g^\alpha|^{(1-m)/m}(\tau^\alpha/g^\alpha) \qquad (7.31)$$

as a generalization from the dislocation dynamics curve of Stein and Low, see equation (4.88). The absolute sign introduced in (7.31) guarantees the sign consistency between $\dot{\gamma}^\alpha$ and τ^α. This consistency is required by the second law of thermodynamics (in a way parallel to the development in subsection 2.8.2) when applied to crystal slip deformation. The symbol $\dot{\gamma}^\alpha_o$ in (7.31) denotes a material constant with the meaning of reference shear strain rate. The exponent m characterizes the rate sensitive exponent and varies from zero to one. Two extremes of $m = 0$ and $m = 1$ relate to the cases of rate independent and visco-elastic response, respectively. The function g^α in (7.31) characterizes the "hardness" in the slip system α, with its evolution governed by a rate independent law

$$\dot{g}^\alpha = \sum_{\beta=1}^{n} \hat{h}^{\alpha\beta} |\dot{\gamma}^\beta| \qquad (7.32)$$

where $\hat{h}^{\alpha\beta}$ is the hardening matrix for the rate dependent materials. Several remarks can be made to the above formulation

(1) The effect of latent hardening has been taken into account in equation (7.32).
(2) Because all coefficients in the hardening matrix $\hat{h}^{\alpha\beta}$ are positive, the deployment of absolute shears such as $|\dot{\gamma}^\beta|$ in (7.32) implies that plastic deformation along any orientation would always cause strain hardening.
(3) The possibility of Bauschinger effect is excluded by the form of equation (7.32). Modification should be devised if Bauschinger effect in the single crystal level is considered.

For the case of equal strain hardening considered by Taylor and Elam[1,2], see equation (6.5), (7.32) can be further simplified to

$$g^\alpha = g^\alpha(\gamma), \quad \gamma = \sum_{\beta=1}^{n} |\gamma^\beta| \qquad (7.33)$$

namely the hardening response is only determined by the total absolute amount of shears in various slip systems.

7.1.5 Constitutive Relations of Single Crystals

We next derive the rate independent constitutive equations for a ductile single crystal. Because the formulation will be cast in a rate form, the following expression for the time derivative of Schmid resolved shear stress

$$\dot{\tau}^\alpha = \mathbf{R}^\alpha : \mathbf{D}^* \qquad (7.34)$$

is required, where $\mathbf{R}^\alpha$ was given in (7.21). The derivation for the above formula is rather tedious, involving the definition of Schmid shear stress (7.27), formula (2.46) to replace the material derivative by a corotational rate in a double contraction expression, and especially a "general pivotal relation"

$$\check{\mathbf{P}}^{*\alpha} : \tau = \beta^\alpha : \mathbf{D}^*. \qquad (7.35)$$

The latter expression can be derived from equation (3.3) in the paper by Hill and Havner[10], by choosing the current configuration as the reference configuration for further hypothetical incremental deformation, and taking Kirchhoff stress and logarithmic strain as the primitive work conjugate pair.

The flow consistency condition (7.30) for the critical slip systems becomes

$$\mathbf{R}^\alpha : \mathbf{D}^* \begin{array}{c} = \\ < \end{array} \sum_{\beta=1}^{n} h^{\alpha\beta}\dot{\gamma}^\beta \longrightarrow \dot{\gamma}^\alpha \begin{array}{c} > \\ = \end{array} 0 \qquad (7.36)$$

by substituting the expression (7.34) for the rate of Schmid shear stress. If the term $\mathbf{R}^\alpha : \mathbf{D}^p$ is further added to the both sides of (7.35), the flow rule proposed by Hill and Rice[4] is obtained

$$\mathbf{R}^\alpha : \mathbf{D} \; \substack{= \\ <} \; \sum_{\beta=1}^{n} g^{\alpha\beta} \dot{\gamma}^\beta \longrightarrow \dot{\gamma}^\alpha \; \substack{> \\ =} \; 0 \tag{7.37}$$

for any slip system α, where

$$g^{\alpha\beta} = h^{\alpha\beta} + \mathbf{R}^\alpha : \mathbf{P}^\beta. \tag{7.38}$$

From equation (7.37), a sufficient condition for the unique determination of various shear rates $\dot{\gamma}^\alpha, \alpha = 1, \ldots, n$, consists of the positive-definiteness of matrix $g^{\alpha\beta}$, a requirement more attainable than the positive-definiteness of $h^{\alpha\beta}$ itself. If $g^{\alpha\beta}$ is positive definite, equation (7.37) can be inverted to give

$$\dot{\gamma}^\alpha = \sum_{\beta=1}^{n} g^{\alpha\beta-1} \mathbf{R}^\beta : \mathbf{D}. \tag{7.39}$$

The elastic-plastic constitutive relation can be obtained by substituting the above expression for the shear rates into equation (7.20)

$$\overset{\triangledown}{\tau} = \mathbf{L}^t : \mathbf{D} \tag{7.40}$$

where the tangential stiffness tensor $\mathbf{L}^t$ is given by

$$\mathbf{L}^t = \mathbf{L} - \sum_{\alpha=1}^{n} \sum_{\beta=1}^{n} \mathbf{R}^\alpha g^{\alpha\beta-1} \mathbf{R}^\beta. \tag{7.41}$$

The Voigt symmetry of $\mathbf{L}^t$ preserves only if $g^{\alpha\beta}$ is symmetric, which requires a symmetric arrangement of slip systems as indicated from (7.38).

Generally speaking, the possibility of a non-positive definite matrix $g^{\alpha\beta}$ cannot be excluded, due to the fact that $h^{\alpha\beta}$ (which forms a considerable contribution to the $g^{\alpha\beta}$ matrix) is usually negative definite. Therefore, the uniqueness on the single crystal deformation, as well as the relative spin between the lattice frame and material frame, cannot be assumed a priori. The conditions by which the deformation uniqueness can be assured are connected closely to the hardening characterization, the stress state, and the configurations of slip systems.

7.1.6 Latent Hardening Measurement

As the first illustration of present theoretical framework on crystalline plasticity, let us consider the planar double slip model as shown in Fig. 7.2, where only the primary and the conjugate slip systems, with subscripts p and c denoting the quantities related to them, are permitted to be operative in the single crystal. These slip planes form angles ϕ_p and ϕ_c with the loading axis (chosen as along the unit base vector $\mathbf{e}_2$), respectively. The angles ϕ_p and ϕ_c obviously vary during the deformation, and their initial values, designated as ϕ_p^o and ϕ_c^o, are assumed as known quantities. For simplicity, the lattice distortion is omitted so that $\mathbf{F}^*$ is solely composed of lattice rotation, which leads to the following expressions for the current values of ϕ_p and ϕ_c

$$\phi_p = \phi_p^o - \beta, \quad \phi_c = \phi_c^o + \beta \tag{7.42}$$

where β signifies the rigid rotation angle for the single crystal specimen.

For a rate independent single crystal, the strain hardening law under the present double slip model could be written as the following form

$$\begin{bmatrix} d\tau_p \\ d\tau_c \end{bmatrix} = h \begin{bmatrix} 1 & q \\ q & 1 \end{bmatrix} \begin{bmatrix} d\gamma_p \\ d\gamma_c \end{bmatrix} \tag{7.43}$$

where $h = h(\gamma)$ is a function of the total deformation amount γ. In written down (7.43), assumptions such as the reciprocal equivalence of self-hardening coefficients and the symmetry of hardening matrix are adopted. The value q in (7.43) is termed the *latent hardening ratio*, and is usually treated as a constant. The attention is then focused on the determination of q from the measurement of *overshoot angle*.

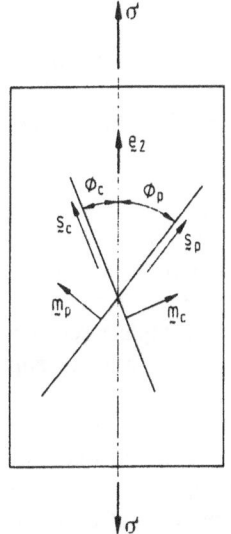

Fig. 7.2. Planar double slip model of a single crystal

In the first stage of tensile deformation of the single crystal, only the primary slip system is operative with $|45° - \phi_p^o| < |45° - \phi_c^o|$. The eliminition of $d\gamma_c$ term provides an easy integration of equation (7.43) for the stresses resolved along the primary and conjugate slip systems. Especially, the ratio between stress τ_p on the primary slip system and stress τ_c on the conjugate slip system is

$$\tau_p/\tau_c = [\tau^o + H(\gamma_p)] / [\tau^o + qH(\gamma_p)] \tag{7.44}$$

where τ^o is the initial yield stress and the hardening function H by slip in the primary slip system is

$$H(\gamma_p) = \int_0^{\gamma_p} h(\gamma_p)\,d\gamma_p. \tag{7.45}$$

We emphasize here that τ_p and τ_c are the *flow resistances* along primary and conjugate slip systems, respectively, under an accumulated shear amount of τ_p along the primary slip system.

From a stress equilibrium analysis, however, the ratio between the driving stresses on primary and conjugate slip systems is

$$\tau_p/\tau_c = \sin 2\phi_p / \sin 2\phi_c = \sin\left[2(\phi_p^o - \beta)\right] / \sin\left[2(\phi_c^o + \beta)\right]. \tag{7.46}$$

Under the prescription of the initial configuration of slip systems, this ratio increases as the rigid rotation of crystal lattice (described by angle β) intensifies. When the two stress ratios listed separately in (7.45) and (7.46) become equal, i.e.

$$[\tau^o + H(\gamma_p)] / [\tau^o + qH(\gamma_p)] = \sin\left[2(\phi_p^o - \hat{\beta})\right] / \sin[2(\phi_c^o + \hat{\beta})] \tag{7.47}$$

the dislocations from the conjugate slip system will begin to operate. This provides an opportunity to determine the particular rigid rotation angle, designated by $\hat{\beta}$ in (7.47), from experimental detection on emission of dislocations from the conjugate slip system. Equation (7.47) contains two kinematic variables γ_p and $\hat{\beta}$. Their interrelationship can be established by the alignment requirement along the loading axis

$$\tan\hat{\beta} = F_{12}^p/F_{22}^p = \left[\gamma_p \sin^2\phi_p^o\right] / \left[1 + \frac{1}{2}\gamma_p \sin 2\phi_p^o\right] \tag{7.48}$$

derived by the observation that the total deformation $\mathbf{F}$, as composed of plastic slip γ_p along primary system and rigid rotation $\hat{\beta}$, would have an eigen vector $\mathbf{e}_2$ parallel to loading axis. Equations (7.47) and (7.48) provide a complete set of a simultaneous system to determine $\hat{\beta}$ and the plastic slip amount in primary slip system at rotation $\hat{\beta}$.

Alternatively, if angle $\hat{\beta}$ is input from experimental data, (7.47) and (7.48) jointly determine the corresponding plastic shear amount prior to the operation of conjugate slip system, and more importantly, the latent hardening ratio q. For the case of equal hardening observed by Taylor and Elam[1,2], the value of q would be equal to one and various characteristic angles at the initiation of double slip are

$$\hat{\beta} = \frac{1}{2}\left(\phi_p^o - \phi_c^o\right) \equiv \beta_o$$

$$\phi_p = \phi_c = \frac{1}{2}\left(\phi_p^o + \phi_c^o\right) \equiv \phi. \tag{7.49}$$

For the non-equal hardening case, an *overshoot angle* is defined as

$$\Delta\beta = \hat{\beta} - \beta_o \tag{7.50}$$

and represents the extent of deviation from a symmetric arrangement of double slip system when the conjugate slip system starts to operate. The overshoot angle can be determined experimentally, which motivates the substitution of the above definitions into (7.47) to give

$$[\tau^o + H(\gamma_p)] / [\tau_o + qH(\gamma_p)] = \sin[2(\phi - \Delta\beta)] / \sin[2(\phi + \Delta\beta)] . \quad (7.51)$$

It is easily observed that when the mean angle of the initial slip system orientations, ϕ, is less than $\pi/4$, a positive (or negative) overshoot angle would be obtained for q greater (or less) than one; and vice versa when ϕ is greater than $\pi/4$. Most experimental data for FCC metals indicate a range of overshoot angle from $0°$ to $3°$ for $\phi < \pi/4$, and consequently predict a value of q from 1 to 1.4.

The hardening law for an individual slip system is conventionally taken as one of the following forms

$$h(\gamma) = h_o [1 - \tanh(\alpha_o \gamma)]$$
$$h(\gamma) = h_o \operatorname{sech}^2(\alpha_o \gamma) \qquad\qquad (7.52)$$

where h_o and α_o are related material parameters. The analytical determination of q-value could be achieved through equations (7.48), (7.51) and (7.52), with the prescription of ϕ_p^o, ϕ_c^o and $\Delta\beta$ data.

7.1.7 Uniaxial Tension of a Single Crystal Bar

Our next example again refers to the uniaxial tension of a single crystal bar, with a further restriction that only the primary slip system, characterized by an arbitrary spatial orientation $\mathbf{m}$ and $\mathbf{s}$, is active, as shown in Fig. 7.3(a). A tensile stress with magnitude of σ is applied along the bar axis designated by a unit vector $\mathbf{T}$. The Kirchhoff stress in the bar is then simply $\sigma\mathbf{TT}$.

For the illustrative purpose, our attention is again confined to a rigid plastic response of the single crystal bar. Then from the decomposition formula (7.1) of deformation gradient $\mathbf{F}$, the total deformation within the bar is composed of a plastic slip $\mathbf{F}^p = 1 + \mathbf{sm}$ as shown in Fig. 7.3(b) and a rigid body rotation $\mathbf{F}^*$ about an axis $\mathbf{r}$ (unit vector) perpendicular to both $\mathbf{s}^*$ and $\mathbf{T}$ which realigns the bar in agreement to the tensile axis, as shown in Fig. 7.3(c). It is straightforward to derive that

$$\mathbf{r} = (\mathbf{s}^* \times \mathbf{T}) / \left[1 - (\mathbf{s}^* \cdot \mathbf{T})^2\right]^{\frac{1}{2}} \qquad\qquad (7.53)$$

and the current slip system $(\mathbf{m}^*, \mathbf{s}^*)$ only rotates an angle β from its initial value in the reference configuration.

The velocity gradient for this deformation can be written as an additive decomposition of a slip deformation rate and a rigid spin as follows

$$\mathbf{l} = \dot{\gamma} [\mathbf{s}^*\mathbf{m}^* + (\mathbf{T} \cdot \mathbf{m}^*)(\mathbf{T}\mathbf{s}^* - \mathbf{s}^*\mathbf{T})] \qquad\qquad (7.54)$$

where the second term, characterizing a correction spin, is expressible by a spin vector $\dot{\beta}\mathbf{r}$, with $\dot{\beta}$ denoting the angular speed around $\mathbf{r}$ axis. Apparently, the current slip system $(\mathbf{m}^*, \mathbf{s}^*)$ spins according to the lattice spin $\dot{\beta}\mathbf{r}$.

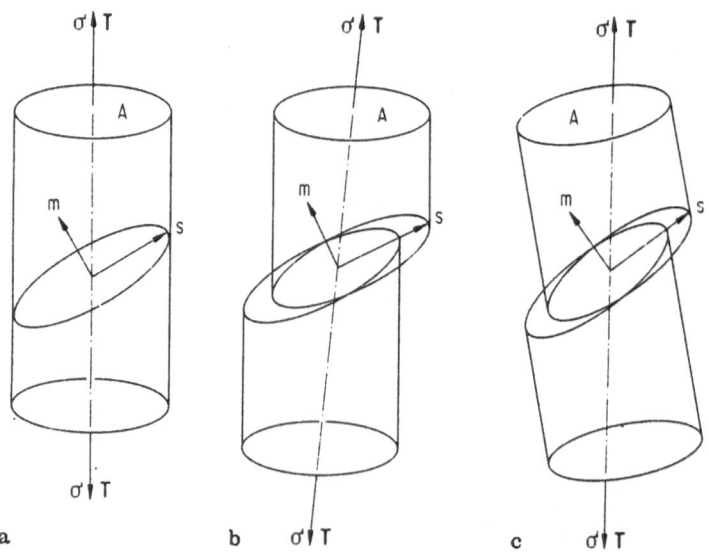

Fig. 7.3. Tension of a single crystal bar. **a** Orientation of primary slip system; **b** plastic slip and induced surface offset; **c** a rotation to realign the tensile axis

For the single crystal bar subject to tension, the material line aligned with tensile axis would remain a line of unchanged orientation, that is

$$\mathbf{l} \cdot \mathbf{T} = \dot{\epsilon}\mathbf{T} \tag{7.55}$$

where $\dot{\epsilon}$ denotes the current elongation rate along the tensile direction $\mathbf{T}$. After substitutions of (7.54) and (7.53) into the above equation, a vectorial expression emerges

$$\dot{\beta}\mathbf{r} \times \mathbf{T} + \dot{\gamma}(\mathbf{m}^* \cdot \mathbf{T})\mathbf{s}^* = \dot{\epsilon}\mathbf{T} \tag{7.56}$$

which relates $\dot{\epsilon}$ and $\dot{\beta}$ to the slip rate $\dot{\gamma}$ in the primary slip system. Explicit expressions for $\dot{\epsilon}$ and $\dot{\beta}$ can be found by taking inner and vectorial products of the above equation with respect to $\mathbf{T}$

$$\dot{\epsilon} = \dot{\gamma}(\mathbf{m}^* \cdot \mathbf{T})(\mathbf{s}^* \cdot \mathbf{T}) = \dot{\gamma}\cos\Theta\cos\phi,$$

$$\dot{\beta} = \dot{\gamma}(\mathbf{m}^* \cdot \mathbf{T})\left[1 - (\mathbf{s}^* \cdot \mathbf{T})^2\right]^{\frac{1}{2}} = \dot{\gamma}\cos\Theta\sin\phi. \tag{7.57}$$

Here the conventional notations in metallurgy are adopted for using Θ and ϕ as the projection angles of vectors $\mathbf{m}^*$ and $\mathbf{s}^*$ along the loading axis $\mathbf{T}$, respectively.

The Schmid resolved shear stress along the primary slip system can be written as

$$\tau = \mathbf{m}^* \cdot \tau \cdot \mathbf{s}^* = \sigma(\mathbf{m}^* \cdot \mathbf{T})(\mathbf{s}^* \cdot \mathbf{T}) = \sigma\cos\Theta\cos\phi \tag{7.58}$$

where $\cos\Theta\cos\phi$ is exactly the Schmid factor. If one substitutes the above expression of τ into the hardening law of primary slip system (assumed as the only operative one), then a uniaxial constitutive relation linking the applied stress σ and the bar elongation ϵ can be derived after lengthy algebra

$$d\sigma/d\epsilon = h/(\cos^2\phi\cos^2\Theta) + \sigma(\cos 2\phi/\cos^2\phi). \tag{7.59}$$

The first term on the right hand side of (7.59) represents the increment of stress with respect to the axial elongation as a result of strain accumulation. This term solely attributes to the intrinsic behavior of the material under consideration, and either *material hardening* or *material softening* may occur by a positive or a negative tangential modulus h. On the other hand, the second term on the right hand side of (7.59) is due to the variation of Schmid resolved shear stress τ caused by lattice rotation during the elongation process, signifying the geometrical change on the slip system configuration induced by deformation. This change of slip system configuration could produce either hardening or softening effect on the tensile response of a single crystal bar in accordance to the current sign of $\cos 2\phi$. If ϕ is greater than $\pi/4$, the *geometrical softening* will take place. Actually, when $\phi > \pi/4$ and $h/\sigma < |\cos 2\phi| \cdot \cos^2\Theta$, the tangent of the stress versus total strain curve of single crystal bar becomes negative and indicates the occurrence of global softening. The important message from this simple demonstration lies in that a single crystal bar could become strain-soften by purely geometrical effect, even if the material behavior is strain hardening. Apparently, under geometrical softening condition, any local perturbation of the lattice rotation would result in further softening in the related portion of single crystal bar, which in turn promotes further deformation concentration and finally the flow localization. The phenomenon of flow localization induced by geometrical softening is possible in both single crystals and the selected grains within a polycrystal aggregate.

7.2 Polycrystals

After the acquisition of constitutive relations for single crystals, attention is then directed to the case of polycrystal aggregates. Two basic problems have to be addressed to resolve the plastic deformation of polycrystals

(1) How to choose slip systems in each single crystals when the polycrystal aggregates are subject to uniform but arbitrary remote loading ?
(2) How to obtain the averaged constitutive relation for a polycrystal aggregate from the knowledge of grain orientation distribution and the constitutive relations for the constituent single crystals ?
    These questions will be dealt with in the subsequent subsections.

7.2.1 Selection of Slip Systems

Let us consider the polycrystal model illustrated in Fig. 7.4, where a central single crystal grain is highlighted from the polycrystal aggregate. Kinematic conditions are prescribed along the aggregate boundary

$$\mathbf{v} = \bar{\mathbf{D}} \cdot \mathbf{x} \qquad \text{for all } \mathbf{x} \text{ in } \mathbf{V}. \qquad (7.60)$$

The model indicated in Fig. 7.4 aims at the following question :

If $\bar{\mathbf{D}}$ is given uniformly along the boundary and a compatible homogeneous deformation prevails in the entire interior of polycrystal aggregate (meaning $\mathbf{D}$ equals to $\bar{\mathbf{D}}$ everywhere), then how to select appropriate slip systems to realize this deformation ?

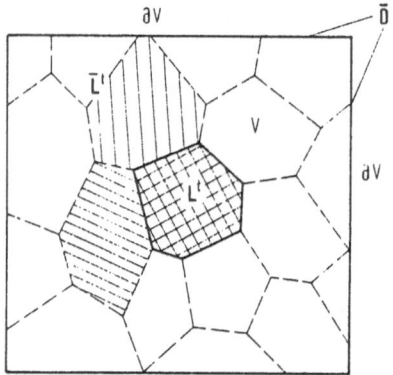

Fig. 7.4. Model of polycrystal aggregate where a single crystal with tangential stiffness matrix $\mathbf{L}^I$ is embedded in a polycrystal surrounding of equivalent tangential stiffness matrix $\bar{\mathbf{L}}^I$

For simplicity, we temporarily focus on the rigid plastic materials. Then a kinematic constraint will be imposed from plastic incompressibility

$$D_{ii} = 0. \qquad (7.61)$$

Thus, there are only five independent components in the symmetric deformation rate tensor. Generally speaking, any given deformation rate can be accomplished by a linear combination of 5 independent slip systems. Accordingly, as many as 792 choices (different combinations for taking 5 elements from 12 possible candidates) are at stake from the twelve potential slip systems for a BCC or FCC polycrystal to realize the given D tensor. Although some of the combinations can be ruled out from further kinematic consideration, the actual selection for the unique combination of slip systems most appropriate to the prescribed deformation and crystallographic orientation would rely on additional physical principles.

7.2.2 Sachs Theory

Sachs appeared to be the first to consider this problem[12]. The basic assumption in his model lies in the uniform equilibrium stress field across every grains of the polycrystal aggregate. This assumption clearly leads to a highly non-uniform kinematic field which contradicts most experimental measurements of polycrystal deformation subject to remote uniform loading. Furthermore, Saches theory underestimates considerably the critical tension/shear stress ratio.

7.2.3 Dual Principles of Plastic Work

Since the rejection of Sachs' hypothesis of uniform stress, scientists searched for new physical principles to guide their selection of slip systems. Some of the subsequently proposed physical principles rooted in the description of plastic work. We would approach this issue by first introducing the following definitions

Definitions
(1) Any combination of possible shearing strain rates $\tilde{\tilde{\gamma}}^{\alpha}, \alpha = 1, \ldots, n$, satisfying

$$\sum_{\alpha=1}^{n} \tilde{\tilde{\gamma}}^{\alpha} \mathbf{P}^{\alpha} = \mathbf{D} \tag{7.62}$$

are called kinematically admissible shear rates.
(2) Any possible stress tensor $\tilde{\tau}$ satisfying

$$\tilde{\tau} : \mathbf{P}^{\alpha} = \tilde{\tau}^{\alpha} \leq \tau_{cr}^{\alpha} \qquad \text{for all } \alpha \tag{7.63}$$

are called statically admissible stress.
After above preparations, the following theorem can be stated

Theorem. The following inequalities hold for all kinematically admissible shear rates or all statically admissible stress

$$\sum_{\alpha=1}^{n} \tau_{cr}^{\alpha} \tilde{\tilde{\gamma}}^{\alpha} \geq \sum_{\alpha=1}^{n} \tau_{cr}^{\alpha} \dot{\gamma}^{\alpha} \geq \sum_{\alpha=1}^{n} \tilde{\tau}^{\alpha} \dot{\gamma}^{\alpha}. \tag{7.64}$$

Proof. It was established by Rice[13] through an application of the second law of thermodynamics that the individual plastic power (work rate) $\tau^{\alpha} \dot{\gamma}^{\alpha}$,in each slip system should be all non-negative, or equivalently, the plastic shear rate along each slip system must take the same direction as the corresponding Schmid resolved shear stress. This result and the definition (7.63) for statically admissible stress lead to the second inequality of (7.64). The same conclusion and (7.63) also give rise to another inequality for any kinematically admissible shear rate $\tilde{\tilde{\gamma}}^{\alpha}$

$$\sum_{\alpha=1}^{n} \tau_{cr}^{\alpha} \tilde{\tilde{\gamma}}^{\alpha} \geq \sum_{\alpha=1}^{n} \tilde{\tau}^{\alpha} \tilde{\tilde{\gamma}}^{\alpha} = \sum_{\alpha=1}^{n} \tilde{\tau}^{\alpha} \dot{\gamma}^{\alpha}$$

where the last equality is based on the definitions of the kinematically admissible shear rates $\tilde{\tilde{\gamma}}^{\alpha}$, see equation (7.62), and of the statically admissible stress, see equation (7.63). Because the above inequality is valid for any statically admissible resolved shear stress $\tilde{\tau}^\alpha$ with the inclusion of the extreme case of $\tilde{\tau}^\alpha$ being equal to τ_{cr}^α. After identifying $\tilde{\tau}^\alpha$ as τ_{cr}^α in the last expression of the above formula, we complete the proof on the first inequality of (7.64). Q.E.D.

The physical significance of (7.64) is similar to that in limit analysis of macroplasticity. That is, the plastic power calculated from the kinematically admissible shear rates or from the statically admissible stress would provide an upper or lower bound of the actual plastic power $\dot{W}^p$. By the historic coincidence, the upper and lower bound theorems for macroplasticity and that just introduced for mesoplasticity were established at the same year period. The true solution should belong to both kinematical and statical admissible solutions, so the respective predictions by the utilization of kinematically or statically admissible fields would provide close bounds on the plastic power $\dot{W}^p$ provided the selections are extensive

$$\min_{\tilde{\gamma}^\alpha} \sum_{\alpha=1}^{n} \tau_{cr}^\alpha \tilde{\gamma}^\alpha = \dot{W}^p = \max_{\tilde{\tau}} \tilde{\tau} : \mathbf{D}. \tag{7.65}$$

A prototype of the first expression in (7.65) was advanced by Taylor[14] in 1938 in his endeavor toward slip system selection. In connection with the equal hardening hypothesis for all slip systems proposed earlier by Taylor and Elam[1,2], the scheme proposed by Taylor was in fact based on the *principle of minimum accumulated shear rate*. The accumulated shear rate is denoted by

$$\tilde{\Gamma} = \sum_{\alpha=1}^{n} \tilde{\gamma}^\alpha$$

and signifies the summation of slip rates in various slip systems. Taylor's proposal is equivalent to a special case of the first expression of (7.65) in which the critical resolved shear stresses are identical in all slip systems. According to the modern terminology, the first expression of (7.65) is called the *principle of minimum plastic dissipation* whereas the second is called the *principle of maximum plastic work*. The formulae in (7.65) suggest the duality nature of these two principles. The establishment of the second principle in its macroscopic and microscopic forms, as well as the duality proof, were due to the landmark work of Bishop and Hill[15].

As the forms of (7.62), (7.63) and (7.65) suggested, the selection of appropriate slip systems can be cast mathematically into the format of classical convex linear programming. That implies either to minimize the first expression of (7.65) under the constraint of (7.62) or to maximize the last expression of (7.65) under the constraint of (7.63) by standard linear programming algorithm. The detailed implementation of the both schemes can be found in reference [16].

7.2.4 Taylor Theory

We return to the rigid plastic polycrystal aggregate model as previously presented in Fig. 7.4, with the boundary condition (7.60) applied. The following averaging theorem for the plastic work

$$<\tau : \mathbf{D}> \; = \; <\tau> : <\mathbf{D}> + <B(\Delta t)>_S \approx \Sigma : \bar{\mathbf{D}} \qquad (7.66)$$

can be established via the stress equilibrium equation. In (7.66), $\Sigma = <\tau>$ denotes macroscopic Kirchhoff stress, $\bar{\mathbf{D}} = <\mathbf{D}>$ is the macroscopic deformation rate, and the term $B(\Delta t)$ represents contribution to the plastic work by unbalanced traction vector Δt along grain boundaries. This unbalanced stress is resulted from the Taylor's hypothesis of a homogeneous deformation rate enforced in the entire polycrystal, and the ignorance of local perturbation on the deformation field near grain boundaries. However, the area average of unbalanced traction along all grain boundaries, denoted symbolically by $<B(\Delta t)>_S$, is usually negligible. This fact justifies the last approximation anticipated in (7.66). By any means, the definition of the macroscopic stress in (7.66) can serve as a very tight upper bound. If the effective (microscopic) plastic deformation rate is defined by

$$\bar{\dot{\epsilon}}^p \equiv [(2/3)\mathbf{D} : \mathbf{D}]^{\frac{1}{2}} \qquad (7.67)$$

and its work conjugate is denoted by $\bar{\sigma}$ (effective stress), then one has

$$\bar{\sigma}\bar{\dot{\epsilon}}^p = \tau : \mathbf{D} = \sum_{\alpha=1}^{n} \tau_{cr}^{\alpha}\dot{\gamma}^{\alpha}. \qquad (7.68)$$

Taylor[14] assumed that critical resolved shear stresses in all slip systems of every crystal grains were identical and all equal to a constant value of τ_{cr}. Under this assumption, the volumetric average of the above equation becomes

$$\bar{\Sigma}\bar{\dot{\epsilon}}^p = \Sigma : \bar{\mathbf{D}} = \tau_{cr}<\dot{\Gamma}>. \qquad (7.69)$$

The ratio between the averaging effective stress of polycrystal aggregate, $\bar{\Sigma} = <\bar{\sigma}>$, and the critical resolved shear stress, τ_{cr}, is termed the *Taylor factor* M. If one first uses the Taylor scheme for slip system selection to obtain the minimum value of $\dot{\bar{\Gamma}}$, namely $\dot{\Gamma}$, and then substitutes this $\dot{\Gamma}$ to equation (7.69) and takes statistical average for different orientations of the constituent single crystals, he would finally arrive at the following results

$$M \equiv \bar{\Sigma}/\tau_{cr} = <\dot{\Gamma}>/\bar{\dot{\epsilon}}^p = \begin{array}{cc} 3.06 & \text{FCC} \\ 2.83 & \text{BCC} \end{array}. \qquad (7.70)$$

The stress-strain curve predicted by Taylor theory is in fairly good agreement with the testing data of coarse grain polycrystals. Systematic deviation, however, exists in comparison to the testing data of fine grain polycrystals. This deviation partially attributes to the ignorance of the dislocation pileup phenomenon near grain boundaries, as discussed earlier in section 6.4. The grain size effect as inferred from the experimental data can be explained by the Hall-Petch relation (6.91).

The following inadequacies are suffered by the Taylor model :
(1) Lattice elastic deformation is ignored.
(2) Requirement on traction balance along various grain boundaries fails to satisfy.
(3) Over-simplification on the latent hardening behavior by the assumption of identical resolved shear stress in all slip systems.

(4) Simultaneous yielding in all slip systems is implied in Taylor's theory.
(5) Possibility of grain boundary sliding (in which case the strain compatibility requirement of Taylor fails to observe) is excluded.

Remedies to overcome some of the above-mentioned inadequacies will be described in the subsections to follow.

7.2.5 Transformation Strain Problem of Eshelby

As a preparation for the self consistent theory of polycrystal aggregates, the famous *transformation strain problem of Eshelby*[17] is addressed in this subsection. For brevity, the discussion hear is confined to the special case of spherical transformation zones. The more general case of ellipsoidal transformation zones, however, can be found from the comprehensive treatment of Mura[18].

The Eshelby transformation strain problem can be stated as follows. A small sphere is removed from an infinite elastic medium characterized by the stiffness moduli **L**. After a transformation strain ϵ' (which is imposed on the tiny sphere under a stress-free condition and without disturbing its material properties), the same tiny sphere is inserted back to the previous cut-off location, as shown in Fig. 7.5. The question is then raised concerning various field characteristics induced by this operation.

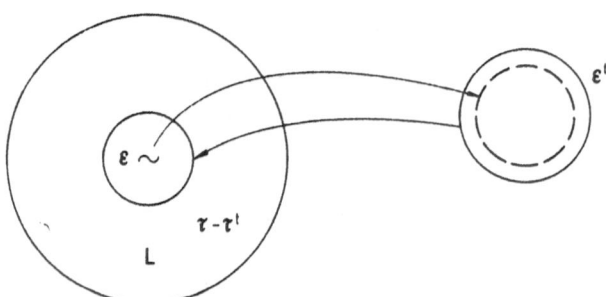

Fig. 7.5. Transformation strain problem for a homogeneous medium

For spherical (or ellipsoidal) transformation zones in an otherwise homogeneous elastic medium, Eshelby was able to show that the strain field ϵ inside a transformation zone is uniform[17]. This uniform strain ϵ relates to the transformation strain ϵ' by

$$\epsilon = \mathbf{S} : \epsilon' \tag{7.71}$$

where the fourth rank tensor **S** is termed *Eshelby tensor* and it only depends on the elasticity tensor **L** for the medium surround the spherical transformation zones.

The evaluation of S can be carried out by the Green function approach described in subsection 4.3.2. The stress applied to the transformation sphere after the insert-back operation is

$$\tau - \tau' = L : (\epsilon - \epsilon') \tag{7.72}$$

where τ and τ' are the imaginary stresses projected by the strains ϵ and ϵ', respectively, in an elastic material with moduli L. Their difference, however, represents the actual stress inside the transformation sphere. For the special case of isotropy, equation (7.71) can be simplified to a two parameter representation,

$$\epsilon = S(\alpha, \beta) : \epsilon' \qquad \text{or explicitly}$$

$$\begin{bmatrix} \epsilon_m \\ \epsilon_{ij'} \end{bmatrix} = \begin{bmatrix} \alpha & 0 \\ 0 & \beta \end{bmatrix} \begin{bmatrix} \epsilon'_m \\ \epsilon'_{ij'} \end{bmatrix} \tag{7.73}$$

see also the discussion in subsection 6.3.2. In (7.73), ϵ_m and ϵ'_{ij} are mean strain and strain deviator, respectively. The volumetric and deviatoric deformations induced by transformation strain are uncoupled for isotropic materials. The constants α and β in (7.73) are

$$\alpha = (1 + \nu)/[3(1 - \nu)]$$
$$\beta = (2/15)(4 - 5\nu)/(1 - \nu) \tag{7.74}$$

respectively.

We next pursue the calculation on non-homogeneous inclusions as delineated in Fig. 7.6(a), where an inclusion with moduli L is embedded, under a stress free configuration, in the surrounding matrix of modulus $\bar{L}$. A remote stress $\bar{\tau}$ (or strain $\bar{\epsilon}$) field is imposed on the composite system. The stress and strain fields in the inclusion are uniform by Eshelby's argument, and they are denoted by τ and ϵ, respectively. If our concern is only limited to field distribution outside the inclusion, we can restate the problem in Fig. 7.6(a) by a superposition of the homogeneous deformation of Fig. 7.6(b) plus a problem described in Fig. 7.6(c) featuring a uniform traction field $(\tau - \bar{\tau}) \cdot \mathbf{n}$ acting on the surface of a spherical void.

The problem depicted in Fig. 7.6(c) can be visualized as that associated with a homogeneous elastic deformation $\epsilon^* = \epsilon - \bar{\epsilon}$ in an imaginary "void material" subject to stress $\tau^* = \tau - \bar{\tau}$. The "constitutive law" of the imaginary void material is described by

$$\tau^* = -L^* : \epsilon^* \tag{7.75}$$

where L^* denotes the *matrix constraint tensor*, in recognition of the constraint of surrounding matrix against the expansion of the "void". The negative sign in (7.75) is introduced to make the coefficients in L^* positive. The analysis on this hypothetical problem is facilitated by the transformation strain problem presented in Fig. 7.5. The traction transmitted through the spherical surface to the outside matrix, designated as τ^* in (7.75), can be identified as $\tau - \tau'$ for the case of transformation strain problem, and the resultant strain of cavity, ϵ^*, is set as the final strain ϵ for the transformation zone. Accordingly, equation (7.75) is converted into

$$\tau - \tau' = \bar{L} : (\epsilon - \epsilon') = -L^* : \epsilon \qquad (7.76)$$

where the first step is due to (7.72) but the matrix moduli of polycrystal aggregate, $\bar{L}$, is used instead of L. With further engagement of Eshelby tensor in (7.71), equation (7.76) can be phrased as the following requirement

$$\bar{L} : (\bar{S} - I_4) : \epsilon' = -L^* : \bar{S} : \epsilon' \qquad (7.77)$$

for all possible transformation strain ϵ', where I_4 again the fourth rank identity tensor, and $\bar{S}$ the Eshelby tensor corresponding to the elastic moduli $\bar{L}$. From the arbitrariness of ϵ' in (7.77), an explicit expression of the elasticity constraint tensor is arrived

$$L^* = \bar{L} : (\bar{S}^{-1} - I_4). \qquad (7.78)$$

For the particular case of isotropy, the above tensorial equation can be simplified to

$$\{L^*\} = 2\bar{G} \begin{bmatrix} 2 & 0 \\ 0 & (1 - \bar{\beta})/\bar{\beta} \end{bmatrix} \qquad (7.79)$$

where $\bar{G}$ is referred to matrix shear modulus and the expression of $\bar{\beta}$ is given by (7.74) provided the mean Poisson's ratio for the material, $\bar{\nu}$, should be used for both quantities.

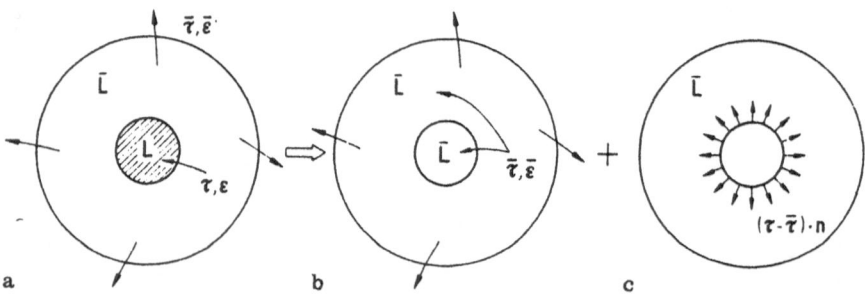

Fig. 7.6. Inhomogeneity problem. **a** An infinite medium with an inclusion is subject to remote loading; **b** homogeneous deformation problem by replacing the inclusion moduli to the matrix moduli; **c** matrix field caused by traction along the void surface

7.2.6 Self Consistent Theory

The *self consistent theory* was first proposed by Hershey[19] in 1954. Its application to polycrystal aggregates was advanced by Kröner[20] and Budiansky and Wu[21], and named after them as the *KBW theory*. The essence of this theory lies in the self-consistent deformation of each single crystal grain when embedded into the overall deformation of surrounding polycrystal aggregate. The mathematical representation of self consistent theory was later perfected by Hill[22] into an elegant format. The theory of Hill will be addressed in this subsection according to the description of Hutchinson[23].

Let us concentrate on the polycrystal model introduced earlier in Fig. 7.4. For simplicity, the single crystal grain under consideration is assumed the shape of a sphere. This assumption, though occasional challenged in the literature for its disability to construct a continuous array of polycrystal, is in fact not far from the theoretical shape of Kelvin tetrakaidecahedron. Consequently, the analysis on a polycrystal aggregate is reduced to that for a spherical single crystal inclusion with designated orientation embedded in a self consistent averaging matrix. The grain boundaries inside the polycrystal (as drawn in dashed line in Fig. 7.4), other than the boundary surrounding the central grain, will be considered effectless in the deformation of polycrystal aggregate. The ductile central single crystal is described by the following constitutive relation

$$\check{\tau} = \mathbf{L}^l : \mathbf{D} \tag{7.80}$$

which is in fact identical to (7.40) with detailed expression for the tangential stiffness tensor $\mathbf{L}^l$ given in (7.41). For the surrounding polycrystal matrix is constituted by the same type of single crystals assembled in different orientations, the global rate form constitutive law should assume a similar form

$$\check{\Sigma} = \bar{\mathbf{L}}^l : \bar{\mathbf{D}} \tag{7.81}$$

where the notation convention of $<\tau> = \Sigma$ and $<\mathbf{D}> = \bar{\mathbf{D}}$ is followed. The local deformation rate $\mathbf{D}$ in (7.80) relates to the global deformation rate $\bar{\mathbf{D}}$ by

$$\mathbf{D} = \mathbf{A} : \bar{\mathbf{D}} \tag{7.82}$$

where the fourth rank *orientation density tensor* $\mathbf{A}$ can be evaluated via the elasticity constraint tensor $\mathbf{L}^*$. Referring to Fig. 7.6(c) for the equivalent cavity problem, the differential stress rate, $\check{\Sigma} - \check{\tau}$, links to the differential deformation rate, $\bar{\mathbf{D}} - \mathbf{D}$, by the matrix constraint tensor $\mathbf{L}^*$ of the aggregate

$$\check{\Sigma} - \check{\tau} = -\mathbf{L}^* : (\bar{\mathbf{D}} - \mathbf{D}). \tag{7.83}$$

Substituting equations (7.80), (7.81) and (7.82) successively into the above equation and making use of the arbitrary choice of $\bar{\mathbf{D}}$, one can get an explicit formula for the orientation density tensor $\mathbf{A}$

$$\begin{aligned}
\mathbf{A} &= (\mathbf{L}^l + \mathbf{L}^*)^{-1} : (\bar{\mathbf{L}}^l + \mathbf{L}^*) \\
&= \left[\mathbf{L}^l - \bar{\mathbf{L}}^l + \bar{\mathbf{L}}^l : \bar{\mathbf{S}}^{-1}\right]^{-1} : \left(\bar{\mathbf{L}}^l : \bar{\mathbf{S}}^{-1}\right).
\end{aligned} \tag{7.84}$$

The last step is due to the application of (7.78).

The self consistent equation can be obtained by taking the volume average of equation (7.80)

$$\check{\Sigma} = <\mathbf{L}^l : \mathbf{D}> = <\mathbf{L}^l : \mathbf{A} : \bar{\mathbf{D}}> = <\mathbf{L}^l : \mathbf{A}> : \bar{\mathbf{D}} \tag{7.85}$$

where equation (7.82) and the orientation indifference of $\mathbf{D}$ is used in various steps of (7.85). The general formula to calculate the overall moduli is then furnished by comparing (7.81) and (7.85)

$$\bar{\mathbf{L}}^t = <\mathbf{L}^t : \mathbf{A}> \tag{7.86}$$

where $\mathbf{A}$ is given in (7.84) as a function involving both overall and single crystal moduli. If the volume average process in (7.86) is executed by integration over all grain orientations, then equation (7.86), in combination to (7.84), would form a nonlinear tensorial integral equation.

Further examination on equation (7.86) would involve details of the averaging integral over all possible grain orientations. As shown in Figure 7.7(a), a local coordinate system with its axes labelled by (1, 2, 3) successively is set up in co-incidence with the three cubic axes of the grain under consideration. The lattice depicted in Fig. 7.7(a) actually corresponds the particular case of a FCC crystal. In Fig. 7.7(b), another coordinate system featured by axes labelled by capital Roman letters (I, II, III) is also set up which coincides with the axes of the testing poly-crystal aggregate. Accordingly, the orientation of a single crystal grain with respect to the polycrystal aggregate is specified by Euler angles Θ, β and ϕ relating the local coordinates (1, 2, 3) to the global coordinates (I, II, III). As demonstrated in Fig. 7.7(b), the volume average in (7.86) will be carried out over all the orientations on a unit sphere under the global coordinate system (I, II, III) centered at point O. The orientation of a given single crystal, as denoted by axes (1, 2, 3), is drawn at a generic point G of the unit sphere. A portion of the unit circle passing through G and the pole of III-axis is also indicated in the figure, which intersects the I-II plane at a point P. The Euler angles Θ, β and ϕ then correspond to the angle from axis-III to axis-3, the angle from axis-I to the line OP, and the angle from the tangent of arc GP at point G to the axis-1, respectively. After the definition of Euler angles, a transformation matrix $\{a_{iK}\}$ denoting the direction cosine of axis-i and axis-K can be established as follows

$$\{a_{iK}\} = \tag{7.87}$$

$$\begin{bmatrix} \cos\Theta\cos\beta\cos\phi - \sin\beta\sin\phi & -\cos\Theta\cos\beta\sin\phi - \sin\beta\cos\phi & \sin\Theta\cos\beta \\ \cos\Theta\sin\beta\cos\phi + \cos\beta\sin\phi & -\cos\Theta\sin\beta\sin\phi + \cos\beta\cos\phi & \sin\Theta\sin\beta \\ -\sin\Theta\cos\phi & \sin\Theta\sin\phi & \cos\Theta \end{bmatrix}$$

where the lowercase and uppercase subscripts are referred to the local coordinates (1, 2, 3) and the global coordinates (I, II, III), respectively. The inner product of the transformation matrix will lead to an identity matrix

$$a_{iK}\, a_{jK} = \delta_{ij}. \tag{7.88}$$

For any elastic-plastic tangential stiffness tensor L^t_{ijkl} of a single crystal grain in its local coordinate system (1, 2, 3), one can always transform it to the global coordinate system (I, II, III) by the transformation matrix

$$L^t_{IJKL} = a_{iI}\, a_{jJ}\, a_{kK}\, a_{lL}\, L^t_{ijkl}. \tag{7.89}$$

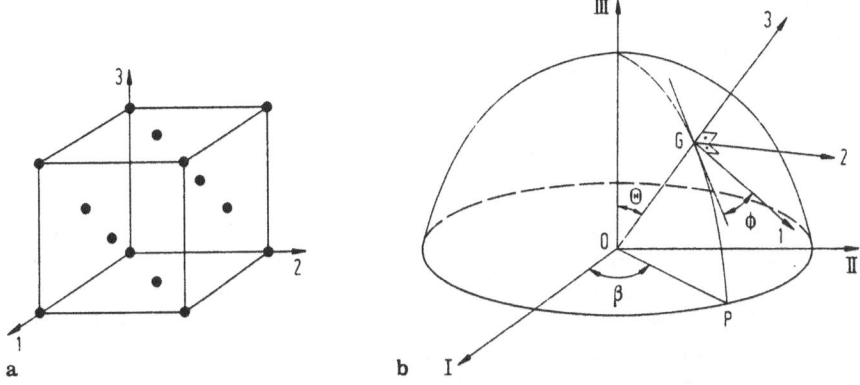

Fig. 7.7. Orientation of a single crystal grain with respect to the polycrystal aggregate. **a** Local coordinate system for a FCC single crystal; **b** Euler angles defining the orientation difference between the local coordinates of the single crystal and the global coordinates of the polycrystal aggregate

Then the volume average of polycrystal behavior can be furnished through an orientation distribution function $D(\Theta, \beta, \phi)$ as

$$\bar{L}^t_{IJKL} = <L^t_{IJMN} A_{MNKL}> = (1/8\pi^2) \left\{ \int_0^{2\pi} d\beta \int_0^{\pi} \sin\Theta d\Theta \right.$$

$$\left. \int_0^{2\pi} a_{iI} a_{jJ} a_{kK} a_{lL} D(\Theta, \beta, \phi) d\phi \right\} L^t_{ijmn} A_{mnkl} \qquad (7.90)$$

where the *orientation distribution function* (ODF), $D(\Theta, \beta, \phi)$, is normalized such that its integration over the unit sphere is one. Please bears in mind that the transformation matrix $\{a_{iK}\}$ in the integrand of (7.90) is also a function of Θ, β and ϕ.

Simplifications on the forgoing expressions will be addressed for specific cases in the next subsection which provide more physical insights to the averaging process of polycrystal behavior.

7.2.7 Examples

The examples conducted in the following correspond to the particular case of elastic (matrix and single crystal) behavior. Parallel results can be obtained for the case of elastic-plastic deformation where the tangential modulus $\mathbf{L}^t$ will be used instead of the elastic modulus $\mathbf{L}$.

Randomly distributed grain orientations. For randomly distributed grain orientations, the global elasticity tensor of polycrystal aggregates is isotropic and can be characterized by a two parameter pair $\mathbf{L}(3\bar{K}, 2\bar{G})$. Accordingly, the results listed in

(7.73) and (7.74) are valid except parameters α, β and ν should be replaced by the corresponding global counterparts with super-imposed bars. The equation (7.79) of the matrix constraint tensor can also be invoked directly. Besides, the randomly distributed orientations reduce the orientation distribution function D in (7.90) to an identity. Further simplifications would rely on the properties of the constituent single crystals.

Isotropic single crystal grains. For isotropic spherical single crystal grains, their stiffness modulus $\mathbf{L}$ only contains two independent parameters, and can be parametralized as $\mathbf{L}(3K, 2G)$, with K or G denoting the stiffness modulus for volumetric or deviatoric deformation. For a polycrystal aggregate composed by arbitrarily distributed spherical isotropic grains, their global moduli, $\bar{K}$ and $\bar{G}$, are in general different to the K and G values for the constituent single crystals. Evaluation of the global moduli, however, is facilitated from the recognition of the parametric form on the fourth rank orientation density tensor $\mathbf{A}$

$$\mathbf{A} = \mathbf{A}\left(\bar{K}/\left[(K - \bar{K})\bar{\alpha} + \bar{K}\right], \bar{G}/\left[(G - \bar{G})\bar{\beta} + \bar{G}\right]\right) \tag{7.91}$$

by the application of the first equality of (7.84) and the simplified expression (7.79) for matrix constraint tensor. When $\bar{\mathbf{L}}, \mathbf{L}$ and $\mathbf{A}$ are all represented in the parametric form, the tensorial averaging equation (7.86) is reduced to the following pair of integral equations for $\bar{K}$ and $\bar{G}$

$$<K/\left[(K - \bar{K})\bar{\alpha} + \bar{K}\right]> = 1$$
$$<G/\left[(G - \bar{G})\bar{\beta} + \bar{G}\right]> = 1. \tag{7.92}$$

For random distribution of isotropic grains, we further have

$$\bar{K} = K, \quad \bar{G} = G, \quad \mathbf{A} = \mathbf{A}(1, 1), \quad D = 1. \tag{7.93}$$

FCC single crystals. The self consistent theory of FCC polycrystal was elaborated by Hutchinson[23] under the general framework laid down by Hill[22]. When the local coordinate axes (1, 2, 3) are chosen to be the cubic axes of a FCC crystal, as shown in Fig. 7.7(a), the FCC symmetry would reduce the elastic modulus to the following three parameter representation

$$\mathbf{L} = \mathbf{L}\left(3\eta_1, 2\eta_2, 2\eta_3\right) \tag{7.94}$$

which symbolizes the following stress-strain relations

$$\sigma_{ii} = 3\eta_1 \epsilon_{ii} = (L_{1111} + 2L_{1122})\, \epsilon_{ii}$$

$$\sigma_{11} - \sigma_{22} = 2\eta_2\, (\epsilon_{11} - \epsilon_{22}) = (L_{1111} - L_{1122})\, (\epsilon_{11} - \epsilon_{22}) \quad \text{etc.}$$

$$\sigma_{12} = 2\eta_3 \epsilon_{12} = 2L_{1212}\epsilon_{12}, \quad \text{etc} \tag{7.95}$$

where the notion η_2 will be equal to η_3 for the particular case of isotropy.

The inversion of $\mathbf{L}$ can be simply proceeded as

$$\mathbf{L}^{-1} = \mathbf{L}(1/3\eta_1, 1/2\eta_2, 1/2\eta_3) \tag{7.96}$$

because the inversion operations can be accomplished from equation to equation in (7.95), without the complication by the coupling effect.

The calculation of the global response, $\bar{L}$, can then be carried out by means of (7.90). For the particular case of random distribution of all orientations, the ODF, D, in (7.90) is equal to identity and gives rise to an isotropic global modulus $L(3\bar{K}, 2\bar{G})$. The values of $\bar{K}$ and $\bar{G}$ can be assessed by taking different contractions of equation (7.90). In particular

$$\bar{L}_{IIKK} = L_{iimn} A_{mnkk}, \quad \bar{L}_{IJIJ} = L_{ijmn} A_{mnij} \tag{7.97}$$

by notifying the property (7.88) of the transformation matrix $\{a_{iK}\}$. The associated constraint tensor is given by the simplified form of (7.79). The orientation density tensor $\mathbf{A}$ can be calculated from (7.94), (7.84) and (7.79) as

$$\mathbf{A} = \mathbf{A}\left(1, \bar{G}/f_2, \bar{G}/f_3\right) \tag{7.98}$$

where

$$f_i = \bar{G}(1 - \bar{\beta}) + \bar{\beta}\eta_i \qquad i = 2, 3. \tag{7.99}$$

Equations (7.94) and (7.97) together give the following expressions of $\bar{K}$ and $\bar{G}$ for a FCC polycrystal aggregate in terms of its single crystal moduli

$$\bar{K} = \eta_1, \quad 2(\eta_2/f_2) + 3(\eta_3/f_3) = 5. \tag{7.100}$$

7.2.8 Miscellaneous Results of Crystalline Plasticity

Besides KBW theory and Hill's self consistent theory, contributions to the formulation of polycrystalline solids were also made by T.H. Lin. In the fifties, Lin modified the rigid plastic model of Taylor to accommodate elastic deformation[24]. He later on proposed a computation scheme[25,26] which could predict the constitutive response of polycrystal aggregate by a deliberate arrangement of single crystal orientations, as shown in Fig. 7.8. Lin assumed that the response of a polycrystalline solid can be closely simulated by the repetition of basic building blocks composed of 64 single crystals, see Fig. 7.8(a). These 64 single crystals in the basic building block are divided into 8 groups in accordance to the symmetry with respect to three mutually perpendicular axes, and consequently 8 single crystals belong to each basic group, referred to Fig. 7.8(b). Thus, the response of a basic building block would appear to be initially isotropic and is capable of symmetric deformation. (The earlier version of the Lin scheme involved only $3^3 = 27$ single crystals, which yielded some unsatisfactory results due to the inability to accommodate symmetric deformation modes.) Among the 8 single crystals in each basic group, 2 have neutral orientations for all coordinate axes. The other 6 single crystals are further divided into 3 pairs with each pair arranged symmetrically in favor of the orientation in a single assigned coordinate axis. T.H. Lin carried out

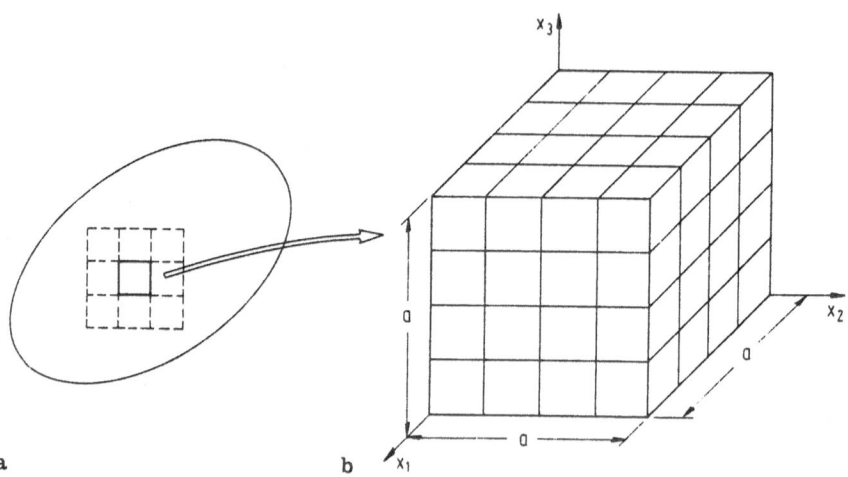

Fig. 7.8. The computational model of T.H. Lin for a polycrystal aggregate. **a** A polycrystal is simulated by the repetition of basic building blocks; **b** a basic building block is composed by 64 single crystals with pre-designed orientation preferences

the numerical computations via this innovative assembly of single crystals to infer the overall response of a polycrystal aggregate.

Various models of polycrystal aggregates were tested numerically by Hutchinson[23] for the simple uniaxial tension case, and results of this calculation is reproduced in Fig. 7.9. The calculation was carried out for the FCC polycrystals with non-hardening flow rules incorporated in various slip systems. Furthermore, the global response of the polycrystal aggregate was assumed to be isotropic and a Poisson's ratio of 1/3 was employed. The abscissa in Fig. 7.9 signifies the extent of plastic deformation, while the ordinate represents the half value of Taylor factor with the horizontal line of 1.53 (as predicted by Taylor) as an upper bound. This horizontal line also indicates an upper bound asymptote to the predictions of other more recent models in the large deformation regime. The upper bound nature of Taylor's estimate is inherent in its foundation, as constructed on the ground of kinematically admissible field. This operation would certainly overestimate the stress response at the initial stage of plastic deformation. Both the self consistent theory and the theory of T.H. Lin predict the same initial uniaxial yielding stress $\sigma = 2\tau_{cr}^{o}$, exactly twice of the initial critical resolved shear stress associated with the most favorable slip system. This value also refers to the minimum value of Schmid factor and indicates that the initial yielding takes place in the most favorably orientated grain in the whole polycrystal for the occurrence of plastic slip. It is noticed that considerable differences still exist among various predictions. T.H. Lin believed that his improved computation[25,26] could best correlate the experimental data of Budiansky and Wu[21].

The forgoing discussions are restricted to the case of rate independent elastic-plastic polycrystal aggregates. A theoretical framework for the alternative situation

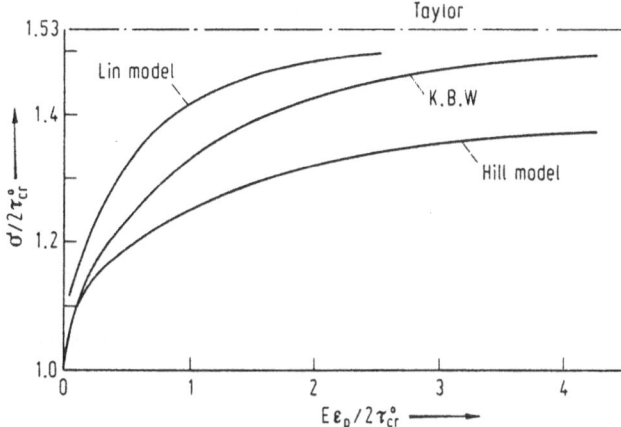

Fig. 7.9. Uniaxial stress strain curves for a FCC polycrystal aggregate, after Hutchinson[23]

of rate dependent deformation was provided by Hutchinson[27]. Another important research topic is the *texture development* from initially isotropic grain orientation arrangement by imposed anisotropic large plastic deformation. In the latter circumstances, the initial random orientations of the constituent single crystal grains would rotate during deformation and become clustered near certain texture groups in a stereographic diagram. This process has been simulated in the finest detail by Asaro and Needleman[28] and by Harren, Deve and Asaro[29]. As an application of the mesoplasticity theory, this issue will be addressed upon later in section 9.3, which grands the significance of mesoplasticity theory to the applications in metal forming processes such as rolling, drawing and forging. The scientists and engineers now are able to monitor the formation and development of deformation texture through a software package interfaced with appropriate mesoplasticity model.

The rapid development of computer technology provide a strong impetus for more detailed modelling of crystalline plasticity. The research on the single crystal behavior is now advanced enough for practical implementations. As long as the hardening matrix $\{h^{\alpha\beta}\}$ of a single crystal is obtained through dislocation interaction calculation, the complete constitutive response of the single crystal can be simulated and an increasing number of applications are accomplished in the recent years. The plastic deformation characteristics, such as necking and cruciform shear band development in a tensile bar, were reproduced[30] by a mesoplasticity simulation based on the double slip model (which means no more than two slip systems can operate at the same time). Success was also reported for the bicrystal calculation by the same mesoplasticity model[31]. With the access of supercomputer, the detailed calculation on polycrystals (up to detailed events of slip operations in any constituent grain) are being actively pursued. These encouraging results indicate that the modelling of plastic constitutive response, at least from the scale of continuous slip has upgraded to a very sophisticated level, and the prediction of versatile plastic deformation characteristics has become reality.

7.3 Macroscopic Responses

7.3.1 Macroscopic Averaging

In mesoplasticity, macroscopic responses are assessed from a macroscopic averaging procedure carried out for microscopic quantities and/or expressions over a sufficiently large macroscopic volume element. We elucidate this process from an example concerning incremental plastic work. As we recall from the derivation towards equation (7.25), for the rate of plastic work expression, $\dot{W}_c^p$, under the dislocation model. If attention is focused on an incremental loading step $d\Sigma$ for macroscopic stress Σ, and consequently an increase df for the dislocation gliding force along various dislocation lines, then the corresponding macroscopic plastic work increment is obtained by integration over all dislocation lines

$$\mathbf{D}_p : d\Sigma = (1/V) \int_{L_{\text{disl}}} v \, df \, dl. \tag{7.101}$$

If the dislocation velocity is only related to its work conjugate force f and dislocation pattern inside the material (termed by Rice as *Pattern of Internal Rearrangement*, PIR), then the integrand in (7.101) can be regarded as a differential of a function w defined by

$$w(f; \text{PIR}) \equiv \int_0^f v(f', \text{PIR}) \, df' \tag{7.102}$$

when PIR keeps fixed. Consequently, the left hand side of (7.101) would also be representable in terms of a differential under fixed PIR, with the implication on the existence of a scalar macroscopic function Ω [15,32] such that

$$\Omega(\Sigma; \text{PIR}) = (1/V) \int_{L_{\text{disl}}} w(f; \text{PIR}) \, dl \tag{7.103}$$

whose differential under fixed PIR would resemble the left hand side of (7.101).

7.3.2 Plastic Strain Rate Tensor

Under current dislocation pattern (meaning PIR is fixed at its current configuration), a comparison between the left hand side of (7.101) and the variation of Ω with respect to the macroscopic stress (named before as the elastic variation) provides us an expression for the macroscopic plastic strain rate

$$\bar{\mathbf{D}}^p = \partial\Omega(\Sigma; \text{PIR})/\partial\Sigma. \tag{7.104}$$

That is, the function Ω characterizes the potential function of the plastic strain rate $\bar{D}^p$. The above reasoning not only establishes the existence of a plastic flow potential Ω, but also provides the normality structure between plastic flow, $\bar{D}^p$, and the constant-Ω surface. The latter is also termed as loading surface in the sequel. We emphasize here that the derivation developed so far is irrelevant to any specific hardening laws and valid for both rate independent or rate dependent response. When Ω only relies on the second invariant of the deviatoric macroscopic stress (namely J_2 theory), equation (7.104) would also imply coaxiality between $\bar{D}^p$ and the macroscopic deviatoric stress.

The above normality proof is based on a Schmid type yield criterion in the corresponding microscopic model. Non-Schmid effect is occasionally recorded for materials capable of microscopic plastic deformation mechanisms other than dislocation glide (such as dislocation climb). The macroscopic consequence for the non-Schmid effect is featured by the lack of normality, or equivalently by a non-associated flow rule. The constitutive framework for materials belong to this category was addressed in reference [6].

7.3.3 Geometry of Ω Surfaces

We recall the dislocation dynamics curves exhibited in Fig. 4.19 between the gliding stress τ (and consequently the gliding force, f, by multiplying the length of Burgers vector, b) and the dislocation velocity v, where the solid and dashed curves refer to the cases of rate independent and rate dependent glide, respectively. The monotonic increase feature shared by all dislocation dynamics curves can be described mathematically as

$$(v^B - v^A)(f^B - f^A) \geq 0 \qquad (7.105)$$

where superscripts A and B denote two arbitrary points on the dislocation dynamics curve. The above inequality will be utilized to depict the geometric characteristics of Ω surfaces in stress space. As shown in Fig. 7.10, a supporting tangent plane can be constructed at any point Σ^A on a Ω surface, and Σ^B represents another stress state outside this supporting tangent plane. Let us consider a loading process from Σ^A to Σ^B along a straight line as drawn in Fig. 7.10, an inequality can be established for this loading process from equation (7.105) and (7.101) as follows

$$\int_{\Sigma^A}^{\Sigma^B} [\bar{D}^p(\Sigma) - \bar{D}^p(\Sigma^A)] : d\Sigma \geq 0. \qquad (7.106)$$

After straightforward integration, it becomes

$$\Omega(\Sigma^B) - \Omega(\Sigma^A) \geq \bar{D}^p(\Sigma^A) : (\Sigma^B - \Sigma^A). \qquad (7.107)$$

For Σ^B locates outside the tangential plane, the difference stress vector $\Sigma^B - \Sigma^A$ should form an acute angle with $\bar{D}^p(\Sigma^A)$ because the latter is perpendicular to the tangent plane passing through Σ^A. Accordingly, the right-hand side of inequality

(7.107) is non-negative. Then a necessary condition for Σ^B lying outside the tangent plane at Σ^A can be phrased as

$$\Omega\left(\Sigma^B\right) \geq \Omega\left(\Sigma^A\right) \tag{7.108}$$

so comes the convexity of the loading surface. As the conclusion, the convexity of Ω surface, regardless any type of rate sensitivity, can be inferred from the monotone non-declining feature of dislocation dynamics curve, which in turn is traced back to basic physical and thermodynamic principles.

Alternatively, if the stress state Σ^B is selected inside the Ω surface, a similar argument like proceeded above would give the following inequality for all possible Σ^B's which do not exceed the current Ω surface

$$\bar{\mathbf{D}}^p\left(\Sigma^A\right) : \left(\Sigma^A - \Sigma^B\right) \geq 0. \tag{7.109}$$

By comparison to (2.9) or (2.40), one immediately recognizes the above inequality represents the macroscopic form of the *principle of maximum plastic work*, used as the cornerstone of classical macroplasticity.

We next discuss the smoothness of the loading surfaces. This question could be phrased through the continuous dependence between the macroscopic plastic strain rate, $\mathbf{D}^p$, and the macroscopic stress, Σ, when they are linked by equations from (7.102) to (7.104) under a fixed PIR.

The case of rate dependent solids described by the dashed curves in Fig. 4.19 is examined first. Then both $v(f)$ and $w(f)$ are smooth functions of the dislocation gliding force f. Furthermore, f is a function of macroscopic stress Σ by the relation of $f = b\mathbf{P} : \Sigma$, and any possible non-smooth transition from one slip system to the other, as anticipated in this expression, is smeared out by the continuous adjustment on contributions from different slip systems. Hence, rate dependent solids are featured by a family of smooth loading surfaces in accordance to different w values, and those loading surfaces are densely allocated in the vicinity of a surface corresponding to critical gliding force f_{cr}.

For the case of rate independent solids dictated by the solid curve in Fig. 4.19, the $v(f)$ function behaves as if it were a Dirac delta function. Thus, the $w(f)$ function, as defined in (7.102), becomes a Heaviside step function with f_{cr} as the transition point. Its corresponding loading surfaces would consequently collapse into a single sheet, termed as the *yield surface*, through the application of (7.103). Moreover, sharp corners on this yield surface cannot be excluded. The development of corner plasticity theory, as described in section 2.7, has been motivated by this fact.

7.3.4 Loading Surfaces of LiF Single Crystals

Fluoride-Lithium, LiF, can be easily produced in the form of single crystal with large size, and the dislocation movement in LiF is also easy to observe. The lattice of LiF is characterized by a NaCl structure, as shown in Fig. 7.11(a). The six

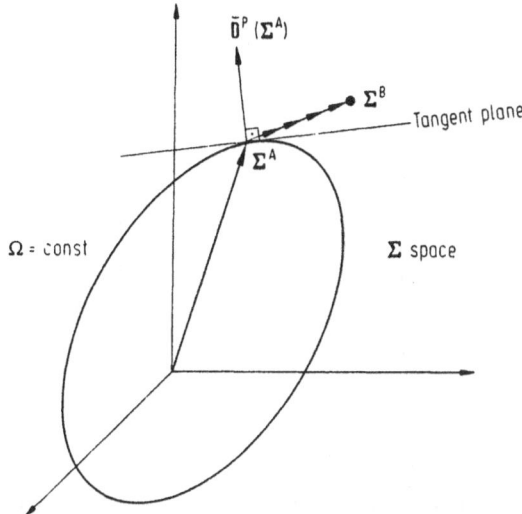

Fig. 7.10. Schematic illustration on the convexity proof of loading surface

slip systems in LiF single crystal are of $\{110\}<110>$ type, and they are listed as follows

No.	Slip plane	Slip direction
1	$(1\bar{1}0)$	$[110]$
2	(110)	$[1\bar{1}0]$
3	$(01\bar{1})$	$[011]$
4	(011)	$[01\bar{1}]$
5	$(\bar{1}01)$	$[101]$
6	(101)	$[\bar{1}01]$

According to the microscopic observations for LiF single crystals, its dislocation dynamics curve can be correlated by a power law description under normal dislocation glide speed regime

$$v = v_o(f/f_o - 1)^n H(f - f_o) \tag{7.110}$$

where $H(.)$ denotes the Heaviside step function. Substituting the above dislocation dynamics curve into (7.102) and carrying out the integration, we have

$$w = [v_o f_o/(n + 1)](f/f_o - 1)^{n+1} H(f - f_o) \tag{7.111}$$

where f refers to the resolved gliding force along a specific slip system under consideration. Equation (7.111) indicates that the surfaces of common w-values

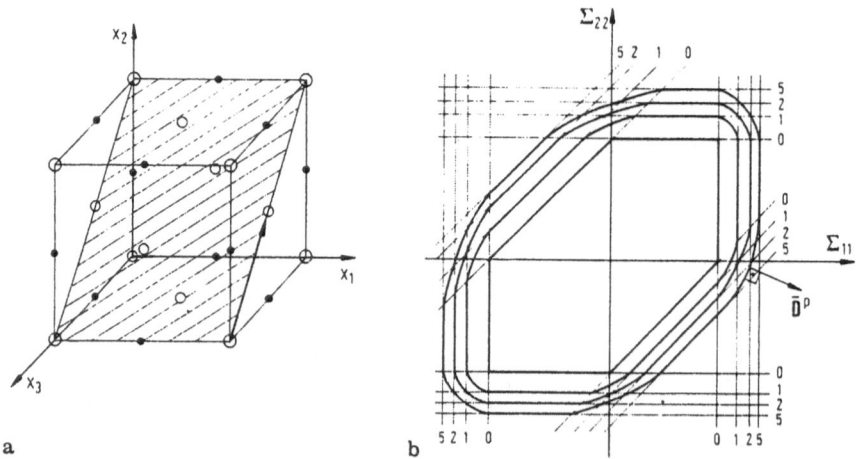

Fig. 7.11. Special case of LiF single crystal. **a** Lattice structure of LiF; **b** Ω surfaces, featuring the transition from the initial yield surface with sharp corners to smooth subsequent loading surfaces

are formed by a family of parallel hypersurfaces on which the dislocation glide force $f = b\mathbf{m}^* \cdot \boldsymbol{\Sigma} \cdot \mathbf{s}^*$ is constant.

For the particular case of LiF, the macroscopic loading surface Ω, as described in (7.103), can be specified as

$$\Omega(\boldsymbol{\Sigma}; \text{PIR}) = \sum_{\alpha=1}^{6} \rho^{\alpha} w(f^{\alpha}) \tag{7.112}$$

where ρ^{α} represents the dislocation density for the α-th slip system, and f^{α} is the resolved gliding force acting upon it. The substitution of (7.111) into (7.112) results in

$$\Omega(\boldsymbol{\Sigma}; \text{PIR}) = [v_o f_o/(n+1)] \sum_{\alpha=1}^{6} \rho^{\alpha} [b\mathbf{m}^{*\alpha} \cdot \boldsymbol{\Sigma} \cdot \mathbf{s}^{*\alpha}/f_o - 1]^{n+1}$$
$$\cdot H(b\mathbf{m}^{*\alpha} \cdot \boldsymbol{\Sigma} \cdot \mathbf{s}^{*\alpha} - f_o). \tag{7.113}$$

At the instant of initial yield, Ω would take the value of zero, the initial yielding surface is consequently given by

$$\mathbf{m}^{*\alpha} \cdot \boldsymbol{\Sigma} \cdot \mathbf{s}^{*\alpha} = f_o/b \qquad \alpha = 1, \dots, 6 \tag{7.114}$$

which resembles the Tresca yield surface enclosed by six hyperplanes, with the appearance of sharp edges and corners. In the subsequent yielding stage, however, Ω defines a smooth and convex closed surface. To visualize this fact, let us consider the following special case

$$\rho^{\alpha} = \rho, \qquad \alpha = 1, \dots, 6$$

$$\Sigma_{ij} = 0 \quad \text{except } \Sigma_{11} \text{ and } \Sigma_{22}. \tag{7.115}$$

When substituting the above simplifications to (7.113), one obtains a two dimensional marcoscopic yield surface in the $\Sigma_{11} - \Sigma_{22}$ plane described by

$$\Omega = [v_o f_o \rho/(n+1)] \left\{ [b\,|\Sigma_{11} - \Sigma_{22}|\,/f_o - 1]^{n+1} H\,(b\,|\Sigma_{11} - \Sigma_{22}| - f_o) \right.$$
$$+ [b\,|\Sigma_{11}|\,/f_o - 1]^{n+1} H\,(b\,|\Sigma_{11}| - f_o)$$
$$\left. + [b\,|\Sigma_{22}|\,/f_o - 1]^{n+1} H\,(b\,|\Sigma_{22}| - f_o) \right\}. \tag{7.116}$$

The initial and subsequent loading surfaces described by (7.116) are plotted in Fig. 7.11(b). The evolution from the initial Tresca type hexagon to the subsequent smooth loading surfaces (which are gradually modulated to resemble a J_2 characterization) is clearly demonstrated there, especially the transition from sharp corners to high curvature contour and then to a low curvature contour, with uniquely defined plastic strain rate.

7.4 Plasticity for Geological Materials

7.4.1 Slip Induced Dilatancy

For geological materials, the additive decomposition of velocity gradient, as phrased in (7.6), is still applicable. But l^p in (7.6) now represents the plastic slip rates for geological materials and l^* relates to the additional elastic strain directly linked to applied stress and the meso-element rotation necessary to make the total deformation compatible. For rock materials, l^* may stands for the elastic distortion rate of the rock matrices; while for granular materials it may represent the elastic deformation rate of the grains themselves plus the changing rate on intergranular contact area. The plastic part of velocity gradient, l^p, can be further decomposed into two parts by a similar approach as previously devised in (7.8)[33]

$$l^p = D^p + W^p = \sum_{\alpha=1}^{n} \dot{\gamma}^\alpha \left(\hat{P}^\alpha + W^\alpha \right). \tag{7.117}$$

The symmetric part, $\hat{P}^\alpha$, in the above formula, however, is different from its metal plasticity counterpart (7.9)

$$\hat{P}^\alpha = P^\alpha + \tan\mu_1\, m^{*\alpha} m^{*\alpha} + \tan\mu_2\, 1 \tag{7.118}$$

where the terms relevant to $\tan\mu_1$ and to $\tan\mu_2$ stand for the expansion in normal direction and the equal-directional dilation, respectively, induced by slip deformation. These two additional terms characterize the *plastic dilatancy* phenomenon inherent by geological materials. There are two cases in which the dilatancy parameters μ_1 and μ_2 vanish. The first case is associated with dilatational free plastic deformation, as discussed in the classical metal plasticity theory. The second case refers to the saturation of dilatational plasticity, termed by Mandel as "profusion plastic shear flow". The application of the latter case can also be found in the formulation of crazed matters developed in thermoplastics.

7.4.2 Matrix Elasticity

Similarly, the elastic part of deformation gradient can also be decomposed into a symmetric part $\mathbf{D}^*$ and a skew-symmetric part $\mathbf{W}^*$, where the former relates to stress from the constitutive law pertinent to matrix elasticity. For geological materials, the normality structure and the Voigt symmetry is nevertheless non-attainable, so Cauchy stress $\boldsymbol{\sigma}$ is frequently adopted as an appropriate stress measure and is also engaged in this section, rather than the Kirchhoff stress $\boldsymbol{\tau}$ used extensively in metal plasticity. The Cauchy stress rate corotational to the slip frame is given by[33]

$$\check{\boldsymbol{\sigma}}^* = \mathbf{L} : \mathbf{D}^* \tag{7.119}$$

where $\mathbf{L}$ is the fourth rank elasticity tensor with Voigt symmetry. When combining with (7.15) and (7.117), the above formula can be rewritten as

$$\check{\boldsymbol{\sigma}} = \mathbf{L} : \mathbf{D} - \sum_{\alpha=1}^{n} \dot{\gamma}^\alpha \left\{ \mathbf{L} : \hat{\mathbf{P}}^\alpha + \mathbf{W}^\alpha \cdot \boldsymbol{\sigma} - \boldsymbol{\sigma} \cdot \mathbf{W}^\alpha \right\}. \tag{7.120}$$

7.4.3 Slip Friction and Pressure Sensitivity

Another feature for the slip deformation of geological materials lies in the existence of Columb type slip friction. By taking into account of its influence, the rate independent constitutive law among active slip systems can be revised to

$$\dot{\tau}^\alpha + \tan \beta_1 \dot{\sigma}^\alpha + \tan \beta_2 \dot{\sigma}_m = \sum_{\beta=1}^{n} h_{\alpha\beta} \dot{\gamma}^\beta \tag{7.121}$$

where τ^α and σ^α are resolved shear stress and normal stress for the α-th active slip system, respectively, while σ_m denotes the mean stress. The definitions of τ^α, σ^α and σ_m in (7.121) are all attached to Cauchy stress. The term relevant to $\tan \beta_1$ represents the Columb type slip friction, whereas the term relevant to $\tan \beta_2$ refers to the pressure sensitivity of constitutive relations to the triaxial stress.

To compute $\dot{\sigma}^\alpha$ and $\dot{\tau}^\alpha$, the changing rates of slip systems, namely $\dot{\mathbf{m}}^{*\alpha}$ and $\dot{\mathbf{s}}^{*\alpha}$, are required. Under the ignorance of elastic distortion, those changing rates are controlled by the spin of the slip systems

$$\dot{\mathbf{m}}^{*\alpha} = \mathbf{W}^* \cdot \mathbf{m}^{*\alpha}, \quad \dot{\mathbf{s}}^{*\alpha} = \mathbf{W}^* \cdot \mathbf{s}^{*\alpha}. \tag{7.122}$$

Then equation (7.121) can be rewritten as

$$\check{\boldsymbol{\sigma}}^* : \mathbf{Q}^\alpha = \sum_{\beta=1}^{n} h_{\alpha\beta} \dot{\gamma}^\beta \tag{7.123}$$

where

$$\mathbf{Q}^\alpha = \mathbf{P}^\alpha + \tan \beta_1 \mathbf{m}^{*\alpha} \mathbf{m}^{*\alpha} + \tan \beta_2 \mathbf{1} \tag{7.124}$$

similar to the $\mathbf{Q}$ tensor defined in the Rudnicki-Rice form of macroplasticity. However, the relation (7.123) only suits for active slip systems. A more complete description can be written as

$$\text{if } 0 < \boldsymbol{\sigma} : \mathbf{Q}^\alpha = \tau_{cr}^\alpha \text{ and } \check{\boldsymbol{\sigma}}^* : \mathbf{Q}^\alpha = \sum_{\beta=1}^n h_{\alpha\beta}\dot{\gamma}^\beta \text{ then } \dot{\gamma}^\alpha \geq 0$$

$$\text{otherwise} \quad \dot{\gamma}^\alpha = 0 \tag{7.125}$$

where τ_{cr}^α still represents the current slip resistance for the α-th slip system.

7.4.4 Elastic-Plastic Constitutive Equations

To obtain the elastic plastic constitutive equations, expressions for the shear rates $\dot{\gamma}^\alpha$ in (7.120) have to be found. If one notices that

$$\check{\boldsymbol{\sigma}}^* : \mathbf{Q}^\beta = \mathbf{D} : \mathbf{L} : \mathbf{Q}^\beta - \sum_{\alpha=1}^n \dot{\gamma}^\alpha \hat{\mathbf{P}}^\alpha : \mathbf{L} : \mathbf{Q}^\beta \tag{7.126}$$

then a comparison between (7.123) and the above expression would lead to

$$\mathbf{D} : \mathbf{L} : \mathbf{Q}^\beta = \sum_{\alpha=1}^n \dot{\gamma}^\alpha N^{\alpha\beta} \tag{7.127}$$

where the matrix N is given by

$$N^{\alpha\beta} = h^{\alpha\beta} + \hat{\mathbf{P}}^\alpha : \mathbf{L} : \mathbf{Q}^\beta \tag{7.128}$$

in contrast to the expression (7.38) for metal plasticity. Therefore, the shear strain rates can be solved from (7.127)

$$\dot{\gamma}^\alpha = \left(\sum_{\beta=1}^n N^{\alpha\beta-1} \mathbf{Q}^\beta \right) : \mathbf{L} : \mathbf{D} \tag{7.129}$$

as long as the matrix N is positive definite. The elastic-plastic constitutive equation is obtained by substituting (7.129) back to (7.120)

$$\check{\boldsymbol{\sigma}} = \mathbf{L}' : \mathbf{D} \tag{7.130}$$

where the elastic-plastic tangential stiffness modulus is given by

$$\mathbf{L}' = \mathbf{L} - \left\{ \sum_{\alpha=1}^n \sum_{\beta=1}^n \left(\mathbf{L} : \hat{\mathbf{P}}^\alpha - \boldsymbol{\sigma} \cdot \mathbf{W}^\alpha + \mathbf{W}^\alpha \cdot \boldsymbol{\sigma} \right) N^{\alpha\beta-1} \mathbf{Q}^\beta \right\} : \mathbf{L}. \tag{7.131}$$

7.4.5 Planar Double Slip Model

Some explicit predictions on the constitutive law presented in the forgoing sub-section can be arrived by the incorporation of a planar double slip model[34] as shown in Fig. 7.12. For simplicity, the coordinates X_1 and X_2 in Fig. 7.12 coincide with the principal stresses σ_{11} and σ_{22}. The only two active slip systems "1" and "2" are symmetric to the X_2-axis with their respective **m** and **s** vectors marked in Fig. 7.12. The angle formed by two symmetrically aligned slip lines is denoted by $\pi/2 - \phi$.

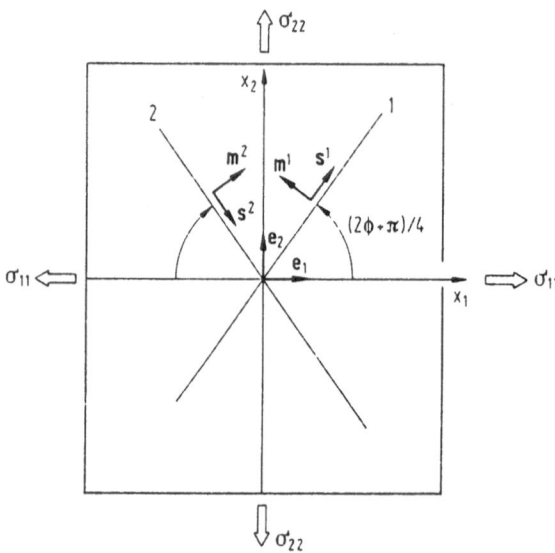

Fig. 7.12. Geometry and slip system orientations in a planar double slip model

As to the material description, we assume the elastic response is described by an isotropic law

$$\mathbf{L} = \lambda \mathbf{1}\mathbf{1} + 2G\mathbf{I}_4 \tag{7.132}$$

where λ and G are the familiar Lamè constants. The instantaneous hardening matrix $h^{\alpha\beta}$ only consists of two slip systems with symmetric and reciprocally equal hardening response assumed, namely

$$h_{11} = h_{22} = h, \quad h_{12} = h_{21} = qh \tag{7.133}$$

where the latent hardening ratio q is presumably a constant. Under the present material behavior and loading condition, the corotational stress rate would degenerate to the conventional stress rate (material derivative). Thus, the constitutive relations under the present idealization becomes

$$\dot{\sigma}_{11} = L_{1111}^{t}D_{11} + L_{1122}^{t}D_{22}$$
$$\dot{\sigma}_{22} = L_{2211}^{t}D_{11} + L_{2222}^{t}D_{22} \tag{7.134}$$
$$\dot{\sigma}_{12} = 2\bar{\mu}D_{12}.$$

The detailed expressions for various coefficients in $\mathbf{L}^{t}$ matrix are given in reference [34] as functions of angle ϕ and various material parameters employed in the previous formulation. These results are lengthy and will not be reproduced here.

7.4.6 Normality, Vertex Modulus and Coaxiality

Several conclusions arrived from this model, however, are worthwhile to address. To begin with, one notices the existence of a rate potential designated by

$$\Phi = \frac{1}{2}\mathbf{D} : \mathbf{L}^{t} : \mathbf{D} \tag{7.135}$$

from which the stress rate can be derived simply as

$$\dot{\sigma} = \partial\Phi/\partial\mathbf{D} \tag{7.136}$$

can be established if and only if the tangential modulus tensor $\mathbf{L}^{t}$ possessing Voigt symmetry, or L_{1122}^{t} equals to L_{2211}^{t} under the present symmetric arrangement between current principal stresses and slip systems. This condition requires

$$[\sin\beta_1 + 2\cos\beta_1\tan\beta_2] / \cos(\phi - \beta_1) =$$
$$[\sin\mu_1 + 2\cos\mu_1\tan\mu_2] / \cos(\phi - \mu_1). \tag{7.137}$$

A sufficient condition for the fulfillment of (7.137) is referred to the case that β_1 and β_2 separately equal to μ_1 and μ_2, then the pressure sensitivity would be balanced by plastic dilatancy. This particular situation, however, is unlikely to occur according to the parameter ranges as reported in section 2.9.

Next, for the present respective orientations between principal stresses and active slip systems, the coaxiality between stress rate and deformation rate will be achieved if

$$[(1 - q)h/2G] (1 - M^*\sin\phi)(1 - B^*\sin\phi) + (M^* - \sin\phi)(B^* - \sin\phi) = 0 \tag{7.138}$$

where M^* and B^* are only related to shear-induced lateral expansion and Columb friction, respectively

$$M^* = \sin\beta_1 / \cos(\phi - \beta_1)$$
$$B^* = \sin\mu_1 / \cos(\phi - \mu_1). \tag{7.139}$$

If condition (7.138) is fulfilled, the tangential modulus $\bar{\mu}$ in (7.134) would become infinity, so that any deviation increment of deformation, namely D_{12}, will be effectively prevented. The modulus $\bar{\mu}$ is termed "vertex modulus" or "Mandel-Spencer non-coaxiality parameter" in the literature. The larger its magnitude, the more likely for coaxiality between stress rate and strain rate.

It is interesting to notice that condition (7.138) is irrelevant to the parameters such as β_2, μ_2 and the average hardening. That is because all these parameters characterize only isotropic, or spherical behavior which has no effect on the coaxiality. For the special cases like equal hardening[1,2] (implying $q = 1$) or weak hardening (when h/G is negligible), the condition (7.138) is reduced to

$$\phi = \beta_1 \text{ or } \mu_1 \tag{7.140}$$

namely coaxiality will be achieved when the orientation difference ϕ of the double slip systems agrees with either friction angle μ_1 or lateral expansion angle β_1. In particular, when μ_1 or β_1 equals to zero (namely free of friction or lateral expansion), the coaxiality between σ and $\mathbf{D}$ would require $\phi = 0$, or the two slip lines perfectly bisect the principal axis of stress rate.

34. References

1. Taylor, G.I. and Elam, C.F., Proc. Roy. Soc. London, A102(1923), p.643.
2. Taylor, G.I. and Elam, C.F., Proc. Roy. Soc. London, A108(1925), p.28.
3. Hill, R., J. Mech. Phys. Solids, 14(1966), p.95.
4. Hill, R. and Rice, J.R., J. Mech. Phys. Solids, 20(1972), p.401.
5. Asaro, R.J., J. Appl. Mech., 50(1983), p.921.
6. Asaro, R.J., in Adv. Appl. Mech., 23(1983), p.1.
7. Asaro, R.J., Mech. Materials, 4(1985), p.343.
8. Lee, E.H., J. Appl. Mech., 36(1969), p.1.
9. Asaro, R.J. and Rice, J.R., J. Mech. Phys. Solids, 25(1977), p.309.
10. Hill, R. and Havner, K.S., J. Mech. Phys. Solids, 30(1982), p.5.
11. Needleman, A., in Plasticity of Metals at Finite Strain, Lee, E.H. and Mallet, R.L. eds., (1981), p.387.
12. Sachs, G., Z. Ver. Dtsch. Ing., 72(1928), p.734.
13. Rice, J.R., in Const. Equat. in Plasticity, Argon, A.S. ed., (1975), p.23.
14. Taylor, G.I., J. Inst. Metals, 62(1938), p.218.
15. Bishop, J.F.W. and Hill, R., Phil. Mag., 42(1951), p.414.
16. Von Houtte, P. and Aernoudt, E., Zeit. fur Metall., 66(1975), p.202.
17. Eshelby, J.D., Proc. Roy. Soc. London, A241(1958), p.376.
18. Mura, T., Micromechanics of Defects in Solids, 2nd ed., Martinus Nijhoff Pub., (1987).
19. Hershey, A.V., J. Appl. Mech., 21(1954), p.236 and p.241.
20. Kröner, E., Acta Metall., 9(1961), p.155.
21. Budiansky, B. and Wu, T.Y., Proc. 4th U.S. Nat. Congr. Appl. Mech., (1962), p.1175.
22. Hill, R., J. Mech. Phys. Solids, 13(1965), p.89 and p.213.
23. Hutchinson, J.W., Proc. Roy. Soc. London, A319(1970), p.247.
24. Lin, T.H., J. Mech. Phys. Solids, 5(1957), p.143.
25. Lin, T.H., in Adv. Appl. Mech., 11(1971), p.255.
26. Lin, T.H., ASME Winter Annual Meeting, New Orleans, (1984).
27. Hutchinson, J.W., Proc. Roy. Soc. London, A348(1976), p.101.
28. Asaro, R.J. and Needleman, A., Acta Metall., 33(1985), p.923.
29. Harren, S.V., Deve, H.E. and Asaro, R.J., Acta Metall., 36(1988), p.2435.
30. Peirce, D., Asaro, R.J. and Needleman, A., Acta Metall., 31(1983), p.1951.
31. Lamonds, J., M.S. Thesis of Brown Univ., June, (1983).
32. Rice, J.R., J. Appl. Mech., 37(1970), p.728.
33. Nemat-Nasser, S., J. Appl. Mech., 50(1983), p.1114.
34. Lance, G.L. and Nemat-Nasser, S., Mech. Materials, 5(1986), p.1.

8 Meso-Damage Theory

8.1 Introduction to Meso-Damage

Spectacular progresses have been achieved by *damage mechanics* during the past decade, which upgrade it into a scientific discipline and one of the most active research frontiers in the failure theories of materials. The scope of nowadays damage mechanics theory encompasses researches originated from macroscopic, meso and microscopic levels, with *meso-damage theory* playing a central role along the interface between solid mechanics and material sciences. The development of meso-damage theory opens a new dimension to mesoplasticity. A brief account is attempted here for this relatively modern topic, with attention focused on the basic, influential theories as well as some important applications to delineate various mechanical behavior of solid materials.

8.1.1 What is Meso-Damage?

Damage is a rather vague term used to characterize the degradation of materials (and structures in some circumstances) by intrinsic mechanical, physical and chemical changes in their internal structures. The geometries, material modeling and research methodologies for microscopic, meso and macroscopic damage theory are summarized in Table 8.1. The microscopic theory studies the physical processes of damage from the length scale comparable to atomic or molecular size based on the methodology of solid state physics. The microscopic data are assembled by (classical or quantum) statistical mechanics. The theoretical completeness and application feasibility of this approach have not yet been achieved. On the other hand, the macroscopic theory provides another vision on the damage. A macroscopic (scalar or tensorial) variable field is introduced to symbolize damage upon which the principles of continuum mechanics and continuum thermodynamics apply, so comes the name of *continuum damage mechanics*. Limited engineering applications have been claimed by continuum damage mechanics. However, it suffers the same limitations as those illustrated in section 2.10 for the case of continuum plasticity.

Meso-damage is referred to a damaged region with certain geometric configuration in a meso-scale, as shown in Fig. 8.1. The geometry and material background of such a region will evolve with respect to loading environment. Meso-damage theory studies the laws governing the formation and evolution of meso-damage, and appropriate criterion by which material or structural failure under prescribed loading environment can be predicted.

The meso-theory of material damage comes as a product of the interdisciplinary studies of solid mechanics and material science. On the one hand, the meso-theory neglects the numerous details during the physical processes of damage so that lengthy calculations in statistical mechanics are avoided. On the other hand, the incorporation of actual geometric images and physical procedures featured in the meso-model provide a realistic background for damage variables and damage evolution. Consequently, damage is substantiated beyond abstract mathematical variables and equations. The meso-damage theory deepens our understanding towards the essential aspects of damage and clarifies the ad hoc phenomenological assumptions engaged in the continuum damage mechanics theory. The meso-model of damage deals with, from geometries as well as thermal-mechanical deformation process, the configurations and distributions of various damage structures and predicts their nucleation, growth and coalescence to macroscopic defects. In essence, meso-damage theory is a continuum theory which links structures of different length scales. As long as the constitutive equations and damage evolution laws for the material element are obtained through a meso-scale analysis, they can be directly invoked to assess the macroscopic damage behavior of specimen and structure components.

Table 8.1. Characterization of microscopic, meso and macroscopic damage theories

	microscopic	meso	macroscopic
geometry	vacancies, atomic bonds, dislocation	voids, grain boundaries, microcracks	specimen, structures, macrocracks
material description	physical equation	matrix constitution and interface modeling	constitutive equations & damage evolution
methodology	solid state physics	continuum mech. and material sciences	continuum mechanics

The study of meso-damage theory was initiated by metallurgists through the study on creep damage of metals at elevated temperature. The void growth mechanism under creep condition was discussed by Robinson[1] and the creep damage controlled by grain boundary diffusion was first discussed by Hull and Rimmer[2]. In the early sixties, the self-consistent method (SCM) was introduced to estimate the stiffness of composite materials and this method led to the famous *Hashin-Shtrikman bound*[3] widely used in composite materials. The participation of solid mechanists in this subject began from the late sixties, following the works of McClintock[4] on the growth equation for a long cylindrical void, and of Rice and Tracey[5] for a spherical void. From a different research front, the pioneer work of Budiansky and O'Connell[6] in 1976 revealed the effect of distributed microcracks on the overall constitutive behavior of brittle-damaged geological materials. Those works shaped up the prototype of meso-damage analysis for ductile and brittle materials. Later on, Gurson[7,8] completed the rate independent ductile damage formulation for void evolution mechanism, as will be described in detail in the next section. This meso-damage framework was subsequently extended by Tvergaard and Needleman, as well as their co-workers, to various sophistications and applications, as summarized in references [9] and [10]. Progresses in this subject also create a sub-discipline called computational micromechanics which grows rapidly with the availability of large computers. On the other hand, following the traditional analyses set by Robinson and by Hull and Rimmer, Cocks and Ashby[11], and later on Ashby and Dyson[12], proposed a spectrum of void growth models to cover various creep damage phenomena. More elaborate analyses on creep damage were furnished by Chuang et al.[13] for diffusion controlled creep mechanisms and by the Hutchinson[14]-Tvergaard[15] theory for the case of creep damage controlled by grain boundary cavitation. The development of meso-damage theory in the eighties was characterized by its applications to various non-metallic material systems, following the early work of Budiansky and O'Connell. Important developments were recorded in the research on rocks, concretes and structural ceramics whose damage behavior is governed by distributed (orderly or randomly oriented) microcracks. The works by Kachanov[16,17] for orderly distributed microcracks, and by Nemat-Nasser et al.[18,19] for contact surface friction, crack array stability and load induced anisotropy, and by Bazant[20] and Kracjnovic[21] for brittle damage characterization highlighted this development. Parallel advances on damage and toughening theories of polymeric solids, e.g., see Argon[22], and especially for composite materials, see Hashin[23], Dvorak et al.[24], have been both intensive and extensive. The meso-damage theory itself was gradually matured from the past decade and was diversified into various related disciplines. It is our objective in this chapter to present the meso-damage theory in a systematic and integrated way as possible, so that our readers can mastery the essential aspects of this newly-developed discipline.

8.1.2 Meso-Damage Structures

Various *meso-damage structures* are demonstrated in Fig. 8.1. They are :
(a) dispersed voids with relatively uniform sizes;
(b) distributed voids with several characteristic sizes;
(c) voids embedded in a macroscopic shear band;
(d) voids (or weak second phases) connected by localized shear bands;
(e) non-intersecting microcracks with random or orderly orientations;
(f) connected orderly aligned microcracks as encountered in the multiple fracture of composite laminates;
(g) craze formation in thermoplastics;
(h) persistent dislocation structures created during cyclic loading, like persistent slip bands and alternating dislocation cells.

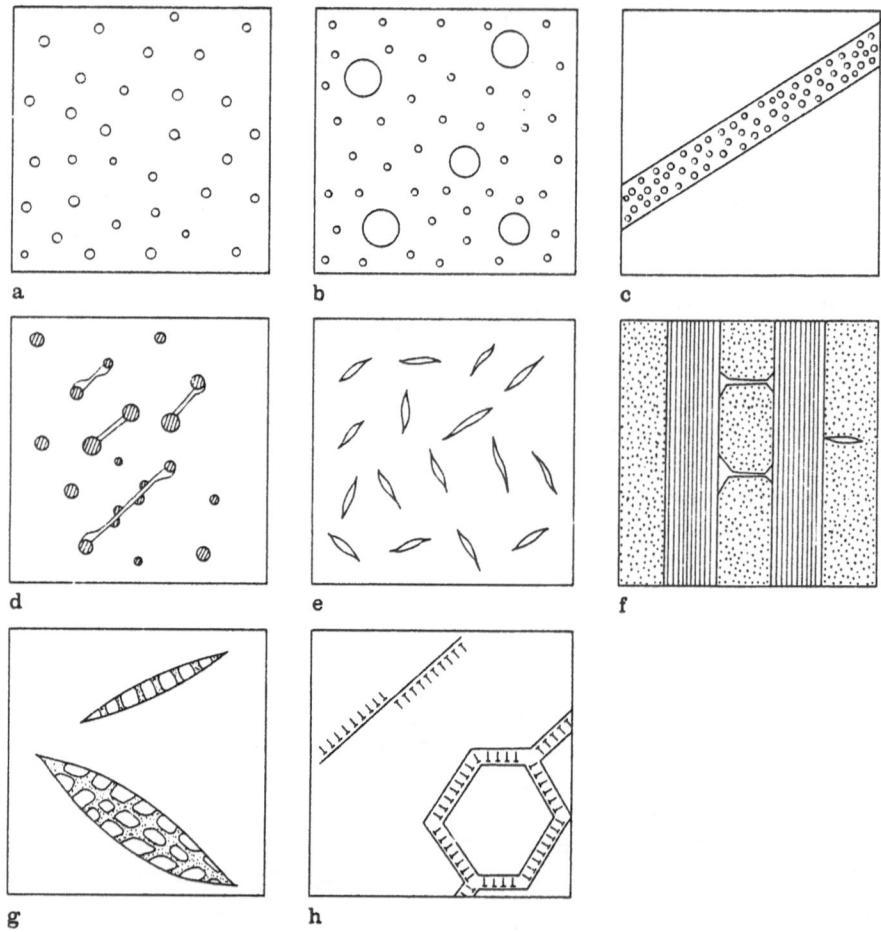

Fig. 8.1. Eight typical meso-damage structures. Graphs **a** to **d** are based on the void mechanism associated with ductile fracture, while **e** and **f** are pertinent to brittle damage manifested by microcracks. Graphs **g** and **h** correspond to the damage in thermoplastics and the damage produced during cyclic plastic deformation

The above listed meso-damage structures provide the research subjects of the meso-damage theory. Due to the limited scope of the present chapter, only the essential aspects for selected meso-damage structures will be treated in the sequel.

8.1.3 Approach to Meso-Damage Study

As we mentioned before, meso-damage theory studies the evolution of various meso-damage structures, including their formation, growth and final coalescence to macroscopic defects. The meso-damage theory regards those meso-damage structures as geometric configurations which could be analyzed by a continuum mechanics approach. Roughly speaking, meso-damage study could be characterized by the continuum mechanics analysis to meso-damage structures reflecting observations from material sciences. The objective for this analysis is to obtain appropriate constitutive equations with the incorporation of damage characterizing variables, and the *averaging procedure* can be carried out once for all. The meso-damage study is featured by

(1) The meso-damage analyses provide links among various length scales and material phenomena through *homogenization* or *heterogenization* technique.
(2) This analysis is characterized by the correlation of the physical description in a relatively rough scale to the computation on certain geometric configuration in a relatively fine scale.
(3) The whole analysis can be decomposed into several steps connecting two adjacent length scales with the pointwise mechanical field of the rough scale serves as the remote boundary conditions for the fine scale.
(4) The transition from one scale to another can be performed once for all.

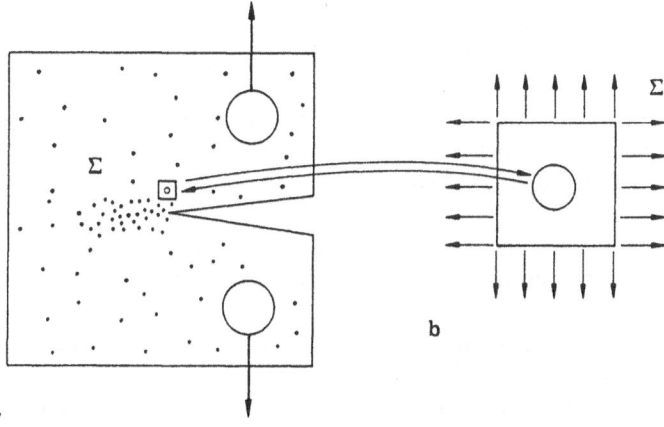

Fig. 8.2. Transition between different research scales. **a** A macroscopic specimen where damage is treated as continuously distributed field; **b** at meso-scale the damage is portrayed by a geometric cavity

To delineate this approach, let us consider the example shown in Fig. 8.2. In the graph (a) of Fig. 8.2, a compact tension specimen is loaded through the pinhole by the testing machine. A continuous stress field Σ is produced by the loading, along with a possible damage distribution across the specimen. From the macroscopic viewpoint, one has to find the constitutive equation and especially the evolution law for the damage. In a meso-damage approach, this task can be accomplished by analyzing a material element (or a basic cell) of the specimen, as examplified in Fig. 8.2(b), where the macroscopic stress field evaluated at that point, namely Σ, serves as the remote boundary condition. In this example, the meso-damage structure underlining ductile fracture could be modeled by distributed voids. The meso-mechanics calculation for the configuration in Fig. 8.2(b) will furnish a relationship between Σ and the averaging remote strain, as well as a law governing the expansion of the void. These results could be substituted back into the macroscopic configuration in Fig. 8.2(a), as the constitutive equation and the damage evolution law, respectively. Furthermore, the macroscopic relations provided by the cell analysis in Fig. 8.2(b) are not only applicable to the calculation in Fig. 8.2(a), but also applicable for any macroscopic configurations as long as the underlying damage process is the same. We thus obtain a macroscopic damage description for a class of problems.

After this brief introduction on the general aspects of meso-damage theory, we will concentrate on the two most common types of meso-damage processes, namely the meso-damage by void evolution and that by progressive cracking, successively in the next two sections.

8.2 Meso-Damage by Void Evolution

8.2.1 Void Nucleation, Growth and Coalescence

As observed by many metallurgists, the ductile failure of metals are usually characterized by *void nucleation, growth and coalescence* process. It is schematically illustrated in Fig. 8.3 that evolution of voids proceeds by three different stages as follows:

(1) Voids nucleate along the surfaces of second phase particles as shown in Fig. 8.3(a), or along grain boundaries in the form of *grain boundary cavitation.*

(2) The void nuclei expand as further load applies, with possible detachment of previously adhered particles resulting in loss of mass.

(3) As the void density exceeds a critical value, void coalescence could occur along a densely populated void sheet, to form a macroscopic penny shape crack.

The above-mentioned three stages of void evolution will be explored in this section under rate independent and rate dependent (creep condition) formulations.

8.2.2 Rate Independent Ductile Damage Formulation

The formulation of *rate independent ductile damage* consists of the following essential aspects

(1) A description of void configuration, including both the density of and (in some circumstances) the geometric shape of the voids.

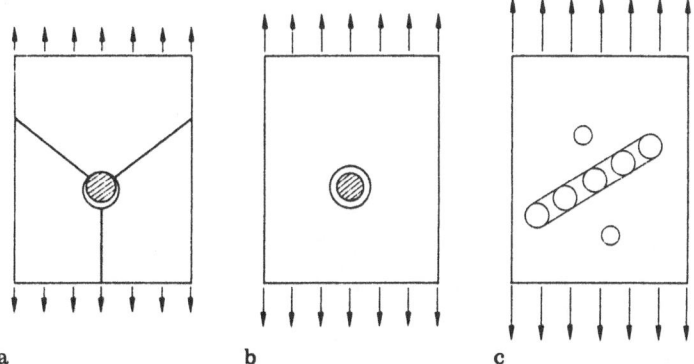

Fig. 8.3. Evolution of voids under load application. a Void nucleation near the second phase boundary; b void growth and detachment of the second phase particle; c coalescence of void sheet

(2) Simulation of the void evolution law.

(3) Characterization of the matrix material, which might encompass both hardening and softening behavior. The latter could be caused by the presence of micro-voids in a finer length scale.

(4) An averaging procedure has to be selected, which could range from cell model, self-consistent method to any other homogenization methods, with the proper account of void interaction.

(5) The pertinence of the model can be evaluated from its resultant macroscopic features, such as dilatational plasticity, pressure sensitivity, non-normality, etc..

A concrete example to demonstrate the above aspects is furnished by the well-known *Gurson theory* as discussed in the next two subsections.

8.2.3 Gurson's Model

Based on the pioneer works of McClintock[4] and of Rice and Tracey[5], Gurson[7,8] proposed the first complete meso-damage formulation to deal with ductile fracture caused by void evolution. The void-containing material is modeled by cell elements demonstrated in Fig. 8.4 whose respective responses would predict the constitutive behavior of the damaged solids. One criticism of Gurson's model is that the cell elements shown in Fig. 8.4 cannot integrate themselves into a solid material without the presence of gaps or overlaps. Nevertheless, the assembly of those cells would be not far from a perfect solid because the spherical cells presented in Fig. 8.4 are quite close to the shape of *Kelvin tetrakaidecahedron*. Among the four cell characterizations depicted in Fig. 8.4, the spherical cell model without the presence of rigid wedges, as shown in Fig. 8.4(b), received most attentions in the subsequent development of the Gurson's theory.

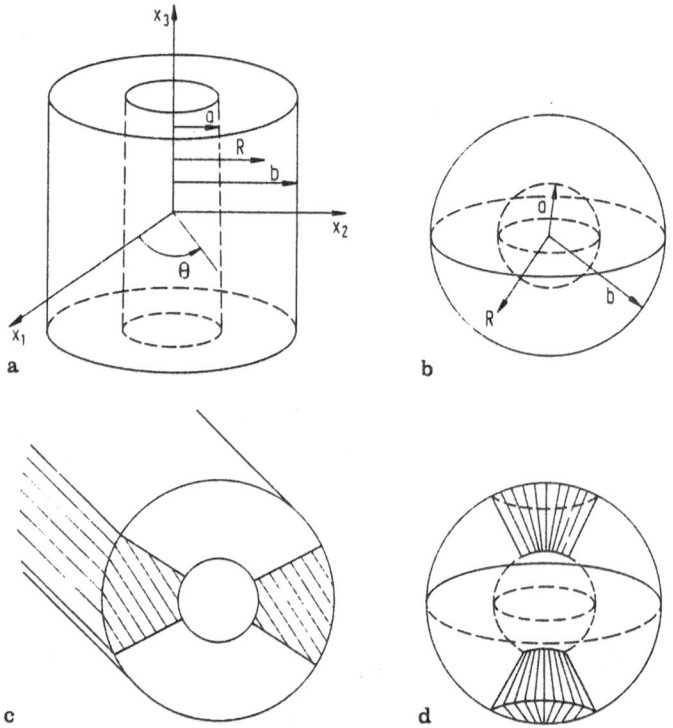

Fig. 8.4. Idealizations of material cells in Gurson's model. **a** Long cylindrical cell; **b** hollow spherical cell; **c** and **d** correspond to **a** and **b**, respectively, but with the presence of rigid wedges

While the void damage is described geometrically in Fig. 8.4, the material matrix surrounding the voids is modeled by a rigid plastic, isotropic hardening J_2 flow theory, implying a convex and smooth von-Mises yield surface for the material matrix. The explorations under a kinematically hardening or mixed hardening formulation will be discussed later. Besides, the employment of a softening matrix response has also been the focus point of several recent studies. If we adhere to the hypothesis that the matrix material (not equivalent to the overall response) strain-hardens, the relation between the effective matrix stress $\bar{\sigma}$ and the effective matrix plastic strain $\bar{\epsilon}^p$ could be described by a single curve as shown in Fig. 8.5 for a rate independent solid, with tangential modulus (both plastic and overall) h given by

$$h = d\bar{\sigma}/d\bar{\epsilon}^p. \tag{8.1}$$

The lowercase bold symbols $\mathbf{v}, \sigma$ and ϵ are used hereafter to represent the velocity, stress and strain tensors inside a material cell where requirement of plastic incompressibility must be satisfied. The macroscopic stress and strain tensors will be denoted by uppercase bold symbols Σ and $\mathbf{E}$, and the latter is compressible provided the *volume fraction of voids*, denoted by f herein, is not negligible.

According to the rigid plastic theory, the velocity field inside a cell is linked to the macroscopic deformation rate $\mathbf{D}$ (or $\mathbf{D}^p$) by the following Bishop-Hill relation[25]

$$\mathbf{D}^p = <\dot{\epsilon}> = (1/2V) \int_S (\mathbf{v}\mathbf{n} + \mathbf{n}\mathbf{v})\, dS \tag{8.2}$$

where V denotes the volume of the material cell, and S the exterior surface with outward unit normal $\mathbf{n}$. If $\mathbf{v}$ and $\dot{\epsilon}$ represent a set of any kinematically admissible fields satisfying kinematic equations, velocity boundary conditions, continuity condition and incompressible requirement, then the macroscopic plastic dissipation $\dot{W}^p(\dot{\epsilon})$ can be defined as

$$\dot{W}^p(\dot{\epsilon}) = <\sigma : \dot{\epsilon}> \equiv \Sigma : \mathbf{D}^p(\dot{\epsilon}). \tag{8.3}$$

Under the definition of $\mathbf{D}^p$ in (8.2), the last equality in (8.3) could be regarded as a definition of macroscopic stress Σ. It is worthwhile to notice that Σ will have the physical significance of $<\sigma>$ only if σ satisfies the equilibrium equations.

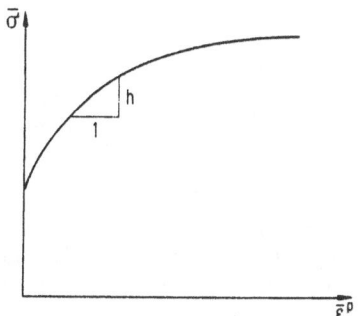

Fig. 8.5. Effective stress vs. effective plastic strain curve in the material matrix inside a cell

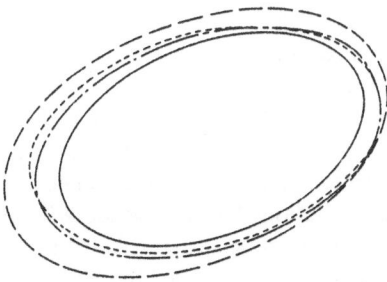

Fig. 8.6. Schematic representation of upper bound nature of approximate yield surfaces

Due to the minimum plastic dissipation principle of Taylor, the actual (marked by a superscript "A") meso-kinematical fields $\mathbf{v}^A$ and $\dot{\epsilon}^A$ should minimize the macroscopic plastic dissipation inside the material cell, namely

$$\dot{W}^p(\dot{\epsilon}^A) = \min_{\dot{\epsilon}} \dot{W}^p(\dot{\epsilon}). \tag{8.4}$$

We emphasize here that the equality sign in (8.4) holds only if $\dot{\epsilon}$ could be selected from all possible kinematically admissible fields, otherwise the estimate on the right hand side of (8.4) only serves as an upper bound for the plastic dissipation. Thus, (8.4) is also called the upper bound theorem of mesoplasticity, referred to the first relation of equation (7.65) in the previous chapter. Gurson was able to prove that when the meso-kinematically admissible fields in accordance to (8.2) and (8.3) are averaged to macroscopic quantities, the resulted approximate stress and deformation rate fields exhibit the following properties:

(1) All approximate macroscopic yield surfaces, $F(\Sigma, \bar{\sigma}, f) = 0$, obtained from this procedure are outside, and at most coincide with, the actual macroscopic yield surface, as shown in Fig. 8.6.

(2) The smoothness and convexity of the yield surface for matrix material could infer the smoothness and convexity of the approximate macroscopic yield surfaces. Furthermore, the following normality flow rule can be applied to obtain $\mathbf{D}^p$ from yield function F

$$\mathbf{D}^p = \Lambda(\partial F/\partial \Sigma) \tag{8.5}$$

where the flow factor Λ should be determined from the consistency condition with the incorporation of specific hardening law. Please notify that (8.5) establishes the existence of a macroscopic potential function.

Gurson[7,8] constructed the respective kinematically admissible fields for the four different cell representations in Fig. 8.4, and then optimized those fields according to (8.3). He finally computed the macroscopic stress by (8.2) and the work conjugate relation

$$\Sigma = \partial \dot{W}^p / \partial \mathbf{D}^p = \langle \sigma : (\partial \dot{\epsilon}/\partial \mathbf{D}^p) \rangle. \tag{8.6}$$

This procedure provides an apparatus to obtain approximate yield surfaces. As illustrated in Fig. 8.6, a yield surface (dashed curve) associated with an arbitrary set of kinematically admissible fields forms an upper bound of the actual macroscopic yield surface (solid curve), whilst the kinematically admissible fields after optimization to a certain degree would give rise of a tighter upper bound (dot-dash curve) to the actual yield surface. Gurson next pursued to obtain approximate, but analytical, representations for the *macroscopic yield surface* (dot curve). For the spherical void with surrounding matrix yielded completely, the following approximate analytical expression was obtained by Gurson

$$F(\Sigma, \bar{\sigma}, f) = (\Sigma_e/\bar{\sigma})^2 + 2f \cosh(3\Sigma_m/2\bar{\sigma}) - 1 - f^2 = 0 \tag{8.7}$$

which defines a macroscopic yield function only relevant to the macroscopic Mises stress Σ_e, the matrix Mises stress $\bar{\sigma}$, the macroscopic hydrostatic stress Σ_m and the void volume fraction f. Numerical calculations[8] showed that the above expression was sufficiently accurate for the actual upper bound yield surface of dilute distributed voids. A pictorial representation of (8.7) is delineated in Fig. 8.7 where the macroscopic hydrostatic stress is served as abscissa and the macroscopic Mises stress is plotted as the ordinate, both normalized by matrix Mises stress. The *plastic volume dilatancy* signified by the void volume fraction f causes a conjugate effect of *pressure sensitivity* on the yield surface, as demonstrated clearly in Fig. 8.7.

Attention is now focused on the evolution laws of the hardening variable $\bar{\sigma}$ and the damage variable f to establish completely the mechanics state of a material cell. The evolution law of the former could be facilitated by the *equivalence principle*

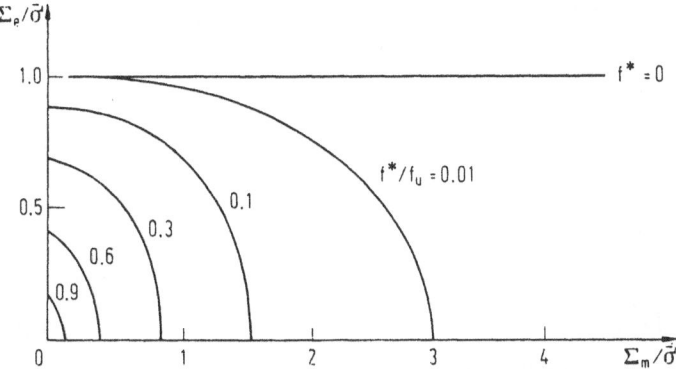

Fig. 8.7. Yield surfaces and pressure sensitivity of Gurson's model

between macroscopic and meso plastic work,

$$(1-f)\bar{\sigma}\dot{\bar{\epsilon}}^p = \Sigma : \mathbf{D}^p \tag{8.8}$$

which also serves as the definition of the matrix effective plastic strain rate, $\dot{\bar{\epsilon}}^p$. From this equality and the definition (8.1) of plastic tangential modulus h, we arrive at an *evolution equation for the matrix effective stress* $\bar{\sigma}$

$$\dot{\bar{\sigma}} = h(\Sigma : \mathbf{D}^p)/[(1-f)\bar{\sigma}] \tag{8.9}$$

where h can be determined from the uniaxial stress strain relation for a void-free material.

Similar to many physical procedures, the *evolution of void volume fraction* is composed of contributions from void nucleation and void growth

$$\dot{f} = \dot{f}_{nu} + \dot{f}_{gr}. \tag{8.10}$$

The plastic incompressibility of the matrix would require the following void growth rate

$$\dot{f}_{gr} = (1-f)tr\mathbf{D}^p. \tag{8.11}$$

Gurson proposed the following kinetic law governing void nucleation

$$\dot{f}_{nu} = A\dot{\bar{\sigma}} + (B/3)\dot{\Sigma}_m \tag{8.12}$$

where A represents the influence on void nucleation by the increment of matrix flow stress, and B describes the same influence arised from the increment of hydrostatic stress. Under rate independent assumption, both A and B are functions irrelevant to stress rates. As readers may have already noticed, equation (8.11) alone would correspond to the conservation of mass during the void growth. The addition of the void nucleation term in (8.10) would imply a loss of material during the formation of void damage structures. This fact should come as no surprise to us because the *detachment of second phase particles* has already been anticipated in Fig. 8.3. The incorporation of a void nucleation term in (8.10) enables us to describe the non-volume conserved deformation of material matrix from the physical viewpoint, and which can also be served to suppress early *damage localization* tendency in Gurson's model as will be addressed in subsection 8.2.8.

We next derive the constitutive response under Gurson's model. The additive decomposition of deformation rate tensor $\mathbf{D}$ into elastic and plastic contributions enables us to write

$$\mathbf{D} = \mathbf{D}^e + \mathbf{D}^p \tag{8.13}$$

where the elastic part is simply governed by the generalized Hooke's law

$$\dot{\Sigma} = \mathbf{L} : \mathbf{D}^e \tag{8.14}$$

with $\mathbf{L}$ being elasticity tensor. Then the overall material response could be obtained by the consistency condition

$$\dot{\Sigma} = \mathbf{L}^t : \mathbf{D} \tag{8.15}$$

where the tangential stiffness matrix $\mathbf{L}^t$ is given in the familiar Rice-Rudnicki form

$$\mathbf{L}^t = \mathbf{L} - (\mathbf{L} : \mathbf{P})(\mathbf{L} : \mathbf{Q})/H \tag{8.16}$$

where

$$\begin{aligned}
\mathbf{P} &= \partial F/\partial \Sigma \\
\mathbf{Q} &= \mathbf{P} + (B/3)(\partial F/\partial f)\,\mathbf{1} \\
H &= \mathbf{Q} : \mathbf{L} : \mathbf{P} - \{(1-f)(\partial F/\partial f)(\partial F/\partial \Sigma_m) + \\
&\quad [h/(1-f)\bar{\sigma}]\,[A(\partial F/\partial f) + \partial F/\partial \bar{\sigma}]\,\mathbf{P} : \Sigma\}.
\end{aligned} \tag{8.17}$$

Obviously, the Voigt symmetry of $\mathbf{L}^t$ will be lost whenever B is nonzero. Special treatment is needed for the numerical implementation of Gurson's theory under a non-trivial B.

To get insights to the function A and B, we have to think about the *controlling mechanisms of void nucleation*. Some metallurgists believe that void nucleation is controlled by the plastic strain at the nucleation sites. The experimental data of Gurland[26] indicated a linear relationship between the void nucleation rate and the effective plastic strain rate and that straight line could be extrapolated to the origin. This proportionality relation also implies that $\dot{f}_{nu}$ is proportional to $\dot{\bar{\sigma}}$. This viewpoint was also shared by Goods and Brown[27] from their study on void nucleation mechanism, which stated that the separation between second phase particles and matrix was controlled by local plastic strain. Under the further assumption that the nucleation rate is described through a normal statistical distribution as given by Chu and Needleman[28]

$$A = \left[f_n / \left(\sqrt{2\pi}\, hS \right) \right] \exp \left\{ -\frac{1}{2} \left[(\bar{\epsilon}^p - \epsilon_N)/S \right]^2 \right\}. \tag{8.18}$$

$$B = 0$$

Namely, the nucleation rate is governed by the value of effective plastic strain $\bar{\epsilon}^p$ over a *profusion strain* ϵ_N, the latter also corresponds to the strain level at the maximum nucleation rate. f_n in (8.18) represents the *saturated volume fraction of void nucleation* as if the deformation could proceed infinitely, and S denotes the standard deviation of normal distribution.

Another group of experimental data, like those proposed by Argon et al.[29], suggest that void nucleation process is controlled by the maximum normal stress transmitted along the interface between matrix and the second phase particles. As pointed out by Needleman and Rice[30], such a nucleation mechanism is controlled by a stress combination of $\bar{\sigma} + \Sigma_m$. Also according to Chu and Needleman[28], the normal statistical distribution formula for such a circumstance becomes

$$A = B = \left[f_n / (\sqrt{2\pi} S) \right] \exp \left\{ -\frac{1}{2} \left[(\bar{\sigma} + \Sigma_m - \sigma_N) / S \right]^2 \right\} \tag{8.19}$$

where σ_N is the *profusion stress* at which the void nucleation occurs at the maximum rate.

We now summarize the basic advantages of Gurson's model as follows:

(1) Gurson's model is proposed under a sound physical basis and has a clear geometric image of meso-damage.
(2) Its formulation is analytic and self-complete.
(3) It is free of curve-fitting parameters.
(4) The numerical implementation of Gurson theory still maintains at the same computer MIPS level as the incremental formulation of conventional plasticity models.
(5) Gurson's model can describe the complete process of ductile fracture, as will be shown in the next subsection.

8.2.4 Improvements and Applications of Gurson's Model

Since the advance of Gurson's model, there have been many research works launched for the improvements and applications of Gurson's theory. Further elaborations on the void nucleation description, as presented in the preceding subsection only consists of one direction of these improvements. The other important improvements on the Gurson's theory will be outlined next.

Interaction among voids. In the original version of Gurson's theory, each individual void is treated separately, with the *effect of void interaction* neglected. To reveal this interaction, detailed finite element calculations were carried out by Tvergaard[31,32] for a rigid plastic body with double periodic void array. He found that if the void volume fraction f in Gurson's yield surface is modified to $q_1 f$

$$F(\Sigma, \bar{\sigma}, f) = (\Sigma_e / \bar{\sigma})^2 + 2q_1 f \cosh(3\Sigma_m / 2\bar{\sigma}) - 1 - (q_1 f)^2$$
$$= 0 \tag{8.20}$$

then the corrected yield surface would fit better to the finite element results with consideration on void interaction. Equation (8.20) would reduce to the original Gurson's yield function when q_1 equals unity. The numerical data of Tvergaard, however, would suggest a q_1 value of 3/2. The issue of void interaction can also be tackled from self-consistent method.

Void Coalescence. Experimental observations indicate that the ligaments between neighboring voids would neck to produce *void coalescence on a densely populated void sheet*, whenever the void growth reaches a critical level. Different approaches to include void coalescence effect were summarized by Tvergaard[9]. One way to accomplish that is to add a void coalescence rate, $\dot{f}_{coal}$, in (8.10) to account for its accelerating effect on the evolution of f. Another effective treatment consists of replacing the void volume fraction f in (8.21) to f^*, Tvergaard and Needleman[33]

$$F(\Sigma, \sigma, f) = (\Sigma_e/\bar{\sigma})^2 + 2f^* q_1 \cosh(3\Sigma_m/2\bar{\sigma}) - 1 - (q_1 f^*)^2$$
$$= 0 \qquad (8.21)$$

where the effect of void coalescence is reflected by the new damage variable f^* defined by

$$\begin{aligned} f^* &= f \qquad f \le f_c \\ f^* &= f_c + (f_u - f_c)(f - f_c)/(f_F - f_c) \qquad f > f_c \end{aligned} \qquad (8.22)$$

where f_c denotes the critical void percentage at which the void coalescence starts. A rough estimate of this value was previously set at 15%, from the argument that the necking of inter-void ligaments occur when the ligament size is equal to the void diameter, under the assumption of periodically distributed void array. The recent analytical, numerical and experimental researches on this subject[34,35,36], however, indicated that the value of 15% overestimated the value of f_c. A more appropriate estimate of f_c may be below 0.1. The other two characteristic values, f_F and f_u, in (8.22) represent the volume fractions for f and f^* at ultimate fracture, respectively. It is assessed from (8.21) that the ultimate value of f^* will be $f_u = 1/q_1 (= 2/3)$ at which the ability of a material cell to withstand macroscopic stress is permanently lost, as shown in Fig. 8.7. The value of f_F, the actual void volume fraction at failure, was selected as 0.25 in Ref. [33]. But recent studies[34,35,36] suggest a smaller value.

Mixed hardening formulation. Although the predictions on flow localization phenomena by Gurson's model are improved remarkably from those assessed by conventional incremental plasticity model with smooth yield surfaces, they still overestimate considerably the bifurcation loads. This fact may partially attribute to the isotropic hardening law employed in the preceding development. The latter would consequently lead to low curvature yield surfaces which offer a stiff response to the sudden switch of stress path experienced by the material elements. A remedy for this undesired response lies in formulating Gurson type theory from the framework of mixed hardening. Researches in this direction was initiated by Mear and Hutchinson[37] under a rate independent mixed hardening formulation and was later extended to rate dependent formulation by Becker and Needleman[38]. In the latter analysis, the Gurson's type yield surface (8.21) was further modified to

$$F(S, \bar{\sigma}, f) = (S_e/\sigma_F)^2 + 2f^* q_1 \cosh(3S_m/2\sigma_F) - 1 - (q_1 f^*)^2$$
$$= 0 \qquad (8.23)$$

where $S = \Sigma - b$ is temporarily used to denote the active stress tensor, with S_e and S_m being the counterparts of Σ_e and Σ_m, respectively. The same interpretation as in subsection 2.5.2 can be given to σ_F. Improvements on predictions of shear band formation and necking instability were reported in references [37] and [38] by using mixed hardening formulation.

Applications of Gurson's model have been very successful in the past decade. Here we do not attempt to enumerate those applications, but rather outline the following selected aspects to show our readers on the scope and understanding gained by those applications.

Ductile failure of smooth specimen under monotonic loading. One of the most successful achievements of Gurson's theory is the *simulation of ductile failure* (including failure modes and up-to-failure stress strain curves) for smooth tensile specimen in both plane strain and axisymmetric conditions. As illustrated in Fig. 8.8, the ductile failure modes under plane strain and axisymmetric loadings are distincted by shear band failure and cup-cone fracture, respectively. Quantitative explanations for the above distinction were not provided until the application of Gurson's theory. The numerical simulations by Saje, Pan and Needleman[39] under plane strain condition, and by Tvergaard and Needleman[33] for an axisymmetric round bar revealed in every detail the failure modes as shown in Fig. 8.8. Besides, the interesting "knee type" stress strain curve for a round bar under uniaxial tension was also reproduced, with different damage stages checked correct with the testing results of Bluhm and Morrisey[40].

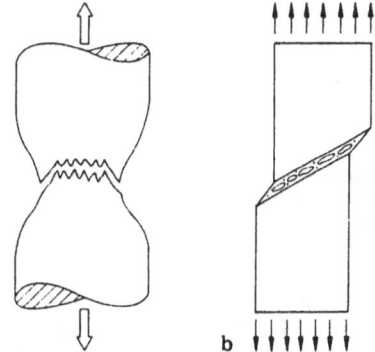

Fig. 8.8. Predictions of ductile failure modes by Gurson model. **a** Cup-cone fracture for a tensile rounded bar; **b** shear bands formation under plane strain tension

Ductile failure of notched specimen. Our next example for the application of Gurson theory consists of the *ductile failure of notched specimen* under plane strain and axisymmetric tensions. This task was accomplished by Needleman and Tvergaard[41], who were able to simulate that the crack formation inside a plane strain specimen would likely start at the notch root whereas cracks would appear from the specimen axis if axisymmetric loading prevails. These predictions agree well with the experimental observations. Gurson's model provides a mean to reveal material failure modes under various geometries and loading conditions. It can also be used to simulate the initiation and growth of cracks.

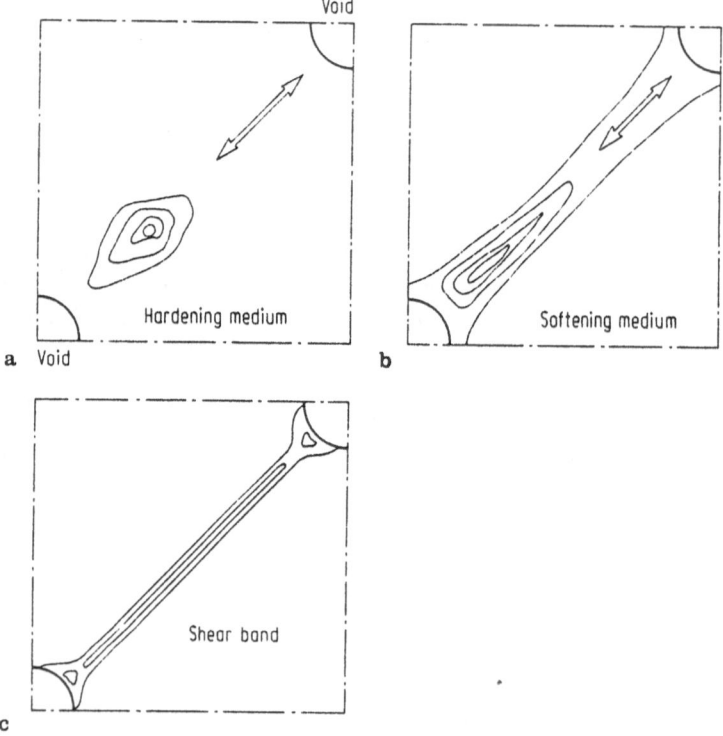

Fig. 8.9. A schematic illustration on the dynamic formation of slip bands. **a** A disturbance pulse bounces between weak spots in a hardening matrix; **b** the gradual intensification of the bouncing disturbance in a softening matrix; **c** formation of a shear band

Formation of shear bands. It is well-known that the critical load for *shear band formation* would be substantially overestimated by classical incremental plasticity theory based on a smooth yield surface. The engagement of the damage-softening behavior, as incorporated in Gurson's theory, would help to reduce the high bifurcation load to a more realistic level. The implementation of this idea was initiated by Yamamoto[42], who clearly demonstrated the reducing effect on bifurcation load for shear band formation by Gurson's formulation. On the other hand, several experimental investigations also revealed a localized high population of void volume fraction within the incipient shear bands. Further elaborations on this aspect were made by Saje, Pan and Needleman[39] where the viscoplastic matrix behavior was incorporated, and by Mear and Hutchinson[37] under a mixed hardening formulation. More exciting case about the dynamic formation of shear band among periodically distributed weak spots was recently reported by Needleman[10] by means of Gurson's model. It is numerically visualized that a disturbance pulse emitting from a weak spot would bounce between the neighboring weak spots if the matrix material is characterized as hardening, as shown in Fig. 8.9(a). If the matrix material softens with respect to deformation, however, the same disturbance pulse initially bouncing between the weak spots would be gradually trapped by the softening matrix, and finally localized to form a shear band, as depicted in Fig. 8.9(b) and (c). The whole procedure of *dynamic formation of shear bands* can be simulated vividly by computer graphics.

Central burst during extrusion. Aravas[43] studied the *central burst* phenomenon occurred during the extrusion process of a steel bar by Gurson's model. His numerical simulations for different processes under various die design parameters agreed well with the experimental observations. Hence, a software package featured by Gurson's model could be used for a closed-loop design study, as well as a central burst prevention check for this particular metal forming process.

Crack-void interaction. *Crack-void interaction* is essential in the understanding of ductile failure, and Gurson's model provides a unique link on how the voids in the material nearby the current crack tip merge onto the advancing crack. A mode I problem for crack interacted with a hole along the crack extension line under remote tensile loading was studied by Aravas and McMeeking[44], where the growth and coalescence of hole onto the long crack was simulated by characterizing the surrounding material by Gurson's model. This study is also helpful in elucidating ductile tearing processes and provides quantitative assessment of fracture parameters. Under mixed mode loading condition, the crack-void interaction was studied by Aoki, Kishimoto and Sakata[45] by Gurson's model where the hole located about 45° above or below the crack extension plane. It was demonstrated by those authors that the hole on one side of the crack will be gradually closed and the hole on the other side will be substantially expanded under a mode II dominated loading. Furthermore, it was revealed that the a linkage between the expanded hole and the running main crack could occur from a location somewhat behind the current crack tip. Which provided explanation for some crack branching phenomena observed in the experiments. Consequently, Gurson's model can be used to explore the *crack propagation path* as well as the *toughness control* of material systems. An extensive computation for distributed weak spots scattering around the main crack was done by and Needleman and Tvergaard[46] on supercomputer. The weak spots are modeled by the same Gurson formulation as in the nearby matrix but with slightly lower profusion stress for void nucleation. Fracture paths of realistic appearances were produced and the vital significance of the distribution pattern of the weak spot was confirmed.

Although signifying an important contribution to the understanding of meso-damage behavior, the Gurson analysis still leaves behind several unresolved theoretical issues and is suffered from some application limitations, as listed in the following:

(1) Only *isotropic damage behavior* is considered in the Gurson's model by employing a scalar damage variable, namely the void volume fraction f. The admissible velocity fields engaged in Gurson's analysis[7,8], however, did not exhibit spherical symmetry, and consequently would result in an anisotropic evolution of void shape. Further analyses indicate that conventional characterization of voids by damage parameters or tensors cannot get the formulation self-closed and certain truncation has to be carried out somewhere. A systematic exploration on this aspect is still to be done.

(2) The Bishop-Hill approach[25] invoked by Gurson was established under a pointwise velocity boundary condition on the exterior of the cell. Whilst in Gurson's analysis the boundary condition of the cell can only be given by the average quantities such as $\mathbf{D}^p$. Therefore, not only the kinematical fields inside the cell but also the pointwise boundary condition are varied during the optimization process represented by (8.3). It is still unclear that if an upper bound theorem could hold for such circumstances. A more pertinent formulation of this problem is still to be found.

(3) Phenomenon of matrix softening has not been explored by the Gurson's theory, although it is applicable in principle.

(4) The damage localization behavior of Gurson model, as will be addressed in subsection 8.2.8, is quite sensitive to the void nucleation condition employed.

(5) The coalescence of voids is over-simplified in the current treatment of Gurson's model. It is better to be treated from statistical considerations.

(6) Some testing data for porous and powder-compacted materials indicate a different void evolution rule from that predicted by Gurson's theory.

8.2.5 Rate Dependent Creep Damage

Rate dependent creep damage happens in metals at elevated temperature when subjected to sustained load. As a preliminary discussion, we first approach this problem from the simple models of Ashby et al.[11,12].

Classification on creep damage. The meso-damage models under creep condition was summarized by Ashby and Dyson[12]. They enumerated the following four classes of creep damage with ten distinct damage models.

A. Damage by loss of external cross-section
    (1) without necking;
    (2) with necking.

B. Damage by loss of internal cross-section
    (3) void formation within grains controlled by power-law creep;
    (4) grain boundary cavitation;
    (5) creep fracture of macroscopic crack and creep damage zone accompany
         with propagating crack tip.
C. Damage induced by degradation of microstructures
    (6) thermal coarsening of precipitated particles;
    (7) increase of mobile dislocation density;
    (8) territory creep induced by material sub-structures.
D. Damage by environment erosion
    (9) damage by internal oxidation;
    (10) damage induced by spalling of oxidation protection films.

The mechanisms underlying the above 10 damage models were given by Ashby
and Dyson[12], along with the one dimensional law for plastic strain accumulation
and damage evolution law for each mechanism. The great variety exhibited by
the above damage mechanisms was invoked by those authors to criticize the universal approach of continuum damage mechanics, where the damage evolution is
always modeled by the Kachanov law[47] regardless the difference in the damage
mechanisms.

Simple computation model of creep damage. As a demonstration of creep damage
analysis, we hereby concentrate on the damage mechanisms characterized by the
loss of internal cross-section. The behavior of material matrix is modeled by the
power-law creep

$$\dot{E}_e^c = \dot{E}_o(\Sigma_e/\Sigma_o)^n \tag{8.24}$$

where $\dot{E}_e^c$ and Σ_e are macroscopic effective creep rate and effective stress, respectively, both defined in a J_2 sense. Parameters Σ_o and n are used to describe
power-law creep. The reference creep rate, $\dot{E}_o$, is proportional to $\exp(Q_c/RT)$,
where Q_c denotes the activation energy of creep, R the gas constant and T the absolute temperature. The general descriptions of creep and damage under sustained
uniaxial tension can be expressed in the following forms

$$\dot{E}_c = \dot{E}_o C(\Sigma, T, f)$$
$$\dot{f} = \dot{E}_o F(\Sigma, T, f) \tag{8.25}$$

where f stands for the *area fraction of voids* in a representative cross-section. The
following characteristic quantities are introduced to measure creep damage
Minimum (or damage free) strain rate $- \dot{E}_m$

$$\dot{E}_m = C(\Sigma, T, 0) \dot{E}_o. \tag{8.26}$$

Fracture time $- t_f$. Under isothermal creep condition, t_f can be evaluated according
to the following formula

$$t_f = (1/\dot{E}_o) \int_{f_i}^{f_c} F(\Sigma, T, f)^{-1} \, df \tag{8.27}$$

where f_i and f_c are the void area fractions at the initial instant and the final rupture of damage evolution, respectively.

Monkman-Grant constant – C_{mg}

$$C_{mg} = \dot{E}_m\, t_f.$$ (8.28)

Fracture strain – E_f. Under isothermal creep condition, E_f becomes

$$E_f = \int_{f_i}^{f_c} \{C(\Sigma, T, f)/F(\Sigma, T, f)\}\, df.$$ (8.29)

Creep damage tolerance – μ, defined by

$$\mu = E_f / C_{mg}.$$ (8.30)

The value of μ can be used to assess the material toughness against creep fracture, as suggested by Leckie and Hayhurst[48].

Simple models of creep damage for this class of mechanisms were discussed by Cocks and Ashby[11]. The creep damage mechanisms are distinguished into the following three types, namely the damage controlled by power-law creep within the grains, the damage controlled by grain boundary diffusion and damage controlled by diffusion along the void surface. The first mechanism is predominant under a high stress level whereas the last two prevail under a low stress level. They could co-exist under an intermediate stress level.

A schematic description on the three creep damage mechanisms are given in Table 8.2 according to the discussion by Cocks and Ashby[11]. The left column correspond to a model of *creep damage controlled by grain boundary diffusion*, which elaborated the early work of Hull and Rimmer[2]. It is assumed that the matter diffusion along void surface is sufficiently rapid so that the void remains spherical and its growth is determined by the matter diffusion along grain boundaries. The simple analysis of Cocks and Ashby[11] gave the following results on creep damage parameters

$$
\begin{aligned}
C &= (2l/d)\{\Phi_o/\ln(1/f)\}(\Sigma/\Sigma_o)\\
F &= f^{-\frac{1}{2}}\{\Phi_o/\ln(1/f)\}(\Sigma/\Sigma_o)\\
C_{mg} &= (4/3d)f_c^{3/2}\\
\mu &\approx 2
\end{aligned}
$$ (8.31)

where d and l are grain size and inter-void distance, respectively. Φ_o in (8.31) denotes a dimensionless parameter composed of material diffusion constant along grain boundaries scaled by stress and temperature combination.

If the diffusion along the void surface is relatively slow, the void would no longer hold its spherical shape and the *creep damage* would be *controlled by surface diffusion*, as in the middle column of Table 8.2. In this case, the Cocks-Ashby estimates are

$$C = \left[4\psi_o\sqrt{f}/(1-f)^3\right](\psi_s/\Sigma_o d)(\Sigma/\Sigma_o)^2$$
$$F = \left[\psi_o\sqrt{f}/(1-f)^3\right](\Sigma/\Sigma_o)^3 \tag{8.32}$$
$$C_{mg} = (8\psi_s/\Sigma_o d)\sqrt{f_i f_c}$$
$$\mu \approx 1$$

where ψ_s characterizes the surface free energy and ψ_o signifies another dimensionless combination of various parameters. The low value of μ for this model implies a rather poor fracture toughness and damage may take the form of crack formation.

If the *creep damage* is instead *controlled by* the ability of *matrix power-law creep*, as shown in the right column of Table 8.2, then the creep damage should be characterized by

$$C = \left\{1 + (2\beta r_o/d)\left[(1-f)^{-n} - (1-f)\right]\right\}(\Sigma/\Sigma_o)^n$$
$$F = \beta\left[(1-f)^{-n} - (1-f)\right](\Sigma/\Sigma_o)^n \tag{8.33}$$
$$C_{mg} = -[1/(n+1)]\ln\left[(n+1)f_i\right]$$
$$\mu \approx 1$$

where r_o denotes the initial radius of the void, and β is a dimensionless constant reflecting the effect of triaxiality in the case of multi-axial loading.

Cocks and Ashby[11] also calculated the stress and temperature ranges suitable for different mechanisms, and presented the results in the form of *creep damage maps*. These maps can serve as a guideline for the engineering applications of meso-damage theory under creep condition.

On the other hand, the continuum theory of creep damage was proposed initially by Kachanov[47], and later extended by Rabotnov[49] and by Leckie and Hayhurst[48]. The one dimensional formulation under continuum damage mechanics is also featured by a pair of equations like (8.25), with functions C and F given by

$$C = (1-f)^{-n}(\Sigma/\Sigma_o)^n$$
$$F = \alpha(1-f)^{-m}(\Sigma/\Sigma_o)^m. \tag{8.34}$$

The following remarks could be made from comparison between (8.34) and the previous meso-damage models:

(1) Various creep damage equations can be derived for different types of damage mechanisms, whereas the form of evolution equations under continuum damage theory is prefixed regardless the distinction of various damage mechanisms.

(2) The damage evolution behavior for all of the above meso-mechanisms would be approximated by a power-law characterization when f tends to unity. This coincidence to the continuum model, however, is insignificant because a majority of creep life is compensated in the small damage regime.

(3) Substantial differences occur between various meso-damage models and the Kachanov model under small f. A finite nonzero damage rate is predicted

Table 8.2. Illustration of creep damage by loss of internal cross-section

Control mechanism	Grain boundary diffusion	Surface diffusion	Power-law creep inside grains

under Kachanov model for a damage free solid, whilst this rate would be either infinitely large or infinitely small, as shown in the curves of Table 8.2, for different meso-damage models. This distinction would greatly influence the damage concentration behavior as will be discussed in subsection 8.2.8.

(4) The damage rate predicted by the continuum model is enhanced monotonically with the extent of damage, while for some meso-damage mechanism, such as that shown in the left column of Table 8.2, the damage rate could first decline and then increase as f increases.

8.2.6 Grain Boundary Cavitation and Sliding

The importance of creep damage behavior gives a strong impetus for both solid mechanists and material scientists to pursue more detailed researches on this subject. A comprehensive survey on diffusion controlled creep behavior was given by Chuang et al.[13] via extensive numerical computations. The coupled effect of power-law grain creep and grain boundary diffusion has also received much attention in the past years. The case of dilute distributed voids was treated by

Needleman and Rice[50]. For the difficult problem of densely distributed voids, numerical calculation was also attempted by Anderson and Rice[51]. A topic of special interest in creep damage study is the *cavitation and sliding of grain boundaries* as constrained by matrix creep. We would proceed to model this problem in a more rigorous manner via a three dimensional constitutive formulation.

According to the experimental observation of Dyson[52], an important failure mode of creep damage lies in the cavitation processes along the grain boundaries normal to the direction of *maximum tensile stress* designated by S. These cavitated grain boundaries are constrained by the deformability of power-law creep matrix and can be regarded as *cracked facets* of approximately penny shape. As shown in Fig. 8.10, these grain boundary facets can be treated as cracks subject to linking stresses denoted by $\sigma_l < \beta S$, where $\beta < 1$ is a projection factor of the maximum tensile stress S to the respective grain boundary facets. The constitutive response and the damage evolution law will be determined by the opening and growing of these linked penny shape cracks constrained in a power-law creep matrix.

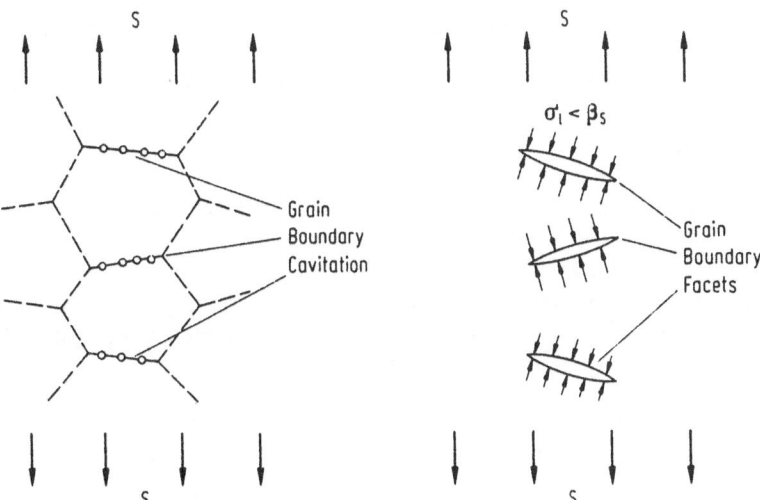

Fig. 8.10. Grain boundary cavitation constrained by power-law creep inside the grains. The crack configuration on the right indicates an idealization of the left graph representing grain boundary facets embedded in power-law creep matrix

8.2.7 Hutchinson-Tvergaard Model

Hutchinson[14] calculated the average response of dilute distributed penny shape cracks constrained in a power-law creep matrix. The approximate solution of He and Hutchinson[53] for a single penny shape crack in an otherwise infinite body was employed in this estimation. According to Hutchinson's calculation, the macroscopic creep rate $\mathbf{D}^c$ is

$$\mathbf{D}^c = \dot{E}_o(\Sigma_e/\Sigma_o)^n \{(3/2\Sigma_e)\mathbf{S} + f[(3(n-1)/2(n+1)) \\ (S/\Sigma_e)^2(\mathbf{S}/\Sigma_e) + (2S/(n+1)\Sigma_e)\mathbf{P}]\} \tag{8.35}$$

where the definition of some of the macroscopic variables are referred to (8.24), and tensor $\mathbf{S}$ denotes the macroscopic stress deviator. The scalar S, however, still conforms to the previous definition of maximum tensile stress, and calculated by

$$S = \Sigma : \mathbf{P}. \tag{8.36}$$

This expression also implies that the *projection tensor* $\mathbf{P}$ in (8.35) and (8.36)corresponds to the orientation of the maximum tensile stress. When the orientation of S coincides to the coordinate direction of X_k, we have

$$P_{ij} = \delta_{i(k)}\delta_{j(k)} \tag{8.37}$$

where the parentheses on index k indicates that no sum should be performed for the quoted index. The damage parameter f in (8.35) represents the "density" of grain boundary cavitation, and can be calculated according to

$$f = 4R^3N(n+1)(1+3/n)^{-\frac{1}{2}} \tag{8.38}$$

where N is the number of microcracks in a unit volume and R denotes the radius of penny shape crack.

The above results due to Hutchinson[14] provide a basic formulation for the grain boundary cavitation problem. But one might notice that the ignorance of unbroken ligaments along the cavitated grain boundary could overestimate the creep strain rate. Rice[54] pointed out that the linking stress on a crack facet should be less than βS, as shown in Fig. 8.10. This average linking stress σ_l across the hypothesized penny shape crack would reduce the maximum tensile stress S to an effective amount of $S - \sigma_l$. According to a modified formulation proposed by Tvergaard[15], the macroscopic creep strain rate is revised to

$$\mathbf{D}^c = \dot{E}_o(\Sigma_e/\Sigma_o)^n \{(3/2\Sigma_e)\mathbf{S} + f[(3(n-1)/2(n+1)) \\ ((S-\sigma_l)/\Sigma_e)^2(\mathbf{S}/\Sigma_e) + (2(S-\sigma_l)/(n+1)\Sigma_e)\mathbf{P}]\} \tag{8.39}$$

where σ_l can be regarded as an internal state variable controlled by an evolution equation. Equation (8.39) forms the basis of *Hutchinson-Tvergaard model* for grain boundary cavitation.

The total creep strain rate is given by a superposition of elastic, creep and thermal deformations, i.e.

$$\mathbf{D} = \mathbf{D}^e + \mathbf{D}^c + \mathbf{D}^T = \mathbf{D}^c + \mathbf{L}^{-1} : \dot{\Sigma} + \alpha\dot{T}\mathbf{1} \tag{8.40}$$

where the expression of $\mathbf{D}^e$ is the same as the one given in (8.14), and the thermal deformation is assumed to be isotropic.

The evolution laws for the internal state variables R and σ_l have to be furnished by calculation on a finer scale, as shown in Fig. 8.11. A set of evolution equations was also provided by Tvergaard[15] based on the previous studies by Needleman

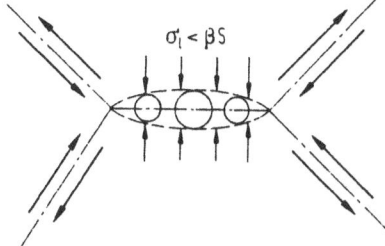

Fig. 8.11. The grain boundary cavitation model with the consideration of linking stress σ_l across the cavitated facets and sliding along nearby inclined grain boundaries

and Rice[50] and by Budiansky[55]. Those studies took into account the effects of *sliding along inclined grain boundaries* and nucleation along the facets, where the latter also controls the evolution of *microcrack density N*. A more detailed study on grain boundary cavitation was later contributed by Anderson and Rice[51] in which the cases of densely populated grain boundary cavitations, as well as the three dimensional grain boundary sliding effect were considered. It is fair to say that the researches on creep damage by grain boundary cavitation forms one of the most active research frontiers of meso-damage theory.

8.2.8 Damage Diffusion, Concentration and Localization

The spatial characteristics of damage evolution plays an important role in the determination of the ductile failure pattern of materials. As shown in Fig. 8.12, the defects are treated geometrically as cracks in a fracture mechanics approach, while they are modeled by a continuously distributed field of damage variables in a damage mechanics methodology. The question now raised is the following: can a distributed damage field be localized to a macroscopic crack under a prescribed damage evolution law? The answer to this problem relies on the possible transition of a slightly non-uniform damage distribution to the concentrated damage pattern as shown in the bottom graph of Fig. 8.12. Precisely speaking, it depends on the behavior of *damage diffusion, concentration and localization*.

A study of this nature could be pursued by the introduction of a small *initial non-uniform damage* distribution, as shown by Yang[36]. The subsequent spatial evolution patterns anticipated under different damage models due to this tiny non-uniform initial damage can be monitored to reveal qualitative damage concentration behavior associated with the respective damage models.

Several simple geometries as shown in Fig. 8.13 can be employed to serve for this purpose. The hydrostatic tension of an initial damage field with spherical symmetry can be used to test the dilatational part of the constitutive equations, as indicated in Fig. 8.13(a). The deviatoric response of the damage model under consideration, on the other hand, can be depicted by the shear model with a slightly

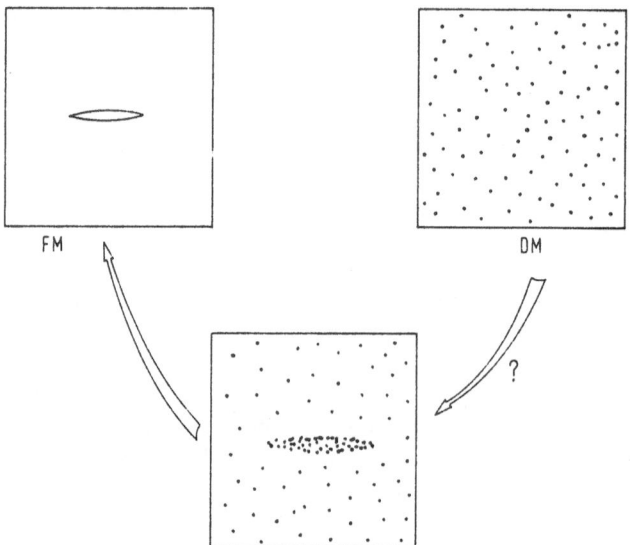

Fig. 8.12. Role of damage concentration on linking the two different defect approaches of damage mechanics and fracture mechanics

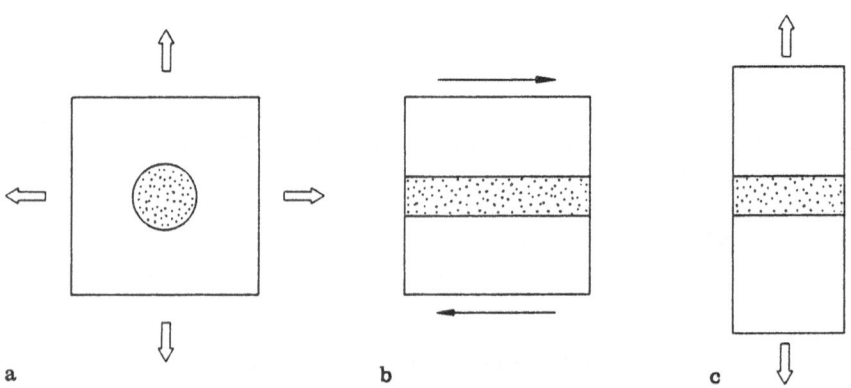

Fig. 8.13. Geometries for testing damage concentration behavior. **a** Hydrostatic tension; **b** simple shear; **c** uniaxial tension

non-uniformed damage band, see Fig. 8.13(b). A combination of the above two damage aspects can be tested under a uniaxial tension model with small nonuniform initial damage along the tensile axis, as shown in Fig. 8.13(c). A thorough check for the last case was reported by Yang[36], and his results are shown in Table 8.3. It is interesting to notice that a spectrum of damage concentration behavior is exhibited by those damage models. The result on the Gurson model suggests

Table 8.3. Predictions of damage concentration features for various common damage models

Damage Models	Damage Concentration Features
Broberg[57]	Damage localization at $D_{cr} < 1$. Global concentration of both damage and plastic strain at failure.
Roussiler[58]	Damage localization at $D_{cr} < 1$. Global concentration of both damage and plastic strain at failure.
Gurson[7,8]	Damage localization at $D_{cr} \ll 1$ if void nucleation is suppressed. Global concentration of damage at failure.
LeMaitre – Dufailly[59]	Less pronounced damage concentration. Infinite failure strain. Extremely strong initial damage sensitivity.
Kachanov[47]	A variety of damage and plastic strain concentration phenomena for different value of m, n and initial damage distribution.
Creep damage controlled by grain boundary diffusion[11]	Diffusion of damage and non-uniform plastic strain.
Creep damage controlled by surface diffusion[11]	Early damage localization. Strong initial defect sensitivity.

the significance of the void nucleation term in suppressing an early and premature damage localization. Quantitative results of this evaluation for seven commonly-used damage models (from literatures of both continuum damage mechanics and meso-damage theory) were given in references [56] and [36].

8.3 Meso-Damage by Progressive Cracking

8.3.1 Crack Damaging Configuration

Another major mechanism of meso-damage is characterized by microcracks distributed in brittle materials, such as rocks, coals, ceramics; or in the brittle matrices of toughened structural materials, such as structural ceramics, polymers and composite materials. A schematic illustration for crack configurations in various material systems are given in Fig. 8.14. They can be classified into the following categories

A. *Simple fracture*
   (1) randomly distributed cracks;
   (2) partially aligned cracks;
   (3) orderly aligned crack arrays.
B. *Multiple fracture*
   (4) random multiple fracture;
   (5) orderly multiple fracture.

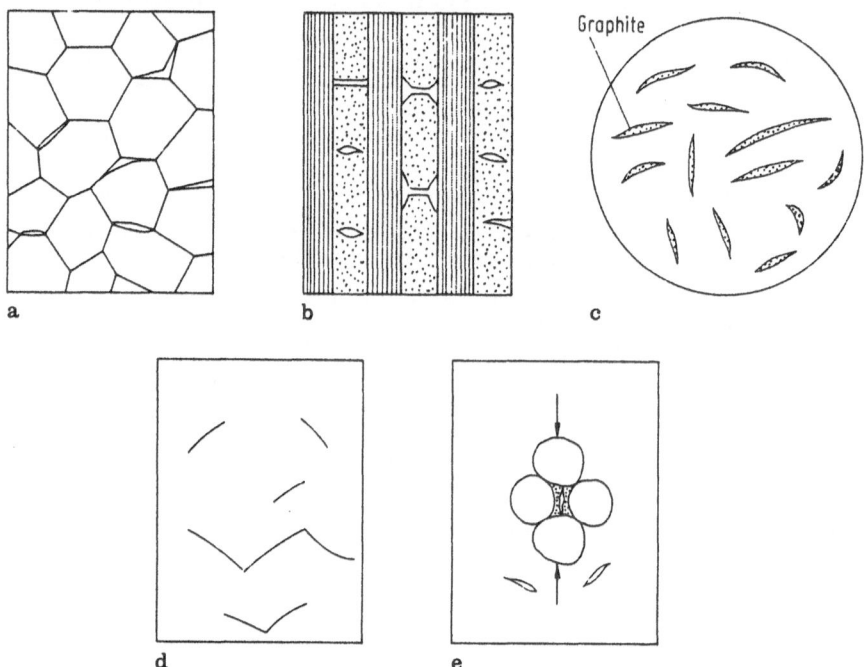

Fig. 8.14. Crack damage configurations in various material systems. **a** Ceramics, **b** composites, **c** cast iron, **d** rock, **e** concrete

Limited by the scope of the present book, here we would rather concentrate in *randomly distributed cracks* and *orderly aligned crack arrays*.

8.3.2 Single Crack Solution

We start from the case of a single crack embedded in a linear elastic anisotropic matrix. The results developed, however, could be extended to the rate type formulation of a nonlinear matrix. The linear elastic constitutive relation can be expressed by the following tensorial form

$$\epsilon = \mathbf{M} : \sigma \tag{8.41}$$

where $\mathbf{M}$ is the fourth rank compliance tensor with Voigt symmetry. Alternatively, one can cast the above equation in a matrix form as commonly proceeded in the literatures. In doing so, we regard both strain ϵ and stress σ as vectors of a dimension of six, and $\mathbf{M}$ as a 6 times 6 matrix. The numbering sequence for this correspondence is shown schematically as follows

In the current treatment, we assume that the compliance matrix $\mathbf{M}$ would at most correspond to an orthotropic medium. The most general form of $\mathbf{M}$ under this assumption is

$$\mathbf{M} = \begin{bmatrix} M_{11} & M_{12} & M_{13} & 0 & 0 & 0 \\ & M_{22} & M_{23} & 0 & 0 & 0 \\ & & M_{33} & 0 & 0 & 0 \\ & & & M_{44} & 0 & 0 \\ & \text{Symmetric} & & & M_{55} & 0 \\ & & & & & M_{66} \end{bmatrix} \tag{8.42}$$

The nine independent compliance coefficients for orthotropic material would be reduced to five if transverse isotropy with respect to $X_1 - X_2$ plane is observed,

$$\begin{aligned} M_{11} &= M_{22} & M_{13} &= M_{23} \\ M_{44} &= M_{55} & M_{66} &= 2(M_{11} - M_{12}). \end{aligned} \tag{8.43}$$

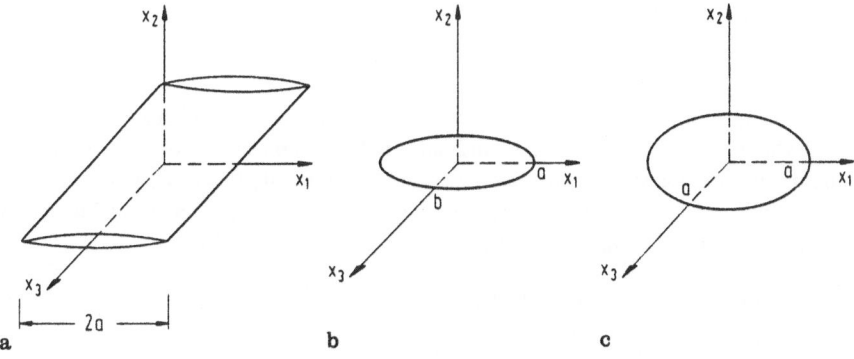

Fig. 8.15. Typical crack configurations. a Plane strain slit; b elliptical crack; c penny shape crack

As beautifully summarized by Laws and Brockenbrough[60] that the energy released due to the formation of a single crack in a matrix characterized by $\mathbf{M}$ as shown in (8.42) can be phrased as a quadratic form of the remote stress tensor σ

$$U = \frac{1}{2} V_{\text{crack}} \, \sigma : \Omega : \sigma \tag{8.44}$$

where U is the energy released by a single crack, V_{crack} the influential volumetric domain of the crack and Ω the fourth rank *energy release tensor*. In the matrix notation, the stress σ in (8.44) can be regarded as a vector of dimension six and Ω a 6 times 6 matrix. The discussion to follow will be exclusively restricted to the three crack configurations shown in Fig. 8.15, namely, a plane strain slit, an elliptical crack and a penny shape crack. The last one represents a special case of the elliptical crack. For those three cases, the influential volumes of a single crack are defined by[60]

$$V_{crack} = \begin{array}{ll} \pi a^2 & \text{plane strain slit} \\ (4\pi/3)ab^2 & \text{for} \quad \text{elliptical crack} \\ (4\pi/3)a^3 & \text{penny shape crack.} \end{array} \tag{8.45}$$

The determination of the *energy release matrix* Ω is rather cumbersome, and we will give its expressions for three cases in Fig. 8.15 one by one. For a plane strain slit, the only non-trivial coefficients of Ω can be evaluated by well-established method of two-dimensional fracture analysis in orthotropic media as

$$
\begin{aligned}
\Omega_{22} &= \left[(M_{22}M_{33} - M_{23}^2)/M_{33} \right] \left(\sqrt{\alpha_1} + \sqrt{\alpha_2} \right) \\
\Omega_{44} &= \sqrt{M_{44}M_{55}} \\
\Omega_{66} &= \left[\left(\sqrt{M_{22}M_{33} - M_{23}^2} \sqrt{M_{11}M_{33} - M_{13}^2} \right) / M_{33} \right] \left(\sqrt{\alpha_1} + \sqrt{\alpha_2} \right)
\end{aligned} \tag{8.46}
$$

where eigenvalues α_1 and α_2 are two roots of the following algebraic equation

$$
\begin{aligned}
\left(M_{22}M_{33} - M_{23}^2 \right) \alpha^2 &- [M_{33}M_{66} + 2 \left(M_{12}M_{33} - M_{13}M_{23} \right)] \alpha \\
&+ M_{11}M_{33} - M_{13}^2 = 0.
\end{aligned} \tag{8.47}
$$

For an elliptical crack as shown in Fig. 8.15(b), the energy release matrix Ω for a general orthotropic medium would take a less transparent form. A numerical scheme was developed instead, see Hoenig[61], to handle this case. For the special case of an isotropic medium, however, the following results were established by Budiansky and O'Connell[6] for the nonzero components of Ω

$$
\begin{aligned}
\Omega_{22} &= \left[2 \left(1 - \nu^2 \right) / E \right] / E(k) \\
\Omega_{44} &= \left[2 \left(1 - \nu^2 \right) / E \right] Q(k, \nu) \\
\Omega_{66} &= \left[2 \left(1 - \nu^2 \right) / E \right] R(k, \nu)
\end{aligned} \tag{8.48}
$$

where ν denotes Poisson's ratio, $k = [1 - (b/a)^2]^{\frac{1}{2}}$, and function Q and R are defined by

$$
\begin{aligned}
R(k, \nu) &= k^2 \left\{ \left(k^2 - \nu \right) E(k) + \nu \left(1 - k^2 \right) K(k) \right\}^{-1} \\
Q(k, \nu) &= k^2 \left\{ k^2 + \left(1 - k^2 \right) E(k) - \nu \left(1 - k^2 \right) K(k) \right\}^{-1}
\end{aligned} \tag{8.49}
$$

where $K(k)$ and $E(k)$ are complete elliptical integrals of the first and the second kinds.

For the special case of penny shape cracks, the following explicit results can be obtained for matrix with transverse isotropy, see Laws[62]

$$\Omega_{22} = (2\eta_1\eta_2/\pi)\,(\eta_1 + \eta_2)\left[(M_{11}^2 - M_{13}^2)\,/M_{11}\right]$$

$$\Omega_{44} = \Omega_{66} = (4/\pi)\,(\eta_1 + \eta_2)\left[(M_{11}^2 - M_{13}^2)\,\sqrt{2M_{44}}\right]/ \tag{8.50}$$

$$\left\{M_{11}\sqrt{2M_{44}} + (\eta_1 + \eta_2)\,(M_{11} + M_{13})\,\sqrt{M_{11} - M_{13}}\right\}$$

where eigenvalues η_1 and η_2 are the two roots of the following algebraic equation

$$(M_{11}^2 - M_{13}^2)\,\eta^2 - [M_{11}M_{44} + 2M_{12}\,(M_{11} - M_{13})]\,\eta$$
$$+ M_{11}M_{22} - M_{12}^2 = 0. \tag{8.51}$$

Thereby we complete the solution for a single crack.

8.3.3 Randomly Aligned Crack Arrays

Attention is now focused on the case of randomly aligned crack arrays under the following assumptions:

(1) Every crack is assumed to be open, and consequently, phenomena such as crack closure and crack surface friction are temporarily omitted.

(2) The crack orientations are randomly distributed so that preference of any crack orientations will not be considered.

(3) The interaction among neighboring cracks are weak in the sense that a self-consistent method could be used to derive the macroscopic response.

Under the above assumptions, we will proceed by self-consistent method to a domain of reference volume V as shown in Fig. 8.16. The left graph of Fig. 8.16 represents a block with zero crack density. It has a compliance of M and is loaded by a remote (macroscopic) stress Σ. The *total potential energy*, Π_o, stored in the block under the application of remote stress Σ is then

$$\Pi_o = \frac{1}{2}\int_V \sigma : \epsilon\, dV - \int_S t \cdot u\, dS = -\frac{1}{2}V\Sigma : M : \Sigma. \tag{8.52}$$

We next consider the case illustrated on the right graph of Fig. 8.16, in which N non-intersected cracks appear in an otherwise identical material block of reference volume V. The presence of distributed cracks would increase the compliance to $\bar{M}$. The total potential energy, designated by Π, possessed by this cracked block is

$$\Pi = \frac{1}{2}\int_V \sigma : \epsilon\, dV - \int_S t \cdot u\, dS = -\frac{1}{2}V\Sigma : \bar{M} : \Sigma. \tag{8.53}$$

From the energy balance, the *total released potential energy* by the presence of N cracks is

$$U_{\text{tot}} = \Pi_o - \Pi = \frac{1}{2}V\Sigma : (\bar{M} - M) : \Sigma. \tag{8.54}$$

This formula serves as the basis for the self-consistent analysis in the sequel.

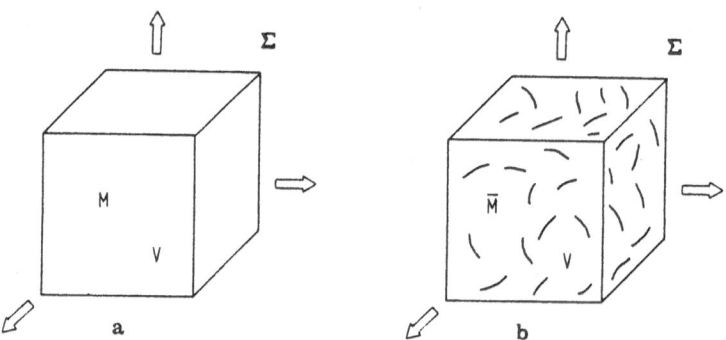

Fig. 8.16. Response of a material block. a Without any cracks; b with cracks

8.3.4 Self Consistent Scheme

We now proceed to estimate the compliance increment, namely $\bar{\mathbf{M}} - \mathbf{M}$, by means of *self-consistent method*. The essence of self-consistent method is to treat crack array by a single crack situated in a self-consistent matrix with compliance $\bar{\mathbf{M}}$. Therefore, according to (8.44) and the volume average notation introduced in equation (7.2), the total released potential energy is

$$U_{tot} = \frac{1}{2} N \Sigma : <V_{crack} \, \Omega(\bar{\mathbf{M}})> : \Sigma \qquad (8.55)$$

where $\Omega(\bar{\mathbf{M}})$ refers to the fact that the energy release matrix Ω as defined in (8.46), (8.48) or (8.50) should be evaluated at the self-consistent compliance $\bar{\mathbf{M}}$. Both equations (8.54) and (8.55) hold true for any macroscopic stress Σ. A comparison between these two equations yields

$$\bar{\mathbf{M}} = \mathbf{M} + <f \Omega(\bar{\mathbf{M}})> \qquad (8.56)$$

where the second term on the right hand side represents the *compliance increment by microcracking*. The symbol f in (8.56) denotes the *crack density*

$$f = N V_{crack} / V. \qquad (8.57)$$

If the statistical distribution of crack size is independent of that of the crack orientation, (8.56) is reduced to

$$\bar{\mathbf{M}} = \mathbf{M} + <f> <\Omega(\bar{\mathbf{M}})> \qquad (8.58)$$

where

$$<f> = N <V_{crack}> / V \qquad (8.59)$$

is the average crack density, with $<V_{crack}>$ being the average crack volume. Under a prescribed average crack density, equation (8.58) constitutes a tensorial nonlinear integral equation to determine the effective compliance $\bar{\mathbf{M}}$. A methodology based on the orientation distribution function (ODF), similar to that developed in subsections 7.2.6 and 7.2.7, can be employed for this problem.

For completely randomly distributed microcracks (where ODF is identical to unity), considerable reduction can be achieved on equation (8.58). For example, for elliptical cracks distributed randomly in an isotropic elastic matrix, its overall moduli are given by

$$\bar{E}/E = 1 - f\left[2\left(1 - \bar{\nu}^2\right)/15E(k)\right]\{3 + [Q(k,\bar{\nu}) + R(k,\bar{\nu})]E(k)\}$$
$$\bar{G}/G = 1 - f\left[8(1 - \bar{\nu})/15E(k)\right]\{4 + 3[Q(k,\bar{\nu}) + R(k,\bar{\nu})]E(k)\} \quad (8.60)$$

where $\bar{\nu}$ is the effective Poisson's ratio as defined by Budiansky and O'Connell[6]. Usually, $\bar{\nu}$ is not far from the matrix Poisson's ratio.

For the penny shape cracks in which a is equal to b, equation (8.60) could be further reduced to

$$\bar{E}/E = 1 - f\left[4(1 - \bar{\nu}^2)/15\pi\right]\{(10 - 3\bar{\nu})/(2 - \bar{\nu})\}$$
$$\bar{G}/G = 1 - f\left[8(1 - \bar{\nu})/15\pi\right]\{(5 - \bar{\nu})/(2 - \bar{\nu})\}. \quad (8.61)$$

Both equations (8.60) and (8.61) claim a steady linear declination of the overall modulus with respect to the crack density f, as shown in the dashed line of Fig. 8.17 from the prediction of self-consistent method. Though this behavior might be appropriate for small crack density f, it is certainly not true for large f especially when f exceeds f_{cr} as marked in Fig. 8.17, after that the self consistent method would break down and predict a negative Young's modulus. A reasonable $\bar{E}/E$ versus f curve should gradually approach the null stiffness for large f, as indicated by the solid curve in Fig. 8.17.

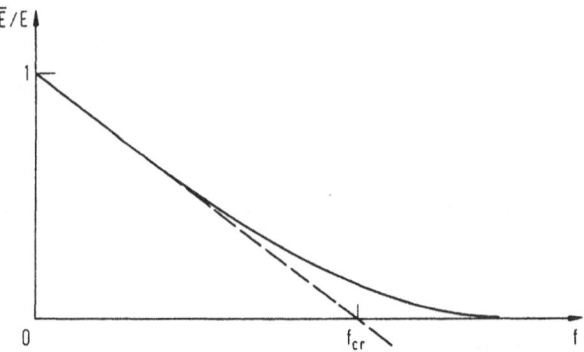

Fig. 8.17. Normalized stiffness versus crack density curve, where dashed curve represents the self-consistent prediction and solid curve indicates a conjectured response

8.3.5 Fabric Tensor

We next turn to the discussion of more general crack arrays where distributions on *crack sizes* (parametrized by r) and *crack orientations* (designated by a unit vector $\mathbf{n}$ which represents the average outward normal of the crack face) are statistically prescribed. A scalar *crack distribution density function* $D(\mathbf{n}, r)$, similar to orientation distribution function employed in subsection 7.2.6, is defined as the distribution density of cracks of outward normal $\mathbf{n}$ and of size r. Obviously, the upper and lower crack faces have the same density

$$D(\mathbf{n}, r) = D(-\mathbf{n}, r) \tag{8.62}$$

then we only need to consider the upper half of the orientation sphere. The distribution density $D(\mathbf{n}, r)$ is normalized as follows

$$\int_0^\infty \int_\Omega D(\mathbf{n}, r)\, d\Omega dr = 2 \int_0^\infty \int_{\Omega/2} D(\mathbf{n}, r)\, d\Omega dr = 1 \tag{8.63}$$

where $d\Omega$ is the solid angle element and Ω is the entire orientation sphere of surface area 4π. If the crack orientation distribution is statistically independent of crack size r, then

$$D(\mathbf{n}, r) = D(\mathbf{n})R(r) \tag{8.64}$$

where $D(\mathbf{n})$ and $R(r)$ are normalized by

$$\int_0^\infty R(r)\, dr = 1 \qquad 2 \int_{\Omega/2} D(\mathbf{n})\, d\Omega = 1, \tag{8.65}$$

respectively. A *fabric tensor* can be defined as, see Oda[63]

$$\mathbf{A} = (\pi N/V) \int_0^\infty \int_\Omega r^3 D(\mathbf{n}, r)\, \mathbf{nn}\, d\Omega dr. \tag{8.66}$$

The integrand in (8.66) is composed of four parts, namely $r\mathbf{n}$, $\pi r^2 \mathbf{n}$, N/V and $D(\mathbf{n}, r)$, representing the statistical weight, the area vector of a single crack, number of crack in unit volume and crack orientation density, respectively. Furthermore, the product of the last three terms represents the area vector of cracks with size r and orientation $\mathbf{n}$. Consequently, $\mathbf{A}$ in (8.66) corresponds to a fabric tensor of crack area distribution weighted by $r\mathbf{n}$, in a sense to formulate the crack volume. Oda[63] argued that the increase on compliance created by anisotropic microcracking can be expressed approximately by the fabric tensor $\mathbf{A}$ as

$$\bar{M}_{ijkl} - M_{ijkl} = (1/4D)(\delta_{il}A_{jk} + \delta_{jl}A_{ik} + \delta_{jk}A_{il} + \delta_{ik}A_{jl}) \tag{8.67}$$

where D relates to the extent of deformation.

Studies on fabric tensors have brought considerable progresses in the recent years. Various forms of fabric tensors have been proposed, such as those by Kachanov[64] and by Kanatani[65]. The most general form of fabric tensors can be deduced from the general tensor representation theory[66]. Their testing principles and techniques are also in the verge of development[63].

8.3.6 Orderly Aligned Crack Arrays

Attention is then focused on *damage caused by orderly aligned crack arrays* , such
as those encountered in composite laminates. The damage behavior for composite
laminates with two families of plies laminated alternatively is discussed hereinafter
as an example of the damage modeling. As revealed in Fig. 8.18, two families of
intralaminar cracks, as well as a family of interlaminar cracks could occur in such
a composite laminate under in-plane loading.

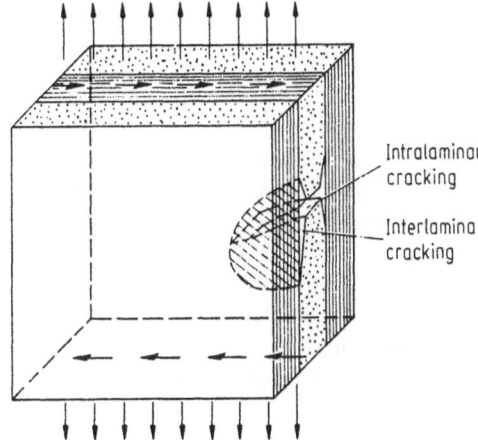

Intralaminar
cracking

Interlaminar
cracking

Fig. 8.18. Damage of a composite laminate under in-plane loading by orderly aligned crack arrays

Under this circumstance, the damage tensor is defined as the additional inelastic
strain caused by microcrackings.

$$\epsilon^D = \epsilon - \epsilon^e. \tag{8.68}$$

It can be established, see Yang[67], that the damage strain defined above could be
expressed by

$$\epsilon^D = \frac{1}{2} \sum_{i=1}^{3} [\mathbf{U}_i \mathbf{n}_i + n_i \mathbf{U}_i] \tag{8.69}$$

where

$$\mathbf{U}_i = (1/V) \int_{S_i^c} [[\mathbf{u}]] \, dS \tag{8.70}$$

is the volume average of displacement discontinuities $[[\mathbf{u}]]$ across defect surface of
type i, denoted by S_i^c. Symbols $\mathbf{n}_1, \mathbf{n}_2$ and $\mathbf{n}_3$ represent the unit normals of two
types of intralaminar cracks and the normal of interlaminar cracks, respectively.

After some manipulations, (8.69) can be written more explicitly as

$$\epsilon^D = \sum_{i=1}^{3} f_i \left[p_i \mathbf{n}_i \mathbf{n}_i + \frac{1}{2} q_i (\mathbf{n}_i \mathbf{s}_i + \mathbf{s}_i \mathbf{n}_i) \right]. \tag{8.71}$$

The first term represents the damage process causing volumetric dilatancy with non-negative p_i, whereas the second term corresponds to the volume preserving deviatoric damage by sliding along microscopic defect surfaces. The definition of crack density f_i was provided by Horii and Nemat-Nasser[68]

$$f_i = (2/V) \sum_{k=1}^{N_i} a_i^{(k)} A_i^{(k)} \tag{8.72}$$

where N_i denotes the number of cracks of type i, and $a_i^{(k)}$ and $A_i^{(k)}$ the size and area of the k-th crack of type i, respectively. The damage evolution in composite laminates will be extremely intricate if interlaminar slip is introduced. Complicated history dependence as well as possible non-normality structure could occur. However, if the following assumptions hold
(1) suppression of localized slip;
(2) damage produced only by cracking;
(3) omission of intralaminar-interlaminar interaction,
one is able to show that the history dependence of damage evolution could be solely represented by

$$g_i^{cr}(t) = \max_{0 \le s \le t} g_i(s) \tag{8.73}$$

signifying the *maximum energy release* by type i cracks. In (8.73), the energy release g_i is defined by

$$g_i = \mathbf{t}_i \cdot \mathbf{Q} \cdot \mathbf{t}_i \quad \text{(no sum on } i) \tag{8.74}$$

where $\mathbf{t}_i$ is the traction vector on S_i^c and $\mathbf{Q}$ is a three times three matrix concerning elastic constants. If the traction vector $\mathbf{t}_i$ on the type i cracks is decomposed into

$$\mathbf{t}_i = \sigma_i \mathbf{n}_i + \tau_i^{II} \mathbf{s}_i^{II} + \tau_i^{III} \mathbf{s}_i^{III} \quad \text{(no sum on } i) \tag{8.75}$$

where $\mathbf{n}_i, \mathbf{s}_i^{II}$ and $\mathbf{s}_i^{III}$ form an orthonormal bases with $\mathbf{s}_i^{II}$ and $\mathbf{s}_i^{III}$ denoting the directions of mode II and mode III crack sliding, respectively. For plane and anti-plane problems, the general quadratic form in (8.74) can be diagonalized to

$$g_i = M_i^I \sigma_i^2 + M_i^{II} \tau_i^{II^2} + M^{III} \tau_i^{III^2} \quad \text{(no sum on } i) \tag{8.76}$$

when $\sigma_i \ge 0$, otherwise the first term should be discarded because compressional damage is neglected. $M_i^{I,II,III} (i = 1, 2, 3)$ will only depend on elasticity constants, and their detailed expressions are

$$M_i^I = (1/2 M_{11} M_{22})^{\frac{1}{2}} \left[(M_{22}/M_{11})^{\frac{1}{2}} + (2M_{12} + M_{66})/2M_{11} \right]^{\frac{1}{2}}$$

$$M_i^{II} = \left(M_{11}/\sqrt{2} \right) \left[(M_{22}/M_{11})^{\frac{1}{2}} + (2M_{12} + M_{66})/2M_{11} \right]^{\frac{1}{2}} \tag{8.77}$$

$$M_i^{III} = 1/2 (M_{44} M_{55})^{\frac{1}{2}} .$$

Using $g_i^{cr} (i = 1, 2, 3)$ as internal variables, the damage evolution equation can be simplified to

$$\epsilon^D(t) = \mathbf{D} \left(g_i^{cr}(t), \, \sigma(t) \right). \tag{8.78}$$

Consequently, one can proceed to classify the possible damage states into 4 different categories, namely
(1) *total damage state*, in which the current values of $g_i, i = 1, 2, 3$, all assume their maximum values;
(2) *partial damage state*, in which part of the current values of $g_i, i = 1, 2, 3$, assume their maximum values;
(3) *pseudo-elastic state*, in which the current values of $g_i, i = 1, 2, 3$, are all below their maximum values. However, at least one of the $g_i^{cr}(t), i = 1, 2$ or 3, is larger than the respective critical value for crack initiation;
(4) *perfect elastic state*, in which no cracks have been initiated.
  As implied by the recent works of Laws and Brockenbrough[60] and by Hashin[23], history independence and normality structure would become inevitable for any total damage state defined by

$$g_i = g_i^{cr} \quad \text{and} \quad \dot{g}_i \geq 0 \quad i = 1, 2, 3. \tag{8.79}$$

This conclusion, however, does not agree with most of the experimental observations. This contradiction provides a strong impetus to pursue interaction calculations which take into account the interlaminar slip, as proceeded by Yang and Boehler[69]. We complete this subsection by providing a pseudo-elastic constitutive relation for composite laminates which observe the above assumptions

$$\epsilon = \mathbf{M}^s : \sigma \tag{8.80}$$

where the secant compliance moduli are given by

$$\mathbf{M}^s = \mathbf{M} + \sum_{i=1}^{3} \{ P_i \left(g_i^{cr} \right) \mathbf{n}_i \mathbf{n}_i \mathbf{n}_i \mathbf{n}_i + (1/4) R_i \left(g_i^{cr} \right) (\mathbf{n}_i \mathbf{s}_i + \mathbf{s}_i \mathbf{n}_i) \tag{8.81}$$

$$\cdot (\mathbf{n}_i \mathbf{s}_i + \mathbf{s}_i \mathbf{n}_i) \}.$$

$\mathbf{M}$ is the moduli for undamaged composite laminates. Functions P_i and R_i have to be furnished by microcrack computation.

8.3.7 Some Related Topics of Microcrack Damage

A brief outline is provided in the following on some related topics of microcrack damage.

Load induced anisotropy. The crack growth law, as provided from fracture mechanics, has orientation preference for distributed microcracks. Furthermore, microcracks respond differently to tensile or compressive stress. The occurrence of crack closure and crack surface friction under compressive loading distinguishes it from the crack open response during tensile loading. In fact, the so-called *load induced anisotropy* behavior could be caused by

(1) Rotation of mean crack face normal **n** toward the maximum tensile stress during the evolution of microcracks.
(2) Uneven response between crack open and crack closure as well as the presence of friction under compressive normal stress.

Loading path dependency. The response of a microcrack damaged solid is load path dependent. As shown in Fig. 8.19, the initial and terminate states of the path I and Path II are identical, whereas their respective intermediate states are different. The shear deformation in the path I is not effected by the application of compressional stress, thus eliminating the possible participation of the friction effect. The shear deformation in the path II, however, is effected by the compressional normal stress through a Columb type law. Therefore, distinct material anisotropic responses will be exhibited under those two load paths, as shown by Kachanov[16] and by Nemat-Nasser[68].

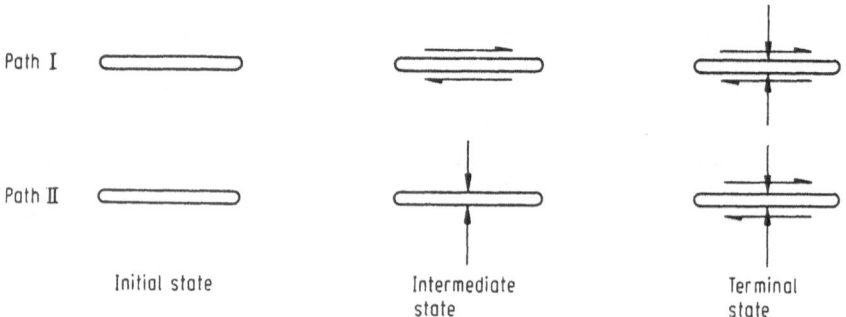

Path I

Path II

Initial state   Intermediate   Terminal
                          state   state

Fig. 8.19. Load path dependency of constitutive responses

Cross effect. For the heterogeneous materials like concretes and structural ceramics, the compressional fracture as shown in Fig. 8.20 is likely to occur. Accordingly, the inelastic deformation can be produced not only by tensile stress normal to the crack surface but also by the compressional stress parallel to the crack surface,

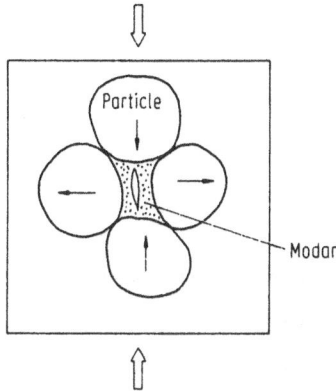

Fig. 8.20. Illustration of compressional fracture and cross effect

as discussed by Ortiz[70] and by Bazant[20]. This phenomenon is termed *cross effect*. Several examples with the incorporation of the cross effect were discussed by Ortiz[70] via numerical computation.

8.4 References

1. Robinson, E.L., Trans. Amer. Inst. Min. Engrs., 74(1952), P.777.
2. Hull, D., and Rimmer, D.E., Phil. Mag., 4(1959), p.673.
3. Hashin, Z. and Shtrikman, S., J. Mech. Phys. Solids, 11(1963), p.127.
4. McClintock, F.A., J. Appl. Mech., 35(1968), p.363.
5. Rice, J.R. and Tracey, D.M., J. Mech. Phys. Solids, 17(1969), p.201.
6. Budiansky, B. and O'Connell, R.J., Int. J. Solids Struct., 12(1976), p.81.
7. Gurson A.L., Plastic flow and fracture behavior of ductile materials incorporating void nucleation, growth and interaction, Ph.D. Thesis, Brown Univ., (1975).
8. Gurson, A.L., J. Engng. Mater. & Tech., 99(1977), p.2.
9. Tvergaard, V., Material failure from void growth to coalescense, The Danish Center for Applied Math. and Mech., Rep. No. S45, (1988).
10. Needleman, A., Computational micromechanics, Keynote Speech at ICTAM XVII, Grenoble, France, Aug. (1988).
11. Cocks, A.C.F. and Ashby, M.F., Progr. in Mater. Sci., 27(1982), p.189.
12. Ashby, M.F. and Dyson, B.F., Proc. of ICF6 Symp., vol.1, (1984), p.3.
13. Chuang, T.-J., Kagawa, K.I., Rice, J.R. and Sills, L.B., Acta Metall., 27(1979), p.265.
14. Hutchinson, J.W., Acta Metall., 31(1983), p.1079.
15. Tvergaard, V., Acta Metall., 32(1984), p.1977.
16. Kachanov, M.I., Mech. Mater., 1(1982), p.19.
17. Kachanov, M.I., Mech. Mater., 1(1982), p.29.
18. Nemat-Nasser, S., Sumi, Y. and Keer, L.M., Int. J. Solids Struct., 16(1980), p.1017.
19. Nemat-Nasser, S., J. Appl. Mech., 50(1983), p.1114.
20. Bazant, Z.P., Appl. Mech. Rev., 39(1986), p.675.
21. Kracjnovic, D., J. Appl. Mech., 50(1983), p.355.
22. Argon, A.S., Sources of Toughness in Polymers, MIT Report, (1988).
23. Hashin, Z., J. Appl. Mech., 54(1987), p.872.

24. Dvorak, G.J., Laws, N. and Hejazi, M., J. Composite Mater., 19(1985), p.216.
25. Bishop, J.F.W. and Hill, R., Phil. Mag., 42(1951), p.414.
26. Gurland, J., Acta Metall., 20(1972), p.735.
27. Goods, S.H. and Brown, L.M., Acta Metall., 27(1979), p.1.
28. Chu, C.C. and Needleman, A., J. Engng. Mat. and Tech., 102(1980), p.249.
29. Argon, A.S., et al., Metall. Trans., 6A(1975), p.815.
30. Needleman, A. and Rice, J.R., in Mechanics of Sheet Metal Forming, Koistinen, D.P. and Wang, N.-M. eds., (1978), p.237.
31. Tvergaard, V., Int. J. Fracture, 17(1981), p.389.
32. Tvergaard, V., Int. J. Fracture, 18(1982), p.237.
33. Tvergaard, V. and Needleman, A., Acta Metall., 32(1984), p.157.
34. Becker, R., Needleman, A., Richmond, O. and Tvergaard, V., J. Mech. Phys. Solids, 36(1988), p.317.
35. Koplik, J. and Needleman, A., Int. J. Solids Struct., 24(1988), p.835.
36. Yang, W., J. Mech. Phys. Solids, (1990), p.725.
37. Mear, M.E. and Hutchinson, J.W., Mech. Mater., 4(1985), p.395.
38. Becker, R. and Needleman, A., J. Appl. Mech., 53(1986), p.491.
39. Saje, M., Pan, J. and Needleman, A., Int. J. Fract., 19(1982), p.163.
40. Bluhm, J.I. and Morrisey, R.J., Proc. 1st Int. Conf. Fracture, vol.3, (1966), p.1739.
41. Needleman, A. and Tvergaard, V., J. Mech. Phys. Solids, 32(1984), p.461.
42. Yamamoto, H., Int. J. Fracture, 14(1978), p.347.
43. Aravas, N., J. Mech. Phys. Solids, 34(1986), p.55.
44. Aravas, N. and McMeeking, R.M., J. Mech. Phys. Solids, 33(1985), p.25.
45. Aoki, S., Kishimoto, K. and Sakata, M., J. Mech. Phys. Solids, 35(1987), p.371.
46. Needleman, A. and Tvergaard, V., J. Mech. Phys. Solids, 35(1987), p.151.
47. Kachanov, L.M., Izv. Akad. Nauk. SSSR, No. 8, (1958), p.26.
48. Leckie, F.A. and Hayhurst, D.R., Acta Metall., 25(1977), p.1059.
49. Rabotnov, Yu.N., Proc. XII IUTAM Congr., Stanford, eds. by Hetenyi, M. and Vincenti, W.G., Springer, (1969), p.342.
50. Needleman, A. and Rice, J.R., Acta Metall., 28(1980), p.1315.
51. Anderson, P.M. and Rice, J.R., Acta Metall., 31(1983), p.919.
52. Dyson, B.F., Metal Sci., 10(1976), p.349.
53. He, M.Y. and Hutchinson, J.W., J. Appl. Mech., 48(1981), p.830.
54. Rice, J.R., Acta Metall., 29(1981), p.675.
55. Budiansky, B., Hutchinson, J.W. and Slutsky, S., in Mechanics of Solids, the Rodney Hill 60th Anniversary Volume, eds. by Hopkins, H.G. and Sewell, M.J., Pergamon Press, (1982), p.13.
56. Yang, W. and Cheng, L., in Role of Plasticity and Damage Mechanics in Fracture of Solids, Proc. of FEFG, Nov. (1988), Tokyo, Japan, p.153.
57. Broberg, H., J. Appl. Mech., 41(1974), p.809.
58. Roussiler, G., Thesis of Ecole Polytech. Univ. Paris, (1979).
59. LeMaitre, J. and Dufailly, J., Third French Congr. on Mechanics, Grenoble, France, (1977).
60. Laws, N. and Brockenbrough, , Int. J. Solids & Struct., 23(1987), p.1247.
61. Hoenig, A., Int. J. Solids Struct., 15(1979), p.137.
62. Laws, N., Mech. Mater., 4(1985), p.209.
63. Oda, M., Mech. Mater., 2(1983), p.163.
64. Kachanov, M.I., Proc. of ASCE, Engng. Mech. Div., 106(EM5)(1980), p.1039.
65. Kanatani, K., Tech. Rep. of Comp. Sci., Gunma Univ., (1982).
66. Cowin, S.C., Mech. Mater., 4(1985), p.137.
67. Yang, W., Proc. of ICAM, Beijing, (1989), p.959.
68. Horii, H. and Nemat-Nasser, S., J. Mech. Phys. Solids, 31(1983), p.155.
69. Yang, W. and Boehler, J.P., Int. J. Solids & Struct., 29(1992), p.1307.
70. Ortiz, M., Mech. Mater., 4(1985), p.67.

Part II
Engineering Applications

9 Modelling of Sheet Metal Textures

Sheet metals are being used increasingly to produce packaging materials or structural components by cold forming. For a given tool profile and friction conditions, the formability can be influenced by the metallurgical structures of the sheet. Metal sheets produced by a combination of rolling and annealing bring about dramatic changes in the mechanical properties and are characterized by the crystallographic textures (or preferred orientations of grains, see Sec. 3.5) which give rise to anisotropic properties. Such anisotropy may be desirable as in the improvement of through thickness of sheet metals and in the permeability of magnetic materials, or it may be harmful as in the case of planar anisotropy which leads to the formation of ears in sheet products. There has been world-wide research interest in understanding and obtaining sheet metal textures by thermal-mechanical processing and relating these to the performance of sheets in forming operations. Research is being done actively in the automobile industry to achieve superior formability and lighter gauge which contributes to the reduction in automobile weight and improve energy-saving performance. Extensive reviews on textures in sheet metals exist in the literatures. In this chapter, a brief background needed on the subject is given and the emphasis is put on the application of mesoplasticity theories which has led to a better understanding of the texture formation in sheet matals.

9.1 Development of Rolling Textures in FCC Metals

Among other deformation textures, the rolling texture by far is the most intensively studied. Early experimental studies on the development of textures in FCC rolled metals and alloys led to the conclusion that pure metals possess a texture similar to copper, while alloys possess the brass-type texture as shown in Fig. 9.1.

The "copper-type" rolling textures is fairly complex. In more modern terms of ODF, the pure metal texture is best represented by a continuous line of preferred orientations extending from $\{110\}<112>$ orientation to an orientation approximately $3°$ from $\{4,4,11\}<11,11,8>$[1]. A slight maximum exists at $\{112\}<111>$. Compared with the copper-type rolling texture, the brass type rolling texture is simple and can be described by the ideal orientation $\{110\}<112>$ with orientation spread linking $\{011\}<100>$. An increase in the solute content (e.g. Zn in Cu) or a de-

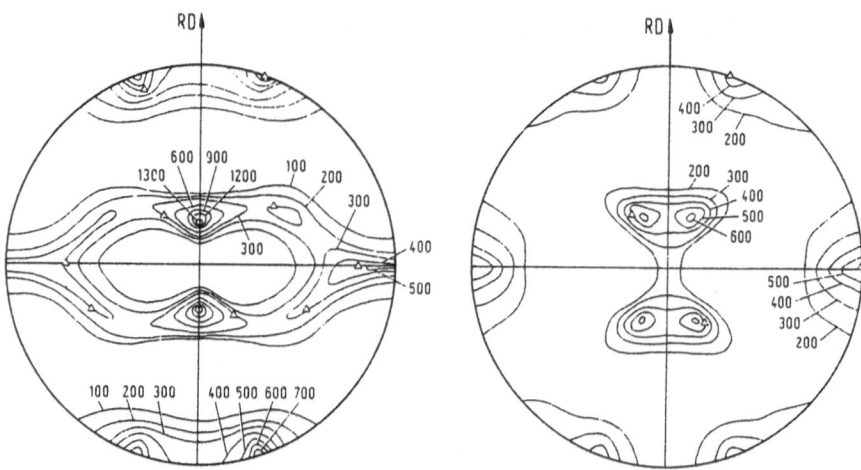

Fig. 9.1. {111} pole figures of (a) copper-type and (b) brass-type rolling texture. (From Hu, H., Sperry, P.R. and Beck. P.A., J.Metals, 4; 1952, p.26

crease in the rolling temperature will shift the copper type texture to the brass type texture and vice versa[2]. Different theories such as overshooting due to unequal hardening rates of the active slip systems and latent slip systems [3], cross-slip theory [4], non-octohedral slip[5], mechanical twinning [6] has been put forward to explain the texture transition in FCC metals.

The rolling texture of copper sharpens with increasing rolling reduction while the brass texture is developed from a complex pattern of changes. The early stages of texture development as shown from X-ray pole figures are essentially identical in both copper and brass. For rolling reductions beyond 40%, the copper texture component $\{112\}<111>$ starts to decrease and its twin $\{552\}<115>$ emerges with a gradual build up of the $\{111\}<uvw>$ components which reach a maximum at $\sim$ 80-85% reduction and thereafter decrease. The $\{110\}$ orientations increase initially but change very little from 40% to 80% reduction. At rolling reduction greater than 90%, there is progressively sharpening of the $(110)<112>$ orientations and the characteristics is well developed. On the other hand, there is very little material left with the $\{112\}<111>$ texture components at such high strain. Further treatment of the subject may be found in review articles written by, Dillamore and Roberts [7], Sowerby and Johnson[8], and Gil Sevillano et al.[9].

9.2 Prediction of Lattice Rotation

When a polycrystalline metal is deformed plastically, a change of shape of the grains occurs. This deformation is described by the sum of a pure strain and rotation. Individual grains rotate to give a more stable configurations determined by the initial grain orientation and the imposed strain tensor. Plastic deformation occurs

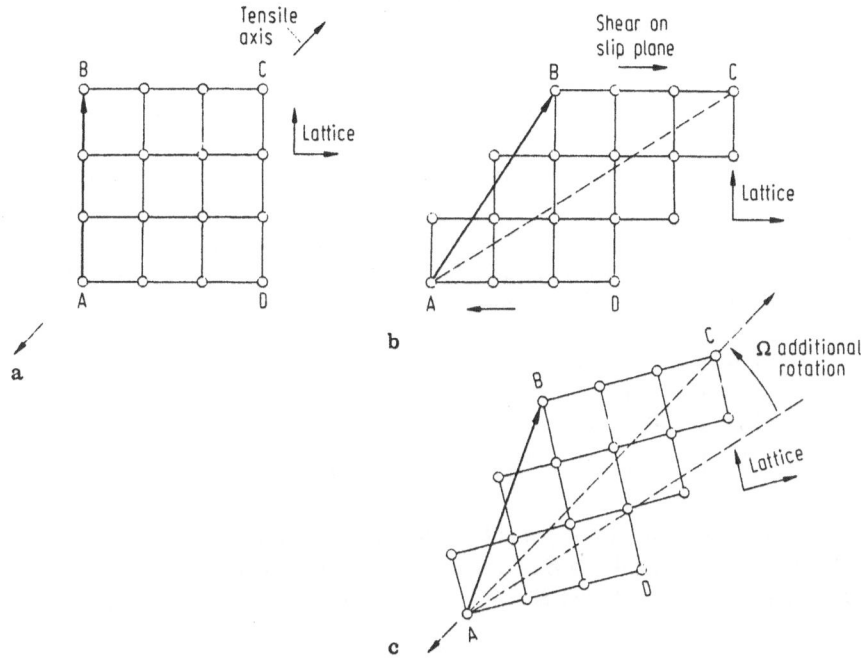

Fig. 9.2. Rotation of a crystal subjected to an uniaxial tension along the tensile axis AC. (From Gil. Sevillano, J., van Houtte, P and Aernoudt, E., Progress in Materials Science, 25; 1981, p.293.)

generally by crystallographic slip processes (as well as twinning in low stacking fault energy materials). As there is only a limited number of slip systems available, the grains will rotate towards a limited number of stable end configurations, thus creating a deformation texture. Fig. 9.2 illustrates how a crystal reorientates itself under a simple uniaxial tension.

Under simple slip, the shear τ on the slip plane will not cause the crystal lattice to rotate although a material vector such as AB may rotate. Since the tensile axis AC will remain fixed in space during the test, an additional lattice rotation Ω will be needed to bring the crystal in a position to the strain force upon it in the direction of elongation AC. Mathematically, the initial orientation matrix T will rotate to a new orientation T' given by

$$T' = (I - \Omega)\,T \tag{9.1}$$

where

$$T = \begin{bmatrix} r_1 & u_1 & n_1 \\ r_2 & u_2 & n_2 \\ r_3 & u_3 & n_3 \end{bmatrix} \tag{9.2}$$

and the components of T are related to the Miller'indices of the crystal as given in equation (5.22) of Chapter 5. I is the unit matrix and the lattice rotation tensor

Ω is given by

$$\Omega = d\omega_{ij} = (1/2)\sum_{s=1}^{12}(a_j^s b_i^s - a_i^s b_j^s)\,\tau^s \tag{9.3}$$

where a_i^s and b_i^s are the components of the unit vectors denoting the slip plane normal and the slip direction respectively and the superscripts denote the slip system in question. All crystallographic data will refer to an orthogonal co-ordinate system which is fixed on major crystal directions (i.e. the cube axis of the crystal). τ^s is related to the shear strain increment of the five independent slip systems as follows:

$$\begin{bmatrix} \epsilon_{22} \\ \epsilon_{33} \\ \epsilon_{23} \\ \epsilon_{13} \\ \epsilon_{12} \end{bmatrix} = \mathbf{E} \begin{bmatrix} \tau^1 \\ \tau^2 \\ \tau^3 \\ \tau^4 \\ \tau^5 \end{bmatrix} \tag{9.4}$$

where ϵ_{ij} are the strain components of the imposed strain tensor referred to the cube axes of the crystal and $\mathbf{E}$ is a square matrix that denotes the direction cosines of the slip systems such that $\mathbf{E} =$

$$\begin{bmatrix} b_2^1 a_2^1 & b_2^2 a_2^2 & b_2^3 a_2^3 & b_2^4 a_2^4 & b_2^5 a_2^5 \\ b_3^1 a_3^1 & b_3^2 a_3^2 & b_3^3 a_3^3 & b_3^4 a_3^4 & b_3^5 a_3^5 \\ b_2^1 a_3^1 + b_3^1 a_2^1 & b_2^2 a_3^2 + b_3^2 a_2^2 & b_2^3 a_3^3 + b_3^3 a_2^3 & b_2^4 a_3^4 + b_3^4 a_2^4 & b_2^5 a_3^5 + b_3^5 a_2^5 \\ b_1^1 a_3^1 + b_3^1 a_1^1 & b_1^2 a_3^2 + b_3^2 a_1^2 & b_1^3 a_3^3 + b_3^3 a_1^3 & b_1^4 a_3^4 + b_3^4 a_1^4 & b_1^5 a_3^5 + b_3^5 a_1^5 \\ b_1^1 a_2^1 + b_2^1 a_1^1 & b_1^2 a_2^2 + b_2^2 a_1^2 & b_1^3 a_2^3 + b_2^3 a_1^3 & b_1^4 a_2^4 + b_2^4 a_1^4 & b_1^5 a_2^5 + b_2^5 a_1^5 \end{bmatrix}. \tag{9.5}$$

For a crystal to undergo any arbitrary shape change, 5 independent slip systems would be necessary. The set of equations (9.3) contains 5 linear equations with n unknowns, where n is the number of potential slip systems (and potential twinning systems in some metals). There are many ways of choosing the 5 independent slip systems out of the 12 possible ones in cubic crystal, and the calculations would become formidable. Most predictions are now based on the Taylor model[10] of the minimum deformation work for each crystal, as modified subsequently by Bishop and Hill[11,12]. Based on these models, each grain undergoes the same prescribed strain as the polycrystalline aggregates. Five independent slip systems are required in general, and they must satisfy the single crystal yield criteria and flow rule. With fully constrained deformation and equal hardening on all slip systems, a yield surface with 56 corners in 6-dimensional space can be constructed and the problem is to discover which of the corners actually causes yielding. There are two types of stress states. One of them exerts equal shear stress on 8 active slip systems, while the other stress state stresses 6 slip systems equally. The selected stress state or corner will correspond to either 8 or 6 slip systems. Each of these corners, or stress state, has co-ordinates which in the Bishop and Hill's notation is a combination of the terms

$$A = (\sigma_{22} - \sigma_{33})/\sqrt{6}\tau \qquad (9.6)$$

$$B = (\sigma_{33} - \sigma_{11})/\sqrt{6}\tau \qquad (9.7)$$

$$C = (\sigma_{11} - \sigma_{22})/\sqrt{6}\tau \qquad (9.8)$$

$$F = \sigma_{23}/\sqrt{6}\tau \qquad (9.9)$$

$$G = \sigma_{31}/\sqrt{6}\tau \qquad (9.10)$$

$$H = \sigma_{12}/\sqrt{6}\tau. \qquad (9.11)$$

For each crystal, the work done dW for the prescribed strain state ϵ is given by

$$dW = \sigma_{ij} d\epsilon_{ij} \qquad (9.12)$$

which can be expressed as

$$dW = Bd\epsilon_{11} + Ad\epsilon_{22} + 2Fd\epsilon_{23} + 2Gd\epsilon_{31} + 2Hd\epsilon_{12}. \qquad (9.13)$$

The work is calculated for all 56 stress states and the stress state whose work done is the highest becomes the yielding stress state. The deviatoric stress can be determined uniquely, but the number of slip systems under the same stress is generally larger than 5 (e.g. 6 or 8 in fcc metals) such that there is an ambiguity in calculating the slip amplitude and hence the texture. This ambiguity can be removed if the "second order plastic term for the plastic work" is minimized as proposed by Renouard and Wintenberger[13]. According to their method, a small strain is applied to each possible set of the slip systems. This leads to a small crystal rotation. For a given imposed strain state, the second order plastic work dW' for each possible combination of slip system is determined from

$$dW' = \sigma'_{ij} d\epsilon'_{ij} \qquad (9.14)$$

where σ'_{ij} and ϵ'_{ij} are the new stress and strain components respectively due to crystal rotation and strain hardening. The set of slip systems which corresponds to the minimum second order plastic work will then be selected. For the deformation of a polycrystals, the Taylor theory assumes that the plastic strain of each grain is equal to the macroscopic plastic strain. All elastic strains are disregarded and it does not satisfy the stress equilibrium conditions. However, the internal stress field set up by the deforming grains at their boundaries would fulfil the stress continuity condition without influencing strain compatibility appreciably. A Pascal program for the calculation of crystal rotation in FCC metals as described in the above procedures is listed in Appendix 1 of this chapter.

The predictions carried out on this basis generally yield the copper-type texture in plane strain compression except that the resulting textures are much too sharp[14]. Such discrepancy between the prediction and the experimental results has been attributed to the micro-inhomogeneity of deformation which could be taken into account by considering non-unique solution of the Taylor theory [15,16,17,18,]. Different approaches have been proposed to improve the problem of non-unique solution. Honneff and Mecking[19] suggested the relaxed constraint model which relies on reducing the number of imposed strain rate components to four or three for those elongated or flattened grains. Other methods include the averaging method proposed by van Houtte and Aernoudt[20] and the rate sensitive model by Kocks[21], Tòth et al.[22], and Neale et al.[23].

The brass-type rolling texture is predicted by a Sachs-Kochendorfer approach[24]. The Sachs deformation pattern when applied with random stress produces texture close to experimental brass type but without the presence of the intermediate {111} components. Duggan et al.[25] in a survey of microstructural changes occurring in 70/30 α-brass observed that twinning was replaced by shear banding as the major deformation mode at rolling reduction in excess of about 50%, and the orientations within the shear bands contributed to the brass texture components. The crystallites in the shear band are extremely small (i.e. submicron) and the mode of deformation within the band is largely unknown. The origin of the brass-type rolling texture still remains unsolved.

9.3 Effect of Strain Path on Slip Rotation

Most of the simulations of deformation texture are performed assuming linear or proportional strain path. The stability of grain orientation is related to the curvature of strain path. Grain orientations which are unstable under a linear strain path may become stable under a non-linear strain path. The strain path represents a locus of successive states of strain followed by a material element and can be represented by the ratio of two principal strains. If the principal strain components increase in proportion with external loading, the path is said to be linear or proportional. When the successive strain varies, the path would be non-linear. The difference between the loading path and the strain path has to be clearly maintained. While the loading path is described by a stress ratio (i.e. σ_{22}/σ_{11}), the strain path is described by a strain ratio (i.e. $\epsilon_{22}/\epsilon_{11}$). For an isotropic material, the strain ratio ($\epsilon_{22}/\epsilon_{11}$) equals -0.5 in uniaxial tension and equals 1 in equibiaxial tension. With an anisotropic material, however, the strain path may not be linear and the ratio of the principal strains deviates from the isotropic cases under the same loading conditions. In macroplasticity, the stress ratio and strain ratio are connected by experimental parameters defined in the various phenomenological yield equations. Those parameters are determined by measuring the ratio of the width and thickness strains of tensile specimen prepared in the plane of the sheet at various angles to the rolling direction. Usually measurements are taken at one level of axial strain and the parameters are assumed to be constant with time.

The strain components that a material can take are related to the boundary conditions and the plastic anisotropy of the material. The boundary conditions will define the number of prescribed strain components for a given shape change. For example, under plane-strain conditions, the strain in the transverse direction is forbidden and a reaction stress will build up. The plastic anisotropy on the other hand will determine the value of the non-prescribed strain components. The strain path followed by a material should be expressed as a function of the material properties and a function of the external imposed constraints, i.e.,

Strain path $= f_1$ (reaction of material) f_2 (imposed stress)

The practice of assigning a fixed strain state for all subsequent deformation will give a strain path that is completely prescribed beforehand. If the strain path is fixed, the imposed stress has to be vary continuously. Strictly speaking, such a fixed strain ratio would not be valid even for an isotropic material as a crystallographic texture will develop eventually during the course of plastic deformation. The strain components would continuously vary and adjust themselves according both to the imposed stress state and the current anisotropy of the material. The strain path determined in this manner will be called a *non-prescribed path*.

A model for the prediction of the strain path of a polycrystalline sheet metal deformed under equibiaxial tension has been proposed by the Chan and Lee[26] and is outlined here. The strain path is defined by the locus of the successive states of the accumulative strain followed by a material element. Elastic strain is neglected and the large plastic deformation is treated as a succession of infinitesimal plastic strain. Further, principal stress axes are assumed to be coincident with principal strain axes. A small cubic element in a metal sheet with its edges parallel to the rolling, transverse and normal direction will be deformed into a rectangular parallelopid. The problem of locating the strain path for a single orientation or a given texture and imposed stress state of a sheet metal is equivalent to that of determining the strain tensor $\epsilon_{(s)(p)}$ of the polycrystals for each of incremental step of deformation, taking into account the grain rotation after each successive step. The imposed plane stress state of the polycrystal is denoted by

$$\sigma_{(s)(p)} = \begin{bmatrix} \sigma_{11} & 0 & 0 \\ 0 & \sigma_{22} & 0 \\ 0 & 0 & 0 \end{bmatrix} \tag{9.15}$$

where the subscripts s and p refer to the specimen axes and the polycrystals of the specimen, and σ_{11} and σ_{22} coincide with the rolling and transverse directions respectively of the sheet. The imposed stress ratio is assumed to be constant (such as in uniaxial tension or equibiaxial tension), and the stress path is linear during the incremental computations. The corresponding strain tensor for equation (9.15) can be written as

$$\epsilon_{(s)(p)} = \begin{bmatrix} X & 0 & 0 \\ 0 & Y & 0 \\ 0 & 0 & -X - Y \end{bmatrix} \tag{9.16}$$

where X and Y are dimensionless parameters to be determined. To reduce the amount of computation, the shear components ϵ_{12} and ϵ_{21} are assumed to be zero. The sheet is supposed to consist of N group of grains each with volume fraction f_g ($g = 1$ to N) and orientation $\{hkl\}<uvw>$. Each grain is assumed to undergo the same strain state as the polycrystal, i.e.

$$\epsilon_{ij(s)(g)} = \epsilon_{ij(s)(p)} \tag{9.17}$$

where the subscript (g) refers to an ideal grain orientation. Due to the stress equilibrium requirement, the summation of the stresses of the individual grains must be equal to the imposed $\sigma_{(s)(p)}$, giving

$$\sigma_{ij(s)(p)} = \sum_{g=1}^{N} f_g \epsilon_{ij(s)(g)}. \tag{9.18}$$

ϵ_{ij} is solved for each group of grains by applying the flow rule to an anisotropic yield loci of the Continuum Mechanics of Textured Polycrystals (CMTP)[27], which is obtained by fitting a continuous curve to the crystallographic Bishop and Hill yield surface[11,12]. Isotropic hardening of slip systems is assumed in the derivation of the yield surface. Since the crystallographic yield locus refers to the $<100>$ axis of the grain, $\epsilon_{ij(s)(g)}$ has to be transformed into the cube axis (c) by the relation

$$\epsilon_{(c)(g)} = \mathbf{P}\epsilon_{(s)(g)}\mathbf{P'} \tag{9.19}$$

where $\mathbf{P}$ is the transformation matrix and $\mathbf{P'}$ is the transpose of $\mathbf{P}$ (see also equation (5.21) and (5.22)).

The crystallographically based non-integer yield function of the CMTP is expressed as

$$F(S_{ij}) = q_1 \left[|S_{11} - S_{22}|^n + |S_{22} - S_{33}|^n + |S_{33} - S_{11}|^n \right]$$
$$+ 2q_2 \left[|S_{23}|^n + |S_{12}|^n + |S_{13}|^n \right] - 1 = 0 \tag{9.20}$$

where S_{ij} are the components of the deviatoric stress tensors referred to the crystal axes, q_1, and q_2 are coefficients chosen to minimize the root mean square distance between the rounded 'quasi' vertices of the crystallographic yield surface and the continuous function taken along the radii leading to the vertices, and n is an exponent greater than 1 or equal to 1 for convexity requirement and depends on the Gaussian distribution of the scatter width of the orientation[28].

The stress tensor is related to the strain tensor by applying the normality rule to the analytical yield function of equation (9.20) as follows:

$$\epsilon_{ij(c)(g)} = \lambda \partial F\left(S_{ij(c)(g)}\right) / \partial S_{ij(c)(g)}. \tag{9.21}$$

The calculation can be simplified by reducing the stress and strain tensor to one which contains only 5 independent components. The deviatoric stress tensor obtained from equation (9.21) is transformed back into the specimen axes, i.e.

$$S_{(s)(g)} = \mathbf{P'}S_{(c)(g)}\mathbf{P}. \tag{9.22}$$

To calculate $\epsilon_{(s)(p)}$, the parameters X and Y in equation (9.16) are varied until the $S_{ij(s)(p)}$ is equal to the components of the imposed stress tensor of equation (9.15) for an incremental deformation of $d\epsilon$. The selected $\epsilon_{(s)(p)}$ will then yield the first point on the strain path. For an incompressible material, the principal strains $\epsilon_{11}, \epsilon_{22}$ and ϵ_{33} are not independent and the strain path described by the ratio of two principal strains $(\epsilon_{11}/\epsilon_{22})$ can be plotted in a two dimensional coordinate system. A grain with an initial orientation $\mathbf{T}$ will rotate to a new rotation $\mathbf{T'}$ given by equation (9.1). The crystal rotation path depends on initial orientation $\{hkl\}<uvw>$ with

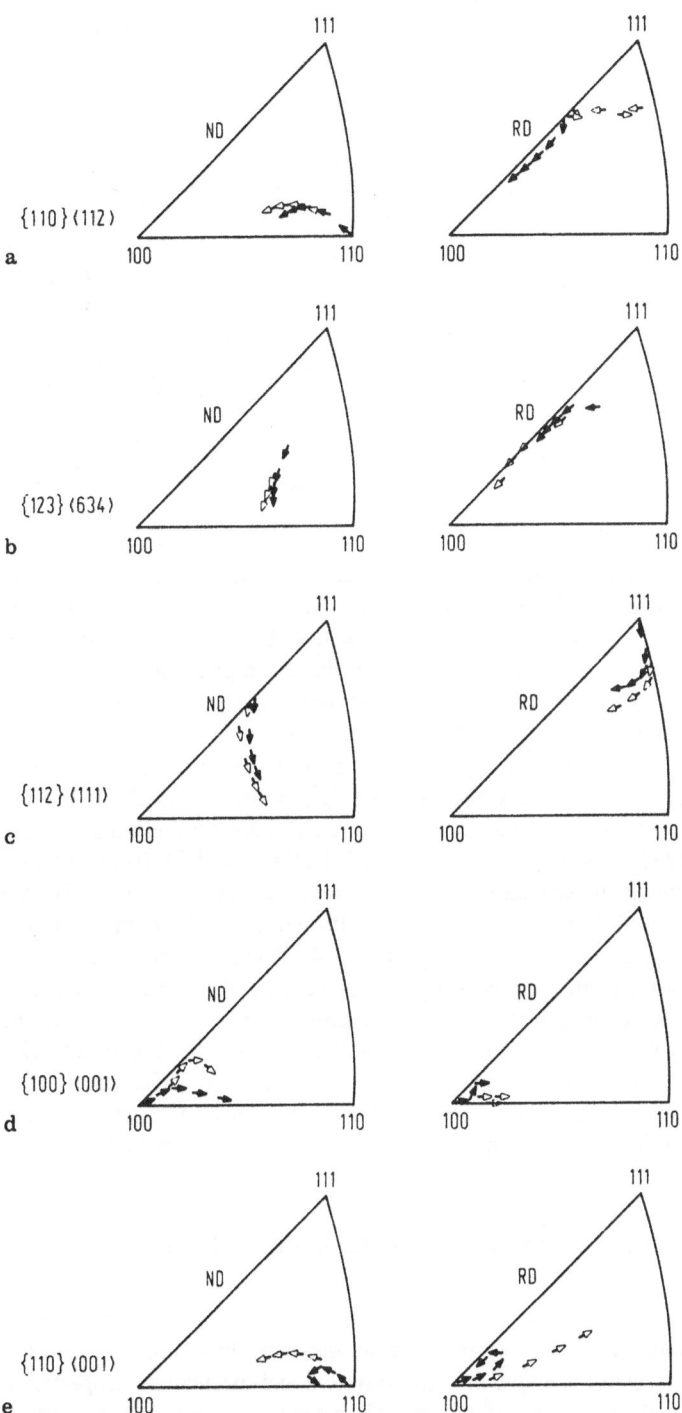

Fig. 9.3. Inverse pole figures for the simulated rotation paths under non-prescribed strain path ($\rightarrow$) and prescribed strain path ($\Rightarrow$) under equibiaxial tension (From Lee and Chan[29])

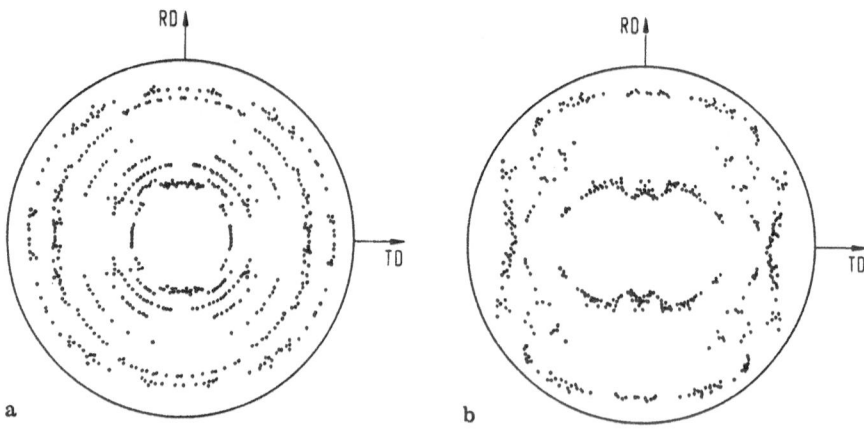

Fig. 9.4. {111} pole figures of simulated equibiaxial texture of C.P. aluminium under (a) prescribed strain path and (b) non-prescribed strain path.(From Lee and Chan[29])

respect to the specimen axes, the critical resolved shear stress of the slip systems and the strain state ϵ calculated from the above iteration procedures. The next strain state is calculated based on the new orientation T'. Repetition of the above procedures will yield a rotation path for each of the strain path followed.

The main difference in the calculation of the rotation path under a non-prescribed strain path and a prescribed strain path is that in the former case ϵ_{11} and ϵ_{22} are varied for each incremental step of deformation $(d\epsilon)$ but in the latter case they are assigned to be fixed. The inverse pole figures for the simulated rotation paths of the main texture components of commercially pure aluminium sheet (i.e., {110}<112>, {123}<634>, {112}<111>, {100}<001> and {110}<001>) under non-prescribed strain path and prescribed strain path are shown in Fig. 9.3, whereas the simulated {111} pole figures of the equibiaxial deformation texture of the aluminium sheet are shown in Fig. 9.4. There is a difference in the predicted deformation texture between prescribed strain paths and non-prescribed strain paths. The degree of clustering of orientations cannot be revealed from the two-dimensional pole figures. From a misorientation analysis, the texture obtained under the non-prescribed strain path is sharper and has more orientations close to {110}<100> and {110}<112>. The equibiaxial deformation texture predicted based on non-prescribed strain path agrees better with the experimental results.

9.4 Generation of New Orientations in Recrystallization

Annealing is an important part of the processing of metals and alloys for engineering applications. Not only the ductility of the work-hardened workpiece is restored, annealing causes changes in many physical and mechanical properties of the materials as well. The study of the annealing behaviour of metal requires an understanding of the structure of the deformed state, interactions between structural defects, and the thermodynamics and kinetics of solid interfaces. Although a lot of research effort and progress were made in areas such as production of extra deep drawing steels and grain oriented magnetic steels, the underlying mechanism of recrystallization remains an area which is less well understood.

Annealing refers to the heating of a cold-worked material and it involves serval stages of internal changes such as recovery, recrystallization and grain growth. Recovery refers to the very early stages of the annealing process during which internal stresses associated with the cold working are removed without significant change in the macroscopic mechanical properties. This stress relaxation is bought about by the reactions of point defects and the rearrangement of dislocations into a low energy configuration such as polygonization[31]. The process of formation and migration of high angle boundaries into cold-worked matrix (nucleation) until all the new grains are in mutual contact (growth stage) is termed recrystallization (Fig. 9.5). Further increase in the size of the grains is called grain growth. A general review of the subject can be found in the work of Beck and Hu [30], and Haessner and Hofmann[31].

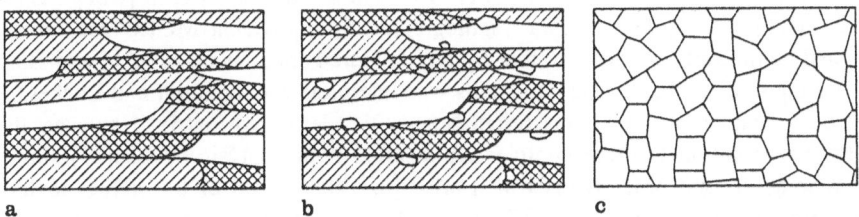

Fig. 9.5. Microstructural changes during recrystallization of metals. a Cold-worked state, b nucleation, c growth stage.

A central question in the recrystallization theories is "How are new orientations generated during recrystallization of the deformed metal?" These forms its recrystallization texture which in many cases differs greatly from the deformation texture shown before annealing. The origin of recrystallization texture is a subject which has been a source controversy for more than thirty years[32,33,34]. There are two main school of thoughts: the theory of *oriented growth* and the theory of *oriented nucleation*. The theory of oriented growth proposes that the nuclei are formed in random orientations and that selection growth decides which of them should grow to give rise to the recrytallization texture. In the theory of oriented nucleation, the new orientation is already present as a special feature of the deformation microstructure, e.g. transition bands, deformation twins, shear bands, etc..

The oriented growth theory is based on the growth selection in aluminium bi-crystals and artificial nucleation in single crystals. The maximum growth rate orientation (MGRO) is related to the matrix by a rotation of 40° around a common <111> axis. In the annealing of polycrystals the fastest growth orientations are those which are able to grow into several different rolling components. For example, the cube orientation is said to be a compromise orientation with respect to the different variants of the copper rolling texture. The oriented growth theory in its original form considers that the orientation dependence of grain boundary mobility is sufficient to account for the observed annealing texture and the role of nucleation is thought to be of no importance[35]. It was realized later by Hu[36] and Schmidt and Lücke[37] that not all 8 possible 40° <111> rotations exist and the selection of the 8 possible ones is determined by the availability of nuclei. The oriented growth theory has special appeal for the computer simulation of recrystallization textures as new textures can be obtained from the deformation textures by a simple matrix transformation.

The abundance of annealing twin formation in FCC metals with a low stacking fault energy has led to the suggestion that twins having the MGRO relationship to the cold-worked matrix formed at the early stage of recrystallization act as nuclei and determine the recrystallization texture[38]. According to Schmidt and Lücke[37], the {326}<835> recrystalization texture can be obtained as second order twins which generate in the spread of high deformation rolling texture at an orientation distance of 22° from {011}<211> in the bridge between {011}<211> and {011}<100>. Not every crystallographically equivalent MGRO can be obtained by multiple twinning so that equally favoured components would be missing from the recrystallization texture. This was interpreted by Schmidt and Lücke[33] to be a lack of nuclei of this orientation in the deformed matrix. However, the physical situation is more complex. A detailed twin analysis by Lee[39] shows that some second order twins which arise from the spread of the rolling texture possess equally favoured MGRO relationship nevertheless fail to appear in the recrystallization texture.

9.5 Lattice Curvature and Recrystallization

The nucleation event is the most critical stage in recrystallization annealing. The concept that nuclei are part of deformed matrix was first proposed by Burgers and Louwerse 1931[40]. Three conditions must be satisfied to generate a successful nucleus in the deformed structure. These are (i) a high local stored energy (small cell or subgrain size), (ii) steep spatial gradient of stored energy (wide cell size distribution) and (iii) sharp lattice curvature. Nuclei are found to form preferentially in regions whether the local degree of deformation is highest. Such sites include deformation bands, twin boundaries, shear bands, and grain boundaries. These special elements of the microstructure correspond to the most strongly deformed ones. In general their orientations deviate widely from the matrix orientation.

The orientation of the recrystallization nucleus is related to the components of the deformation texture as a consequence of textural rotation during slip. A theory of pre-formed nucleus have been proposed by Dillamore and Katoh[41]. By applying the Taylor theory, small lattice curvature is shown to develop where the rotations converge towards stable end orientations and large curvatures occur when there are rapidly divergent rotations. The rotation rates of crystal orientations having $<110>$ or $<111>$ axis parallel to the transverse direction of rolling are shown in Fig. 9.6.

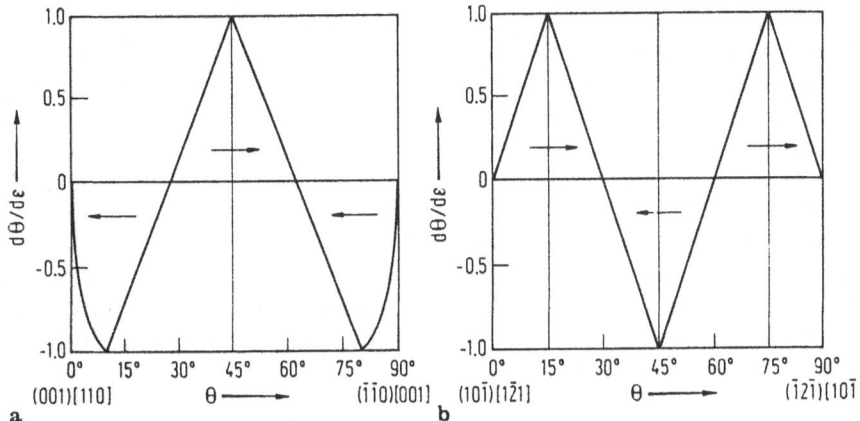

Fig. 9.6. Slip rotation rates for orientation having (a) $<1\ 1\ 0>$ and (b) $<111>$ parallel to the transverse direction of rolling (After Dillamore I.L. and Kotoh, H., Polycrystalline Plasticity and Texture Development in Cubic Metals, BSIRA Open Report, MG/39/71, 1871.)

Stable orientations are those for which the slope of the curve $d\theta/d\epsilon$ against θ is negative at $d\theta/d\epsilon = 0$. Metastable orientations are those for which the slope is positive. A metastable orientation may split into two orientation ranges linked by a transition band which accommodates the misorientation between the adjacent neighbours. The planes containing the stable orientations are called *convergent planes*, and those containing the metastable orientations *divergent planes*. Recrystallization nuclei usually form on these divergent or convergent planes.

The nucleation and growth of a successful grain in a deformed matrix is not an easy process. For a potential nucleus of 1 μm in size to grow into a recrystallized grain of 100 μm, the probability will be 1 in 10^6. The rarity of potential nuclei makes the refinement of grain size by recrystallization alone very difficult. At low strain, new grains are formed from the growth of subgrain and at grain boundaries. With increasing strain, nucleation will form more copiously in transition bands, recovery twins, and shear bands. A knowledge of the evolution of substructural state will be essential for a better understanding of the formation of recrystallization nuclei. Some typical nucleation events in single phase materials are described below:

(a) Nucleation from transition band

One of the simplest model for the nucleation of recrystallization is the growth of a polygonized subgrain[42]. During recovery, excess dislocations of the same sign align themselves into walls perpendicular to slip planes. This process is called polygonization and both slide and climb of edge dislocations are required to form the new walls. Since climb is essential for polygonization, the process will occur most easily in high or intermediate stacking fault energy materials. Such a subgrain, once formed, can grow by absorbing other dislocations in the sub-boundary. Subgrain boundaries thickly populated by dislocations quickly assume a high angle boundary and become a successful nucleus. If a subgrain is of the same size as its neigbours but misoriented by a larger than average amount will tend to shrink by virtue of its boundary energy. A subgrain must have a size advantage over its neighbours if it is to act as recrystallization nucleus. This condition was shown by Dillamore et al.[43] to be satisfied in the larger-than-average subgrain in a transition band (see Sec. 3.6). A quantitative model involving the growth of a subgrain in a transition band is shown schematically in Fig. 9.7.

As shown in Fig. 9.7, the vertical boundaries are parallel to the transition band and would be expected to have large misorientations. On annealing, the structure relaxes as shown in Fig. 9.7(b). The position at which the equilibrium boundary angle ϕ is reached is given by

$$\cos\phi = \sigma_r/\sigma_t \tag{9.23}$$

where σ_r is the energy of the random boundaries (horizontal boundaries) and σ_t is the energy of the transition boundary (vertical boundaries). Nucleation will occur if the points b and c meet each other before the boundary tension in segment ab restricts its movement. The critical length D_r for the subgrain to grow is given by

$$Dr \geq 4/3 \left[d_r + d_t \left(4\sigma_t^2/\sigma_r^2 - 1 \right)^{-1/2} \right] \tag{9.24}$$

where d_t and d_r is the mean spacing of the transition and random boundaries respectively.

Conditions likely to satisfy equation (9.24) would have an increasing dislocation density (a small d_r and a large sub-boundary misorientation Θ) from the centre of curvature to either site of the band. The dislocation density ρ relates approximately to d_r and Θ through

$$\rho = \Theta/bd_r \tag{9.25}$$

where b is the magnitude of the Burger's vector of the dislocation. As ρ increases with total shear strain, the parameters Θ and dr are also function of the Taylor factor M, which is defined as

$$M = \Sigma d\gamma/d\epsilon \tag{9.26}$$

where $\Sigma d\gamma$ is the sum of shears on defined slip systems and $d\epsilon$ is the largest principal strain. The rate of dislocation storage is a function of the M value. The larger the

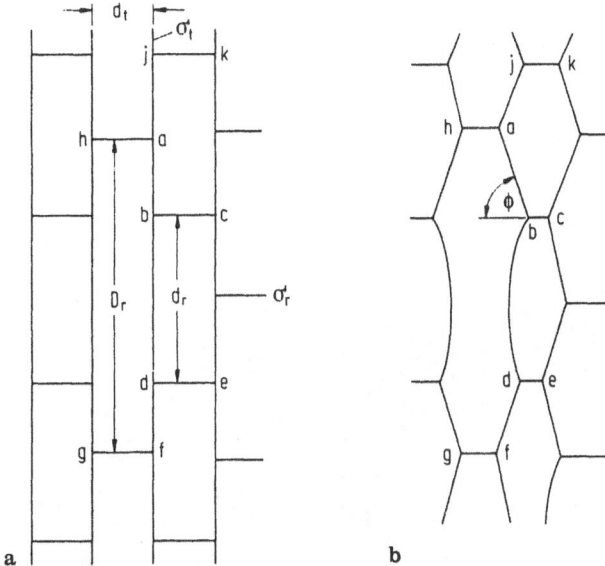

Fig. 9.7. Formation of recrystallization nuclei in a transition band. a Transition band as formed, b relax on annealing (From Dillamore et al.[43])

value of M, the cell boundary misorientation Θ is larger and the cell boundary spacing is smaller. The internal stored energy and the dislocation structures present in the various deformed grains of the cold-rolled sheet represent the driving force for recrystallization on annealing. Dillamore et al.[44] showed that the cell size and cell boundary misorientation measured by transmission electron microscopy in iron cold rolled to 70% vary in a systematic way according to the deformed grain orientation. Such orientation dependence of the deformation structure is responsible for biasing the recrystallization process in favour of certain texture components.

(b) Nucleation in shear bands

The structure of shear band and its formation has been discussed in Sec. 5.2. The dependence of nucleation on lattice curvature in FCC metals can be extended to low stacking fault energy materials to take into account of the occurrence of shear bands. An example of nucleation in shear band is shown in the electron micrograph of Fig. 9.8. The effect of shear band formation in generating lattice curvature is large. Crystal rotation rates $d\Theta/d\gamma$ in shear deformation mode for orientations having $<110>$ or $<111>$ parallel to the transverse direction (TD) of the shear plane have been calculated by Dillamore and Bush[45] as follows:

(i) $<111>$ TD

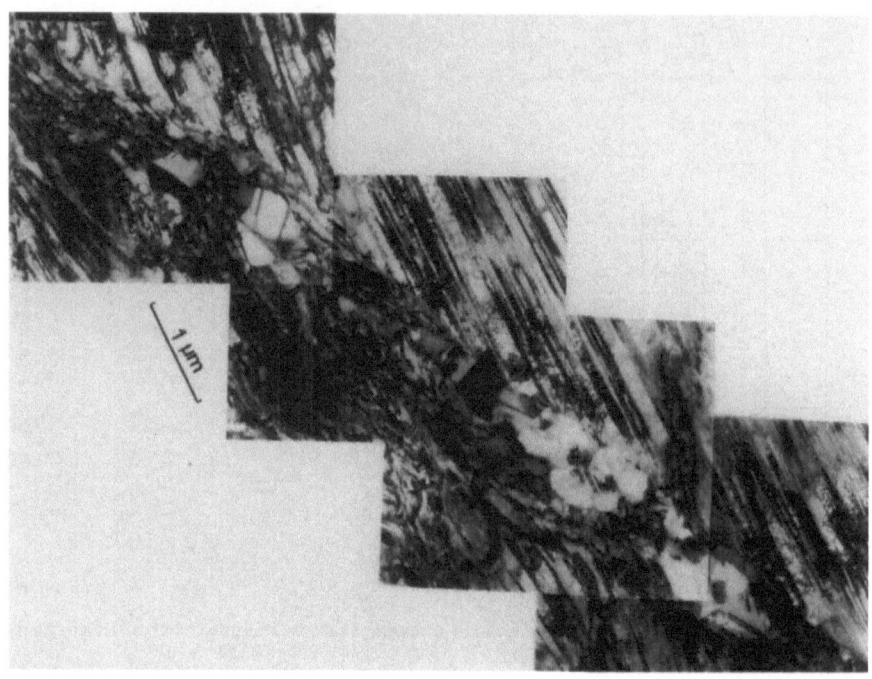

Fig. 9.8. Transmission electron micrograph showing nucleation of a new grain in a shear band in α-brass. Rolling direction is parallel to micron marker

$$d\Theta/d\gamma = \sin(\Theta + \beta)\left[\sqrt{3}\cos(\Theta + \beta) + \sin(\Theta + \beta)\right] \quad 0° \leq \Theta + \beta \leq 30° \quad (9.27)$$

$$= 3\cos^2(\Theta + \beta) - \sin^2(\Theta + \beta) \qquad\qquad 30° \leq \Theta + \beta \leq 60° \quad (9.28)$$

$$= \sin(\Theta + \beta)\left[\sin(\Theta + \beta) - \sqrt{3}\cos(\Theta + \beta)\right] \quad 60° \leq \Theta + \beta \leq 90° \quad (9.29)$$

(ii) *<110> TD*

$$d\Theta/d\gamma = \cos^2(\Theta + \beta)\left[\sqrt{2}\tan(\Theta + \beta) - 1\right] \qquad 0° \quad \leq \Theta + \beta \leq 35°16' \quad (9.30)$$

$$= \cos^2(\Theta + \beta)\left[1 - 2\tan^2(\Theta + \beta)\right] \qquad 35°16' \leq \Theta + \beta \leq 54°44' \quad (9.31)$$

$$= -\cos^2(\Theta + \beta)\left[\sqrt{2}\tan(\Theta + \beta) + 1\right] \quad 54°44' \leq \Theta + \beta \leq 90° \qquad (9.32)$$

where Θ is the angle between the rolling plane and the {110} plane in the transverse direction zone and β is the angle between the rolling plane and shear plane. Alternatively the rotation rate can be calculated from equation (9.1) to (9.14) with an externally imposed shear strain. The rotation of crystals in a shear band contained in grain with a <110> TD after shear strains of 0.5 and 2 are illustrated in Fig. 9.9 for a shear band angle of 35°16'. Θ_i denotes the initial orientation and Θ_f

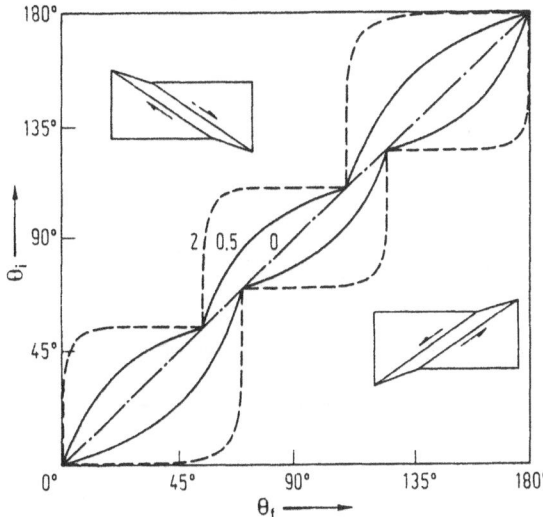

Fig. 9.9. Graph showing initial and final orientation of crystal in a <110> TD zone for different shear strain increments in a 35° shear band. (After Dillamore and Bush[45])

denotes the final orientation after a shear strain increment. The position $\Theta_i = 0°$ corresponds to the {110}<001> orientation.

From the diagram it can be seen that very large crystal rotation are experienced by crystals in the <110> TD zone except the texture component {4 4 11}<11 11 $\bar{8}$> which is stable within an orientation of ±8°. For example a crystal 5° from {110}<001> will be rotated 60° in a 35° shear band at an imposed shear strain of 0.5. {1 1 0}<1 1 2> orientation is similar to {4 4 11}<11 11 $\bar{8}$> and is confined within a small spread of ±5°. The above results hold for a 35° shear band only. A shear band angle of 30°, 27° and 22° will make {110}<1$\bar{1}$2> and {44 11}<11 11 $\bar{8}$> unstable respectively. All curvatures generated in the shear deformation give rise to orientations in shear bands having transverse rolling direction in a {110} plane. Such curvature is important in stimulating recrystallization. This is supported by the experimental evidence[46] that about 45% of 100 recrystallized new grains from the shear bands have their transverse directions parallel to the <110> crystal axis.

(c) Nucleation from recovery twins

Twin formation has been observed at very early stages of recrystallization in low stacking fault materials. The first structural change after a short time anneal in heavily cold-rolled α-brass is the occurrence of recovery twins in strain-free matrix[47]. These twins could be distinguished from the deformation twins and annealing twins and developed in the crystallites of shear bands at an early stage of

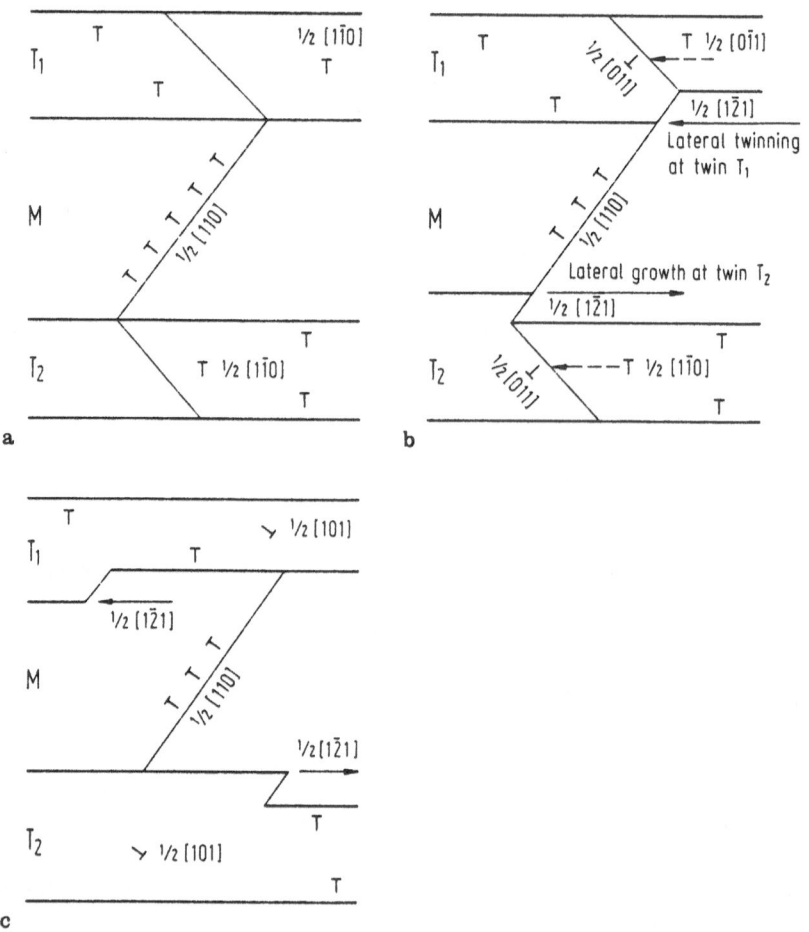

Fig. 9.10. Origin of recovery twin (After Verbraak[48])

heating. The formation of recovery twins corresponds well with the temperature range of pre-crystallization release of stored energy.

Verbraak[48] suggested that formation of a recovery twin is a low stacking fault energy equivalent of polygonization in FCC metals with a high stacking fault energy. A schematic drawing of the possible origin of recovery twin is shown in Fig. 9.10.

Deformation twins are supposed to form on {111} planes. During deformation, pile ups of dislocation exist between two deformation twins. On annealing, the stresses generated by these pile ups cause the dislocations to cross the twin boundaries and react with the dislocations already existed in the deformation twins. These reactions will cause a reduction in the dislocation density. According to this mechanism, some of the deformation twins will be eliminated and some will show a lateral growth.

9.6 Recrystallization Textures of Stretched Sheet Metals

Recrystallization textures in rolled sheet metals have been extensively studied but relatively little has been done in sheet metal deformed under complex states of stress. In the industrial multi-stage forming of sheet metal components, process annealing is carried out to restore the ductility of the metal for further forming. The recrystallization texture at this stage depends on the strain history of the formed component and will affect the subsequent processing step.

There is an orientation dependence of the recrystallized structures or substructures which relates to the amount of dislocation movement, and to the strain history of the deformed material. The effect of strain path (see Sec. 9.3) on recrystallization texture development has been shown by Lee et al.[49] in aluminium alloys. Consider the case of two grains which would deform independently if there is no constraints are now prevented from doing so because of the requirement of material continuity. Elastic stresses will set up between the two grains. i.e.

$$\epsilon_g = \epsilon_s + C(\sigma_g - \sigma_s) \tag{9.23}$$

where ϵ_g = grain deformation tensor
ϵ_s = sample deformation tensor
σ_g = grain stress tensor
σ_s = sample stress tensor
C = a proportionality factor.

Case (i) $C = 0$, both grain orientations stable.

If a grain has a low dislocation activity in accommodating strain, then it has a low value of Taylor factor M. Hence for two neighbouring grains A and B deforming with the same macroscopic shape change as shown by the same deformation path in Fig. 9.11. If $M_A < M_B$, it follows that the respective dislocation densities ρ_A and ρ_B are of magnitude $\rho_A < \rho_B$. Recovery is assumed to be faster in A with lower dislocation storage. Furthermore, since growth of the nucleus arises from the low energy blocks then it is likely that grain A will consume B. Nucleations at strain inhomogeneity such as transition bands and shear bands are ignored as both grain orientations are assumed to be stable.

Case (ii) $C \neq 0$, orientation A stable and B metastable.

In general $C \neq 0$, the difference in strain between grain A and grain B and the macroscopic deformation will be taken up gradually from the grain interior to the

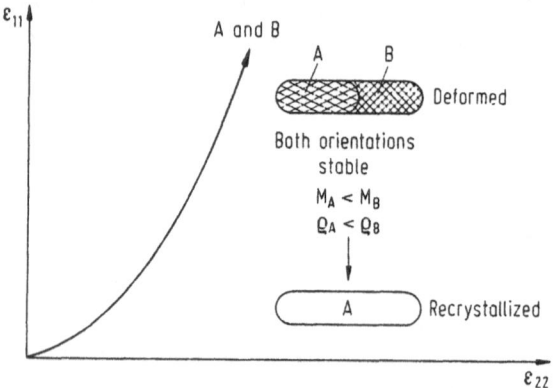

Fig. 9.11. Recrystallization of two grains deformed with the identical strain path. Both grain orientations stable. (From Lee et. al.[49])

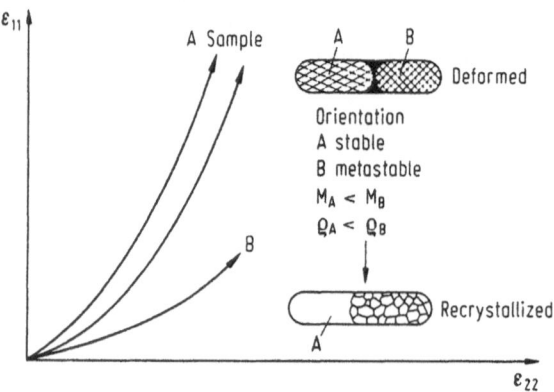

Fig. 9.12. Recrystallization of two grains deforming with different strain path. Orientation A stable and B metastable. (From Lee et.al.[49])

grain boundaries where the deformation will be higher, raising the probability of nucleation in the grain boundary region. Whether the new grain will grow and survive or not depends on the relative dislocation density and grain boundary mobility with its neighbour. If the orientation A is stable, it would recover first and have a size advantage over newly recrystallized grains nucleated from the heavily deformed grain boundary regions.

The different recrystallization behaviours of aluminium sheets deformed in either uniaxial tension or equibiaxial tension are shown in Fig. 9.13. Under uni-axial tension, there is little non-compliance of the major texture components with the macroscopically imposed shape change (Fig. 9.14), i.e. deformation is fairly

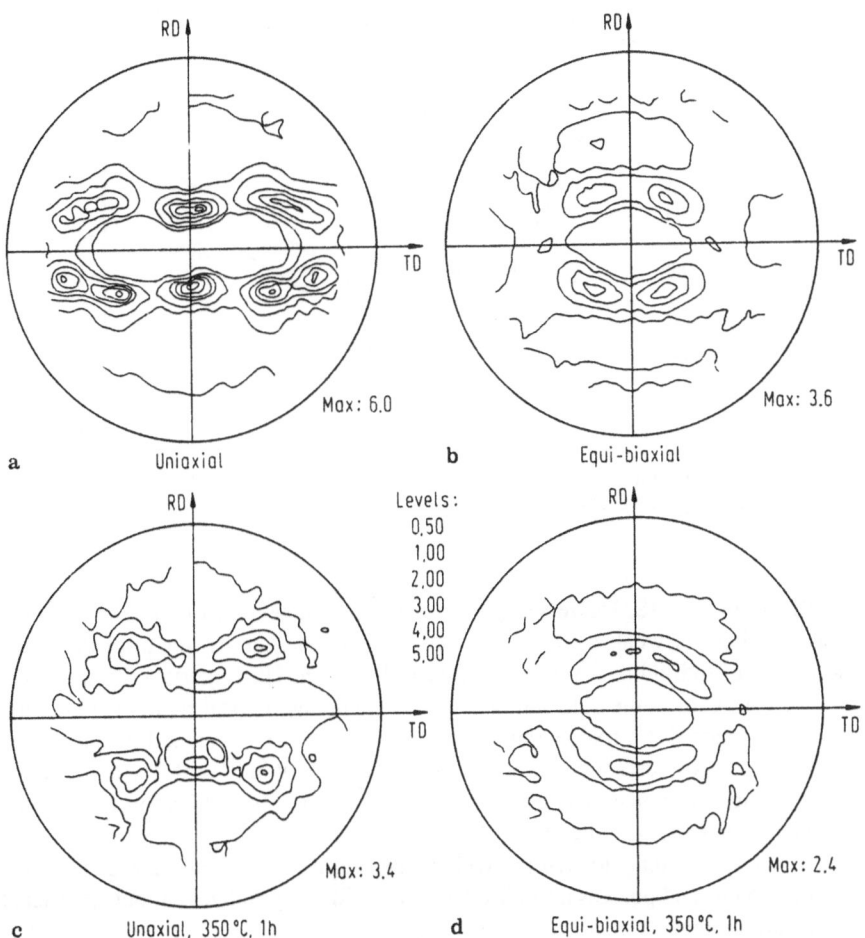

Fig. 9.13. {111} pole figures of deformed and annealed commercially pure aluminium sheets (From Lee et. al.[49])

uniform throughout the sample, and these orientations are relatively stable. Cube grains tend to recover most readily and have some preference over other grains during annealing because among other things, they have the lowest Taylor factor (2.45). In equibiaxial tension, there is larger deviation between the strain paths of the major single orientations and the macroscopic strain path of the whole sheet (denoted by curve C in Fig. 9.14). The largest differences occur in {112}<111> and {100}<001> texture components. Upon equibiaxial stretching, the {112}<111> and {100}<001> components rotate rapidly to other orientations and are destroyed rapidly. The {110}<001> orientation having a lower Taylor factor value over {110}<112> (i.e. 2.57 cf. with 3.47) will be slightly favoured in the final re-

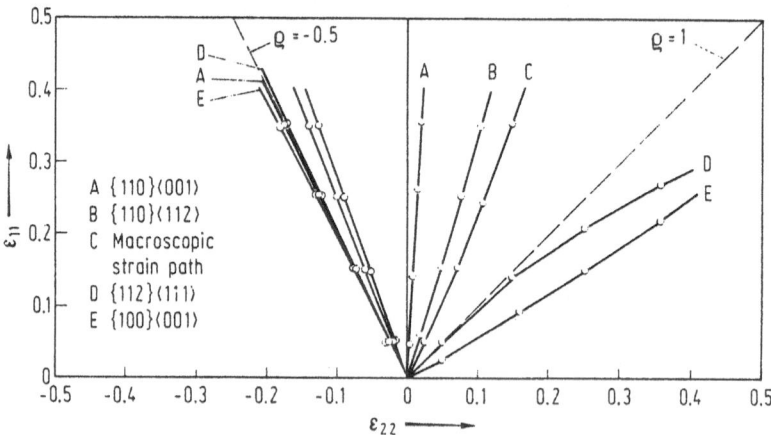

Fig. 9.14. The strain path taken by some of the major texture components in an commercially pure aluminum sheet (From Lee et al.[49])

crystallization texture. However, the recrystallization texture of the equibiaxially deformed sheet is weak as a result of copious nucleation arising from divergent strain paths of individual crystals under biaxial tension. The importance of using the correct strain path in the calculation of the M-value and crystal rotation should not be underestimated. If a fixed strain path (i.e. $\epsilon_{22}/\epsilon_{11} = 1$) is assigned in equibiaxial tension, the M-values calculated for $\{110\}<112>$ and $\{110\}<001>$ orientations will show no difference. The effect of strain path on recrystallization texture can thus be summarized as follows:

(a) Grains which tend to deform with macroscopic shape change (i.e. identical strain path) and possess a low Taylor factor will recrystallize without significant texture changes.

(b) Copious nucleation leading to randomization of the texture will be expected where there is a difference in the strain path taken by the individual grains with the macroscopic strain path.

Appendix 9

Pascal program for the calculation of crystal rotation in FCC metals.
```
{========================================================= }

UNIT ROTUNIT;

{ THIS UNIT DETERMINES CRYSTAL ROTATION.              }
(                                                      }
{ THIS UNIT CONSISTS OF 8 PROCEDURES                  }
{ 'GETARRAYS' AND 'GETSLIPSYS RETRIEVE DATA FROM      }
```

```
{  DATA FILES.                                             }
{ 'DIRECTION' DETERMINES COMPONENTS OF A ORIENTATION       }
{  MATRIX.                                                 }
{ 'STRAINS' DETERMINES STRAIN COMPONENTS REFERED TO        
{  CRYSTAL AXES.                                           }
{ 'SSELECT' SELECT SLIP SYSTEMS BASED ON THE PRINCIPLE     }
{  OF  SECOND ORDER MINIMUM WORK.                          }
{ 'CYCLE' GIVES THE NEW ORIENTATION.                       }


INTERFACE

USES PRINTER,FCSSYS;

{ THE UNIT 'PRINTER' ENABLES THE OUTPUT SEND TO PRINTER  }
{ THE UNIT 'FCSSYS' DETERMINES TAYLOR FACTORS AND SHEAR  }
{ STRAINS OF THE SETS OF SLIP SYSTEMS SELECTED ACCORDING }
{ TO TAYLOR/BISHOP AND HILL METHOD.                      }


TYPE  ROW=ARRAY[1..300,1..3] OF EXTENDED;

VAR SEQ,LL,ZZ,KK,Z,L:INTEGER;
    R,W,ROT:ARRAY[1..3] OF EXTENDED;
    D,G,ER:ARRAY[1..3,1..3] OF EXTENDED;
    RR2,NN2,RR1,NN1,UU1:ROW;
    EMD,MAX,SMD,SMD1,INC,RES:EXTENDED;
    LET:CHAR;


PROCEDURE GETARRAYS;
PROCEDURE GETSLIPSYS;
PROCEDURE ROTATE;
PROCEDURE DIRECTIONS(VAR NN1,RR1:ROW);
PROCEDURE STRAINS;
PROCEDURE LSELECT;
PROCEDURE SSELECT;
PROCEDURE CYCLE;


IMPLEMENTATION


{---------------------------------------------------------}

PROCEDURE GETARRAYS;

{ THIS PROCEDURE READS DATA OF                           }
{ (i)   STRESS COMPONENTS MAT[1..28,1..5] BASED ON       }
{       BISHOP AND HILL'S YIELD SURFACE FROM THE DATA    }
{       FILE BHMAT.DAT                                   }
{ (ii)  DIRECTION COSINES OF THE SLIP SYSTEMS            }
{       MX[1..8,1..12] FROM THE DATA FILE                }
{       STROTE.DAT                                       }


VAR MAT,CAT:FILE OF INTEGER;
BEGIN
ASSIGN(MAT,'BHMAT.DAT');
```

```
RESET(MAT);
FOR K:=1 TO 28 DO
BEGIN
FOR I:=1 TO 5 DO
 BEGIN
 READ(MAT,AR[K,I]);
 END;
END;
CLOSE(MAT);
ASSIGN(CAT,'STROTE.DAT');
RESET(CAT);
FOR K:=1 TO 8 DO
 BEGIN
 FOR I:=1 TO 12 DO
  BEGIN
  READ(CAT,MX[K,I]);
  END;
 END;
CLOSE(CAT);
END;



(------------------------------------------------------}


PROCEDURE GETSLIPSYS;
                                                      }
( THIS PROCEDURE READS DATA OF                        }
{ THE POSSIBLE SLIP SYSTEMS FOR EACH STRESS STATE     }
{ ACCORDING TO THE BISHOP AND HILL'S METHOD.          }


VAR WS1,WS2:FILE OF INTEGER;
BEGIN
ASSIGN(WS1,'WS1F.DAT');
RESET(WS1);
FOR I:=1 TO 12 DO
 BEGIN
 FOR K:=1 TO 8 DO
  BEGIN
  READ(WS1,SX[I,K]);
  END;
 END;
CLOSE(WS1);
ASSIGN(WS2,'WS2F.DAT');
RESET(WS2);
FOR I:=1 TO 16 DO
 BEGIN
 FOR K:=1 TO 6 DO
  BEGIN
  READ(WS2,SY[I,K]);
  END;
 END;
CLOSE(WS2);
END;


{------------------------------------------------------}
```

```
PROCEDURE DIRECTIONS(VAR NN1,RR1:ROW);

{ THIS PROCEDURE GIVES THE ORIENTATION MATRIX              }


BEGIN
SMD:=SQRT(SQR(NN1[SEQ,1])+SQR(NN1[SEQ,2])+SQR(NN1[SEQ,3]));
FOR I:=1 TO 3 DO NN1[SEQ,I]:=NN1[SEQ,I]/SMD;
SMD1:=SQRT(SQR(RR1[SEQ,1])+SQR(RR1[SEQ,2])+SQR(RR1[SEQ,3]));
FOR I:=1 TO 3 DO RR1[SEQ,I]:=RR1[SEQ,I]/SMD1;
UU1[SEQ,1]:=NN1[SEQ,2]*RR1[SEQ,3]-NN1[SEQ,3]*RR1[SEQ,2];
UU1[SEQ,2]:=NN1[SEQ,3]*RR1[SEQ,1]-NN1[SEQ,1]*RR1[SEQ,3];
UU1[SEQ,3]:=NN1[SEQ,1]*RR1[SEQ,2]-NN1[SEQ,2]*RR1[SEQ,1];
D[1,1]:=RR1[SEQ,1]*COS(VDEG)+UU1[SEQ,1]*SIN(VDEG);
D[2,1]:=RR1[SEQ,2]*COS(VDEG)+UU1[SEQ,2]*SIN(VDEG);
D[3,1]:=RR1[SEQ,3]*COS(VDEG)+UU1[SEQ,3]*SIN(VDEG);
D[1,2]:=-RR1[SEQ,1]*SIN(VDEG)+UU1[SEQ,1]*COS(VDEG);
D[2,2]:=-RR1[SEQ,2]*SIN(VDEG)+UU1[SEQ,2]*COS(VDEG);
D[3,2]:=-RR1[SEQ,3]*SIN(VDEG)+UU1[SEQ,3]*COS(VDEG);
D[1,3]:=NN1[SEQ,1];
D[2,3]:=NN1[SEQ,2];
D[3,3]:=NN1[SEQ,3];
END;



{---------------------------------------------------------}

PROCEDURE STRAINS;

{ THIS PROCEDURE DETERMINES COMPONENTS OF A STRAIN TENSOR}
{ REFERRED TO CRYSTAL AXES                               }

BEGIN
FOR I:=1 TO 3 DO
 BEGIN
 FOR J:=1 TO 3 DO
  BEGIN
  RES:=0;
  FOR K:=1 TO 3 DO
   BEGIN
   FOR L:=1 TO 3 DO
    BEGIN
    RES:=RES+D[I,L]*D[J,K]*G[L,K];
    END;
   END;
   ER[I,J]:=RES;
  END;
 END;
E[1]:=ER[1,1];
E[2]:=ER[2,2];
E[3]:=ER[3,2]+ER[2,3];
E[4]:=ER[3,1]+ER[1,3];
E[5]:=ER[1,2]+ER[2,1];
R[1]:=ER[2,3]-ER[3,2];
R[2]:=ER[3,1]-ER[1,3];
R[3]:=ER[1,2]-ER[2,1];
END;
```

```
{---------------------------------------------------------}

PROCEDURE SSELECT;

{ THIS PROCEDURE SELECT SLIP SYSTEMS BASED ON THE        }
{ PRINCIPLE OF MINIMUM SECOND ORDER   WORK               }
                          }

VAR M1:ARRAY[1..16] OF EXTENDED;
BEGIN
IF ZZ=1 THEN KK:=1 ELSE BEGIN
 FOR LL:=1 TO ZZ DO
  BEGIN
   FOR KK:=1 TO 3 DO
    BEGIN
    W[KK]:=ROT[KK];
    FOR I:=1 TO 12 DO W[KK]:=W[KK]+SR[LL,I]*MX[(5+KK),I];
    W[KK]:=W[KK]*INC/2;
    END;
   RR2[SEQ,1]:=RR1[SEQ,1]-W[3]*RR1[SEQ,2]+W[2]*RR1[SEQ,3];
   RR2[SEQ,2]:=RR1[SEQ,1]*W[3]+RR1[SEQ,2]-RR1[SEQ,3]*W[1];
   RR2[SEQ,3]:=-W[2]*RR1[SEQ,1]+W[1]*RR1[SEQ,2]+RR1[SEQ,3];
   NN2[SEQ,1]:=NN1[SEQ,1]-NN1[SEQ,2]*W[3]+NN1[SEQ,3]*W[2];
   NN2[SEQ,2]:=NN1[SEQ,1]*W[3]+NN1[SEQ,2]-W[1]*NN1[SEQ,3];
   NN2[SEQ,3]:=-W[2]*NN1[SEQ,1]+W[1]*NN1[SEQ,2]+NN1[SEQ,3];
   DIRECTIONS(NN2,RR2);
   STRAINS;
   EMVAL;
   M1[LL]:=EM;
   END;
  MAX:=10;
  FOR LL:=1 TO ZZ DO
   BEGIN
   IF M1[LL]<MAX
   THEN
    BEGIN
    MAX:=M1[LL];
    KK:=LL;
    END;
   END;
 END;
END;


{---------------------------------------------------------}

PROCEDURE CYCLE;

{ THIS PROCEDURE GIVES NEW CRYSTAL ORIENTATION           }

BEGIN
DIRECTIONS(NN1,RR1);
STRAINS;
FOR I:=1 TO 3 DO ROT[I]:=R[I];
EMVAL;
EMD:=EM;
SLIPSYS;
```

```
ZZ:=J;
SSELECT;
FOR LL:=1 TO 3 DO
 BEGIN
 W[LL]:=ROT[LL];
 FOR I:=1 TO 12 DO W[LL]:=W[LL]+SR[KK,I]*MX[(5+LL),I];
 W[LL]:=W[LL]*INC/2;
 END;
RR2[SEQ,1]:=RR1[SEQ,1]-W[3]*RR1[SEQ,2]+W[2]*RR1[SEQ,3];
RR2[SEQ,2]:=RR1[SEQ,1]*W[3]+RR1[SEQ,2]-RR1[SEQ,3]*W[1];
RR2[SEQ,3]:=-W[2]*RR1[SEQ,1]+W[1]*RR1[SEQ,2]+RR1[SEQ,3];
NN2[SEQ,1]:=NN1[SEQ,1]-NN1[SEQ,2]*W[3]+NN1[SEQ,3]*W[2];
NN2[SEQ,2]:=NN1[SEQ,1]*W[3]+NN1[SEQ,2]-W[1]*NN1[SEQ,3];
NN2[SEQ,3]:=-W[2]*NN1[SEQ,1]+W[1]*NN1[SEQ,2]+NN1[SEQ,3];
FOR I:=1 TO 3 DO
 BEGIN
 NN1[SEQ,I]:=NN2[SEQ,I];
 RR1[SEQ,I]:=RR2[SEQ,I];
 END;
END;
END.


{========================================================}

UNIT FCSSYS;

{ THIS UNIT CALCULATES TAYLOR FACTORS AND SHEAR STRAINS }
{ OF FIVE INDEDPENDENT SLIP SYSTEMS OF ALL THE          }
{ COMBINATIONS SELECTED FROM THE BISHOP AND HILL'S      }
{ METHOD. THE AMBIGUITIES ARE REMOVED IN THE UNIT       }
{ 'ROTUNIT' BASED THE SECOND ORDER MINIMUM WORK         }
{ PRINCIPLE.                                            }


{ THIS UNIT CONSISTS OF 7 PROCEDURES:                   }
{                                                       }
{ 'GETSYS' GETS ALL POSSIBLE SLIP SYSTEMS BASED ON THE  }
{  SELECTED STRESS STATE ACCORDING TO BISHOP AND HILL'S }
{  METHOD.                                              }
{ 'EMVAL' DETERMINES TAYLOR FACTORS.                    }
{ 'PARTIAL' CARRYS OUT THE PIVOTING PROCESS.            }
{ 'ELIMIN' CARRYS OUT THE TRIANGULARIZATION PROCES.     }
{ 'MATRIX' PERFORMS BACK SUBSTITUTION PROCESS.          }
{ 'JSET' REMOVES ANY REDUNDANT SETS OF SLIP SYSTEMS     }
{ 'SLPIPSYS' DETERMINES THE SHEAR STRAINS OF THE        }
{  SLIP SYSTEMS.                                        }



INTERFACE

USES POWER;

{ THE UNIT 'POWER' CALCULATES THE VALUE OF A VARIABLE   }
{ POW1 TO ITS POWER POW2                                }



CONST RT=2.4494897428;
```

```
VAR LAR,EM:EXTENDED;
    X,COT1,COT,I,K,J,P:INTEGER;
    AA:ARRAY[1..8] OF INTEGER;
    XS,B,E:ARRAY[1..5] OF EXTENDED;
    YY:ARRAY[1..5] OF INTEGER;
    A:ARRAY[1..5,1..5] OF EXTENDED;
    MX:ARRAY[1..8,1..12] OF INTEGER;
    AR:ARRAY[1..28,1..5] OF INTEGER;
    SR:ARRAY[1..16,1..12] OF EXTENDED;
    SX:ARRAY[1..12,1..8] OF INTEGER;
    SY:ARRAY[1..16,1..6] OF INTEGER;


PROCEDURE EMVAL;
PROCEDURE GETSYS;
PROCEDURE PARTIAL;
PROCEDURE ELIMIN;
PROCEDURE MATRICS;
PROCEDURE JSET;
PROCEDURE SLIPSYS;

IMPLEMENTATION


{----------------------------------------------------------}

PROCEDURE EMVAL;

{ THIS PROCEDURE DETERMINES THE TAYLOR FACTOR EM FROM    }
{ THE STRAIN COMPONENTS E[1..5] AND THE STRESS COMPONENTS}
{ AR[1..28,1..5].                                        }

VAR MAX:EXTENDED;

BEGIN
MAX:=0;
FOR I:=1 TO 28 DO
 BEGIN
EM:=AR[I,1]*E[1]+AR[I,2]*E[2]+AR[I,3]*E[3]+AR[I,4]*E[4]+AR[I,5]*E[5];
 IF EM>=MAX THEN
  BEGIN
  MAX:=EM;
  P:=I;
  END;
 IF -EM>=MAX THEN
  BEGIN
  MAX:=-EM;
  P:=-I;
  END;
 END;
EM:=MAX*RT/2;
END;



{----------------------------------------------------------}

PROCEDURE GETSYS;
```

```
{THIS PROCEDURE GETS THE DATA OF THE ACTIVE SLIP SYSTEMS }
{BASED ON THE SELECTED STRESS STATE DETEREMINED IN EMVAL }

BEGIN
IF ABS(P)<13 THEN
 BEGIN
 X:=8;
 FOR I:=1 TO X DO AA[I]:=SX[ABS(P),I];
 IF P<0 THEN
  BEGIN
  FOR I:=1 TO X DO AA[I]:=-AA[I];
  END;
 END;
IF ABS(P)>12 THEN
 BEGIN
 X:=6;
 FOR I:=1 TO X DO AA[I]:=SY[(ABS(P)-12),I];
 IF P<0 THEN
  BEGIN
  FOR I:=1 TO X DO AA[I]:=-AA[I];
  END;
 END;
END;




(-----------------------------------------------------------)


PROCEDURE PARTIAL;

( THIS PROCEDURE PERFORMS THE PROCESS OF PIVOTING BY     }
{ ROWS IN GAUSSIAN ELIMINATION METHOD TO SOLVE 5 LINEAR  }
{ EQUATIONS                                              }


VAR BDUM:EXTENDED;
     RO:INTEGER;
   ADUM:ARRAY[1..5] OF EXTENDED;

BEGIN
LAR:=0;
FOR I:=J TO 5 DO
 BEGIN
 IF ABS(A[I,J])>LAR THEN
  BEGIN
  LAR:=ABS(A[I,J]);
  RO:=I;
  END;
 END;
 IF LAR>0 THEN
  BEGIN
  FOR K:=J TO 5 DO
   BEGIN
   ADUM[K]:=A[J,K];
   A[J,K]:=A[RO,K];
   A[RO,K]:=ADUM[K];
   END;
  BDUM:=B[J];
```

```
  B[J]:=B[RO];
  B[RO]:=BDUM;
 END;
END;


{--------------------------------------------------------}

PROCEDURE ELIMIN;

{ THIS PROCEDURE PERFORMS THE TRIANGULARIZATION OPERATION}
{ AFTER PIVOTING PROCESS PERFORMED IN THE PROCEDURE      }
{ 'PARTIAL'                                              }


VAR MM:ARRAY[1..5,1..5] OF EXTENDED;

BEGIN
FOR J:=1 TO 4 DO
 BEGIN
 PARTIAL;
 IF LAR>0 THEN
  BEGIN
  FOR I:=J+1 TO 5 DO
   BEGIN
   MM[I,J]:=A[I,J]/A[J,J];
   FOR K:=J TO 5 DO
    BEGIN
    A[I,K]:=A[I,K]-A[J,K]*MM[I,J];
    END;
   B[I]:=B[I]-B[J]*MM[I,J];
   END;
  END;
 END;
END;



{--------------------------------------------------------}

PROCEDURE MATRICS;

{ THIS PROCEDURE PERFORMS THE BACK SUBSTITUTION OPERATION}
{ BASED ON THE RESULTS FROM THE PROCEDURE 'ELIMIN'       }

VAR Z:INTEGER;
    SUM:EXTENDED;

BEGIN
FOR I:=1 TO 5 DO
 BEGIN
 FOR K:=1 TO 5 DO
  BEGIN
  IF YY[K]>0 THEN A[I,K]:=MX[I,YY[K]] ELSE A[I,K]:=-MX[I,ABS(YY[K])];
  END;
 END;
B[1]:=E[1];
B[2]:=E[2];
B[3]:=E[3];
```

```
B[4]:=E[4];
B[5]:=E[5];
ELIMIN;
FOR I:=1 TO 5 DO
 BEGIN
 Z:=5+1-I;
 SUM:=0;
 IF A[Z,Z]<>0 THEN
  BEGIN
  FOR J:=Z+1 TO 5 DO SUM:=SUM+A[Z,J]*XS[J];
  XS[Z]:=(B[Z]-SUM)/A[Z,Z];
  END
  ELSE XS[Z]:=999;
  END;
END;



(--------------------------------------------------------}

PROCEDURE JSET;

{ THIS PROCEDURE REMOVES THE REDUNDANT SETS OF SLIP      }
{ SYSTEMS                                                }

VAR ANS,ANS1:INTEGER;
    NEG,POS:EXTENDED;

BEGIN
ANS1:=0;
NEG:=-POW(10,-18);
POS:=999;
IF (XS[1]>NEG) AND (XS[2]>NEG) AND (XS[3]>NEG) AND (XS[4]>NEG) AND (XS[5]>NEG)
THEN
 BEGIN
 IF (XS[1]<POS) AND (XS[2]<POS) AND (XS[3]<POS) AND (XS[4]<POS) AND
(XS[5]<POS)
 THEN
  BEGIN
  FOR I:=1 TO COT DO
   BEGIN
   ANS:=0;
   FOR K:=1 TO 5 DO
    BEGIN
    IF YY[K]>0 THEN
     BEGIN
     IF (SR[I,ABS(YY[K])]-XS[K]>NEG) AND (SR[I,ABS(YY[K])]-XS[K]<-NEG)
     THEN ANS:=ANS+1;
     END
     ELSE
     BEGIN
     IF (SR[I,ABS(YY[K])]+XS[K]>NEG) AND (SR[I,ABS(YY[K])]+XS[K]<-NEG)
     THEN ANS:=ANS+1;
     END;
    END;
    IF ANS=5 THEN ANS1:=ANS1+1;
   END;
   IF ANS1=0 THEN
    BEGIN
```

```
    COT:=COT+1;
    FOR K:=1 TO 12 DO SR[COT,K]:=0;
    FOR I:=1 TO 5 DO
     BEGIN
      IF XS[I]>0 THEN
       BEGIN
       IF YY[I]>0 THEN
       SR[COT,ABS(YY[I])]:=XS[I]
       ELSE
       SR[COT,ABS(YY[I])]:=-XS[I];
       END;
      END;
     END;
    END;
   END;
 END;
```


```
{----------------------------------------------------------}

PROCEDURE SLIPSYS;

{  'SLPIPSYS' DETERMINES THE SHEAR STRAINS OF THE         }
{   SLIP SYSTEMS.                                         }

VAR T,N,M,L,O:INTEGER;
    MIN:EXTENDED;
    MINS:ARRAY[1..12] OF EXTENDED;

BEGIN
GETSYS;
COT:=0;
FOR T:=1 TO X-4 DO
 BEGIN
 YY[1]:=AA[T];
 FOR N:=T TO X-4 DO
  BEGIN
  YY[2]:=AA[N+1];
  FOR M:=N TO X-4 DO
   BEGIN
   YY[3]:=AA[M+2];
   FOR L:=M TO X-4 DO
    BEGIN
    YY[4]:=AA[L+3];
    FOR O:=L TO X-4 DO
     BEGIN
     YY[5]:=AA[O+4];
     MATRICS;
     JSET;
     END;
    END;
   END;
  END;
 END;
MIN:=10;
COT1:=0;
FOR I:=1 TO COT DO
 BEGIN
```

```
 MINS[I]:=0;
 FOR K:=1 TO 12 DO MINS[I]:=MINS[I]+ABS(SR[I,K]);
 IF MIN>MINS[I] THEN MIN:=MINS[I];
 END;
FOR I:=1 TO COT DO
 BEGIN
 IF MINS[I]-MIN<=POW(10,-18) THEN
  BEGIN
  COT1:=COT1+1;
  FOR K:=1 TO 12 DO SR[COT1,K]:=SR[I,K];
  END;
 END;
J:=COT1;
END;
END.


UNIT POWER;
INTERFACE
FUNCTION POW(POW1:EXTENDED;POW2:EXTENDED):EXTENDED;
IMPLEMENTATION
FUNCTION POW(POW1:EXTENDED;POW2:EXTENDED):EXTENDED;
BEGIN
IF POW1=0 THEN POW:=0 ELSE POW:=EXP(POW2*LN(POW1));
END;
END.
```

9.7 References

1. Bunge,H.J., Phys.Stst.Sol., 26(1968), p.167.
2. Smallman, R.E., J.Inst.Metals, 84(1955), p.10.
3. Calnan, E.A. and Clews, C.J.B., Phil.Mag., 41(1950), p.1085.
4. Dillamore, I.L. and Roberts, Acta Metall., 12(1964), p.281.
5. Haessner, F., Z. Metallkunde, 54(1963), p.98.
6. Wassermann, G., Z. Metallkde. 54(1963), p.61.
7. Dillamore, I.L. and Roberts, W.T., Met.Review, 10(1965), p.39.
8. Sowerby, R. and Johnson, W., Met.Sci.Eng., 20(1975), p.101.
9. Gil. Sevillano, J., van Houtte, P and Aernoudt, E., Progress in Materials Science, 25(1981), p.69.
10. Taylor, G.I., J. Inst. Metals, 62(1938), p.307.
11. Bishop, J.F.W. and Hill, R., Phil.Mag. 42(1951), p.414.
12. Bishop, J.F.W. and Hill, R., Phil.Mag. 42(1951), p.1298.
13. Renourad, M and Wintenberger, M., C.r.Acad.Sci.Paris, B292(1971), p.385.
14. Bunge, H.J., Kristall u.Technik, 5(1971), p.145.
15. Dillamore, I.L. and Roberts, W.T., Acta Metall., 12(1964), p.281.
16. Aernoudt, E., Proc.of the 5th Int.Conf.on Textures of Materials, Germnay, 1(1978), 42.
17. Kocks, U.F. and Canova, G.R., in Deformation of Polycrystals (edited by Hansen, N. et al.), Riso National Laboratory, Denmark, (1981), p.35.
18. Bunge, H.J., Textures and Microstructures, 8&9(1988), p.55.
19. Honeff, H. and Mecking, H., Textures of Materials (edited by Gottstein, G. and Lucke, K.), Springer-Verlag, Berlin, 1978, p.265.
20. van Houtte, P. and Aernoudt, E., Z.Metallkde, 66(1975), p.428.
21. Kocks, U.F., in Constitutive Equations in Plasticity (edited by Argon, A.S.), MIT Press, Mass.,1975, p.81.

22. Tòth, L.S., Neale, K.W. and Jonas, J.J., Acta Metall., 37(1989), p.2197.
23. Neale, K.W., Tòth, L.S. and Jonas, J.J., Int.J.Plasticity, 6(1990), p.45.
24. Leffers, T., Phys.Stat.Sol., 25(1968), p.337.
25. Duggan, B.J., Hatherly, M., Hutchinson, W.B. and Wakefield, P.T., Met.Sci., 12(1978), p.293.
26. Chan, K.C. and Lee, W.B., Int.J.Mech.Sci., 32(1990), p.497.
27. Montheillet, F., Gilormini, P. and Jonas, J.J. Acta. Metall., 33(1985), p.705.
28. Bunge, H.J. and Roberts, W.T., J.Appl.Cryst. 2(1969), p.116.
29. Lee, W.B. and Chan, K.C., Textures and Microstructures, 13(1990), p.31.
30. Cahn, R.W., Recrystallization, Grain Growth and Textures, ASM, Ohio, (1965), p.99.
31. Hassner, F. and Hoffmann, S., Recrystallization of Metallic Materials, Stuttgart, (1978), p.63.
32. Hutchinson, W.B., Met.Sci., 8(1974), p.185.
33. Lücke, K., 7th Int. Conf. on Textures of Materials, Holand, p.195.
34. Hatherly, M., Recrystallization '90, Publication of the the Minerals, Metals and Materials Society, (1990), p.59.
35. Beck, P.A., Acta Metall., 1(1953), p.230.
36. Hu, H., Z.Metallkde., 60(1969), p.69.
37. Scmidt, U. and Lücke, K., Texture, 3(1979), p.85.
38. Gottestein, G., Acta Metall., 32(1984), p.1117.
39. Lee, W.B., unpublished work.
40. Burgers, W.G. and Louwerse, P.C., Z.Physik., 61(1931), p.605.
41. Dillamore, I.L., Katoh, H. Met.Sci., 8(1974), p.21.
42. Cahn, R.W., Proc.Phys.Soc., 60A(1950), p.323.
43. Dillamore, I.L., Morris, P.L., Smith, C.E.J., and Hutchinson, W.B., Proc. R.Soc.Lond., 329(1972), p.405.
44. Dillamore, I.L., Smith, C.J.E. and Watson, T.W., Met.Sci. 1(1968), p.49.
45. Dillamore, I.L. and Bush. A.C., 5th Int. Conf.on Textures of Materials, Achen, 1(1978), p.367.
46. Duggan, B.J. and Lee, W.B., in Annealing Processes, Recovery, Recrystallization, and Grain Growth (edited by Hansen, N. et al. eds), Riso National Laboratory, Denmark, (1986), p.297.
47. Huber, J. and Hatherly, M., Met.Sci., Dec., 1979, p.665.
48. Verbraak, C.A., 6th Int.Conf.on Textures of Materials, Tokyo, 1(1981), p.98.
49. Lee, W.B., Chan, K.C. and Duggan, B.J., in Recrystallization'90, Publication of the Minerals, Metals and Materials Society, (1990), p.723.

10 Prediction of Formability

10.1 Structure and Formability

The plastic deformation of a sheet metal into complicated shapes is brought about by the imposition of a combination of stresses. Complex stamping requires varying amount of stretching and drawing, to which bending and buckling are added. Traditionally, prediction of forming behaviour of materials are based on tensile test data, simulative tests and continuum mathematical models in which major input material variables are average macroscopic properties such as elongation, yield strength and work hardening rate. Tensile test results bear little correlation to the more complex stress states commonly encountered in sheet metal forming operations. For example, an aluminium alloy which has excellent formability when used for the deep drawing and ironing of beverage cans may perform very badly if it were stretched formed for an automobile panel.

The large plastic deformation encountered in the forming process can be explored from different length scales. Two approaches have been used in the past for the study of forming problems. These are (1) the macroplasticity approach and (2) the microplasticity approach. The major framework of continuum mechanics is developed at a time modern analytical techniques was not available for the characterization of material structures. A full quantitative characterization of micro- and meso-structures (see Sec. 3.4) becomes now possible due to the advances in modern data acquisition techniques. Examples of modern techniques which give quantitative data about material structures include image analyzer, quantitative crystallographic texture analysis, and quantitative X-ray analysis.

Failures in a wide spectrum of forming process are often associated with the development of strain inhomogeneity or specific fracture events at the microstructural scale. Examples of some typical failure modes in sheet metals are illustrated in Fig. 10.1. Not all failure modes are sensitive to microstructural parameters to the same extent. Some are dictated by the macroscopic properties and the boundary conditions while others are more sensitive to short range variations in the amplitude of the mesol-scale inhomogeneity[1] as shown on the left part of Fig. 10.1. An example of such failure by selective growth of grain scale inhomogeneities is discussed in Sec. 10.4.

Two types of microstructures are differentiated by Parker[2]: the *static microstructures* which exit in the material prior to the forming process and the *dy-*

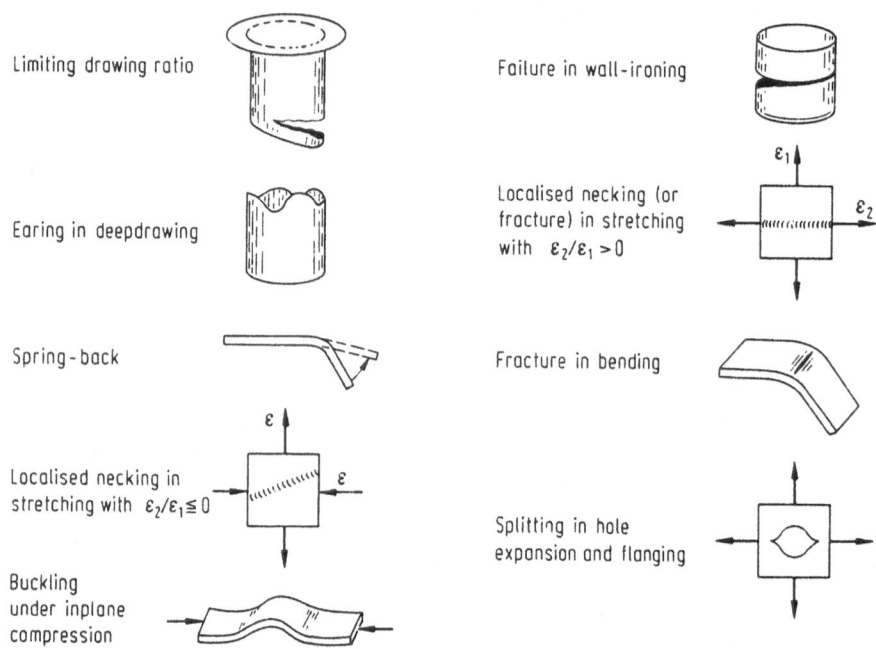

Limiting drawing ratio

Failure in wall-ironing

Earing in deepdrawing

Localised necking (or fracture) in stretching with $\varepsilon_2/\varepsilon_1 > 0$

Spring-back

Fracture in bending

Localised necking in stretching with $\varepsilon_2/\varepsilon_1 \leqq 0$

Splitting in hole expansion and flanging

Buckling under inplane compression

Fig. 10.1. Forming limits in various sheet forming process (After Wilson[1])

namic microstructures which evolve during the forming process (see also Sec. 3.6.). Static microstructures include solute, second phase particles, inclusions, initial grain size and textures, while dynamic microstructures include dislocation configurations, cell walls, slip bands, Lúder bands, microbands and shear bands, etc.. The effects of various microstructures on the mechanical properties which affect formability are listed in Table 10.1

The choice of a particular microstructural features depends on the deformation geometry. For biaxial stretching, a high strain hardening rate is to be prefered so as to spread any local strain concentration. This is achieved in fine-grained alloys which do not recover readily and have strong solute atom/dislocation interaction. In deep drawing a low strain hardening rate is desirable as the undeformed element in the base of the drawn cup may be unable to support the load transmitted by the strained flange. Low strain hardening rate can be obtained by slight cold rolling prior to deep drawing.

While the effect of microstructures on basic mechanical properties are fairly well understood, they are not expressed explicitly in the various mechanics models but through the average properties (such as n, m and r) of the microstructures. Attempts have been made to abandon the phenomenological descriptions of the yield surface by applying the theory of polycrystalline plasticity to calculate the yield surface of textured polycrystals and limit strains in sheet metals[4,5,6]. So far,

Table 10.1. Effect of microstructure on mechanical properties [After Parker, B.A., Formability and Metallurgical Structures. Edited by Sachdev, A.K and Embury, J.D., the Metallurgical Society Inc., 1987, p.73.]

Property features	Strain hardening (n)	Strain rate hardening(m)	Plastic anisotropy(r)
Solute atoms			
-mobile	increase	decrease	little effect
-immobile	increase	little effect	little effect
Particles(μm)			
< 0.05	decrease	little effect	little effect
0.05-3	increase	slightly increase	little effect
3	slightly increase	little effect	little effect
High dislocation density	decrease	slightly increase	little effect
Dislocation cells	decrease	increase	slightly increase
Small grain size	increase	increase	decrease
Texture	little effect*	little effect	increase

* The texture effects on work hardening are overwhelmed by the grain size effect[3]. The strain hardening will decrease for very small grain size about 1 μm.

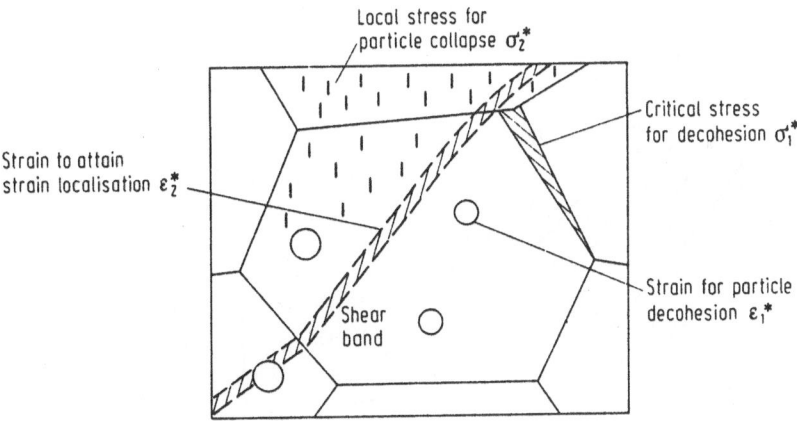

Fig. 10.2. Local events at the microstructural scale which may limit the formability (From Embury, J.D.[7])

these attempts take into account the average properties of the static microstructures such as the grain orientation distributions. The development of strain inhomogeneity and fracture events are influenced to a greater extent by the local properties of the dynamic microstructures. Local events (Fig. 10.2) such as void formation, particle fracture and shear banding limit the formability that can be attained in a given material[7]. Examples of these events are

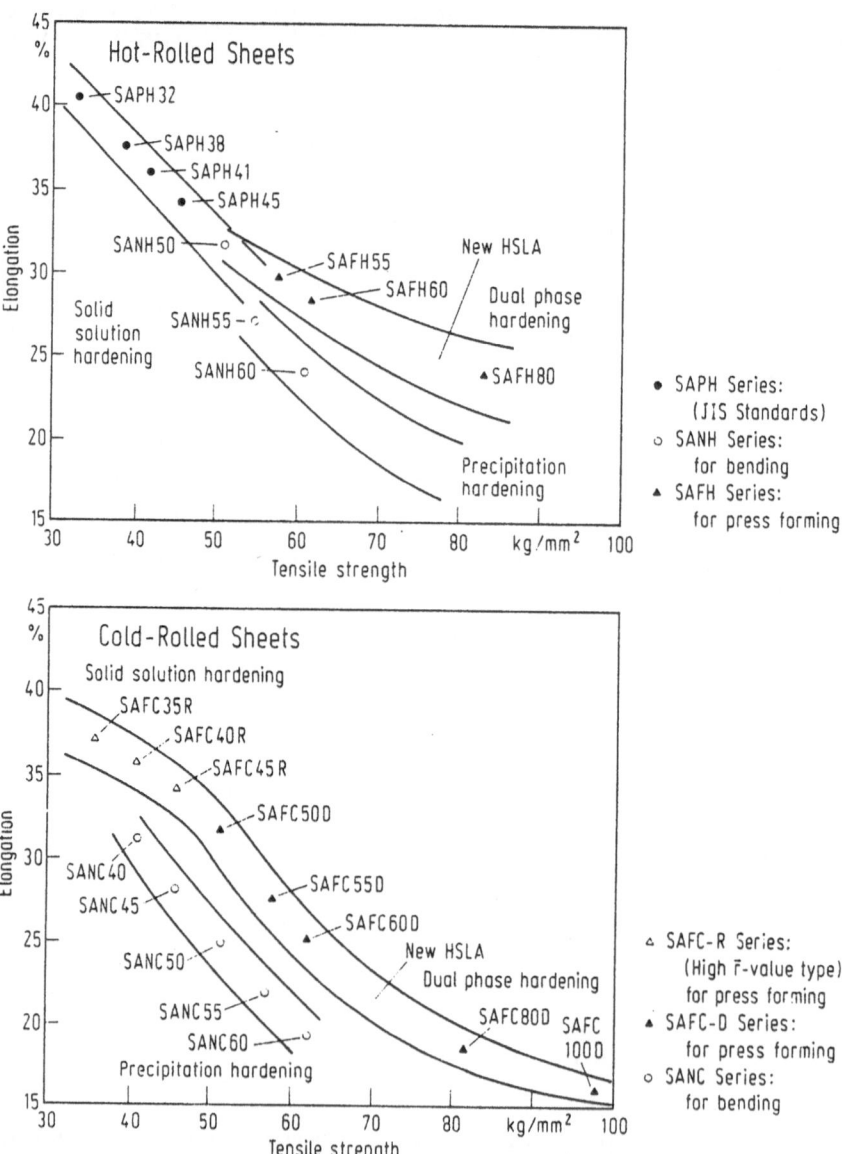

Fig. 10.3. High strength steel sheets for automobile parts (From Nippon Steel News, January, 1980)

(a) dislocation channelling through a dispersion of shearable particles.
(b) void formation and particle fracture caused by plastic strain concentration
(c) texture softening caused by lattice rotation in textured polycrystals leading to shear band formation.

Compared with the forming limit curve (see Sec. 10.4) the fracture limit curve has received relatively little attention. Fracture strains have been observed to be much higher than the limit strains in uniaxial tension. This is not necessary true as fracture limits can approach the limit strains for other deformation geometry. The "mechanics" criteria of fracture are often based on the attainment of a critical level of critical stress or strain, or a critical work per unit volume[8]. The type of McClintock calculations[9] will be needed to provide the link between these fracture criteria and the evolution of the dynamic microstructures through a detailed study of the various failure mode in different alloys. A better understanding of the failure mechanism is of prime importance to provide the physical basis for microstructural and alloy design.

Examples of success in the application of microstructural control in the improvement of commercial alloys are best illustrated in the production of deep drawing steels for automobile application. Traditionally, the best quality of formable steels has been manufactured by box annealing. The {111} texture components which are present in the cold rolled state and retained after annealing possess high R-values (see Sec. 10.2). The good drawability is brought about by slow heating rate and long time of holding at the annealing temperature. For a given steel, the R-values and grain sizes are closely related. Grain growth is permitted as this will sharpen the {111} texture components. Due to the long cycles involved in the box-annealing process, there has been significant movement towards the continuous annealing process on which the processing time can be reduced from a few days to only a few minutes. Production of these deep drawing steels depends critically on the control of the grain sizes, textures and second phase dispersions.

The demand for higher strength, lighter gauge, superior formability steels has led to the development of high-strength low-alloy (HSLA) steels (Fig. 10.3). The stretch-formabilities of solid-solution hardened high-strength low-alloy (HSLA) steels are relatively good. The strengthening of the alloy by a dispersion of non-deformable particles is undesirable to stretch-formability as the Orowan dislocation-loop formation round the particles (see Sec. 6.3.3) leads to a rapid increase in flow stress in the early stages of deformation, and stress concentration at the non-deformable particles can lead to decohesion. Steels with a tensile strength between 35-45 kg/mm^2 are used for press forming of outer and inner panels of automobiles and are produced by adding solute elements such as phosphorous to ultra-low carbon steels of deep drawing quality. Transformation hardening is used to produce steels with higher strength (over 80 kg/mm^2) for frames, structural parts, safety parts, and impact bars, etc. These steels are basically dual phase steels which are characterized by a microstructure of ferrite and martensite[10].

10.2 Calculation of Plastic Strain Ratio

Industrial sheets produced by a combination of cold rolling and annealing bring about directional properties. Sheet metals have different plastic properties in different directions. This is known as *plastic anisotropy*. There are two sources of plastic anisotropy in sheet metals; these are *crystallographic anisotropy* arising from preferred orientations or textures and *microstructural anisotropy* caused by variations in grain size and shape, inclusion stringers, non-uniform distribution of phases, etc.. Common sheet metal forming processes such as deep drawing are sensitive to the crystallographic anisotropy in the sheet. If the sheet has a high deformation resistance in the thickness direction and less in the rolling plane, then it will have a larger limiting drawing ratio (i.e. the ratio of the maximum blank size to the size of the cup that can be drawn successfully). On the other hand, plastic anisotropy results in earing in deep-drawn components and a non-uniform wall thickness. In rolled sheets, there is a marked degree of plastic anisotropy. The plastic properties in the plane of the sheet differs from the through thickness properties (known as *normal anisotropy*) and vary with orientation in the plane of the sheet (known as *planar anisotropy*). In continuum plasticity, the plastic anisotropy of metals can be represented by various yield criteria such as those suggested by Hill[11]. For example, the quadratic yield criterion $f(\sigma_{ij})$ for materials with orthotropic symmetry can be written as

$$2f(\sigma_{ij}) = F(\sigma_{22} - \sigma_{33})^2 + G(\sigma_{33} - \sigma_{11})^2 + H(\sigma_{11} - \sigma_{22})^2$$
$$+ 2L\sigma_{23}^2 + 2M\sigma_{31}^2 + 2N\sigma_{12}^2 = 1 \tag{10.1}$$

where σ_{ij} are the principal stress components The material parameters F, G, H, L, M, and N specify the current state of anisotropy and can be determined by means of tension or shear tests. The *plastic strain ratio r* is defined to be the ratio of the in-plane transverse strain to the through-thickness strain, i.e.

$$r = d\epsilon_{22}/d\epsilon_{33} \tag{10.2}$$

and may depend in any manner on the strain history. In tensile tests, $(\sigma_{22} = \sigma_{33} = 0, \sigma_{11} \neq 0)$, it can be shown that

$$R_o = H/G, \quad R_{90} = H/F, \tag{10.3}$$

where the subscripts denote the angle of the tensile specimen to the rolling direction (Fig. 10.4). For an isotropic material $r = 1$. If the material is isotropic in the plane of the sheet, then $r_0 = r_{90}$. A high r-value indicates a high resistance to thinning and conversely a low r-value implies a low through thickness strength.

The plastic strain ratio expressed above refers to the ratio of incremental strains and is different to the initial definition by Lankford et. al.[12] referred to the ratio of the total strains. In metallurgical literatures, the normal anisotropy (R) is often defined with finite strains (i.e $R = \epsilon_{22}/\epsilon_{33}$). Usually measurement are taken at only one level of axial strain and R is assumed to be constant with strain. However, precautons must be taken as R has been found to vary with the amount of straining. The planar anisotropy is commonly defined as

$$R = (R_0 + 2R_{45} + R_{90})/4 \tag{10.4}$$

Hosford and Backofen[13] showed that R can be correlated with the shape of the yield locus, and is measured by the slope of the tangent to the locus where it

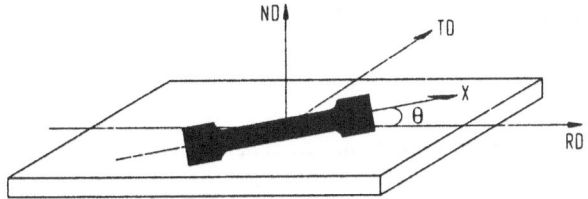

RD Rolling direction
TD Transverse direction
ND Normal direction
X   Tensile axis

Fig. 10.4. Tensile specimen cut from a rolled sheet

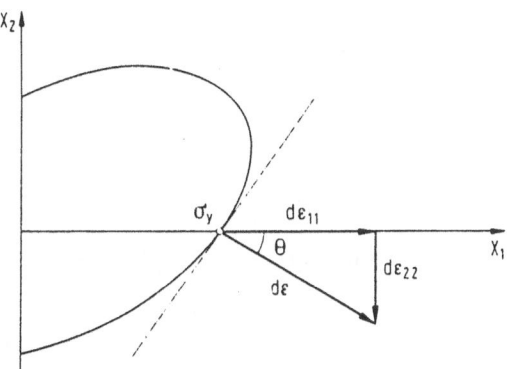

Fig. 10.5. The determination of R from plane stress yield loci

cuts the X_1 or X_2 axis as shown in Fig. 10.5, i.e.

$$\tan \theta = -\frac{d\epsilon_{22}}{d\epsilon_{11}} = R/(1+R). \tag{10.5}$$

When R-value increases, θ decreases and the yield locus is stretched into the first and third quadrants. The stress state which produces plane strain state towards the balanced biaxial stretching is constrained.

Continuum plasticity models are able to represent the plastic anisotropy of metals in a relatively simple way. These continuum yield criteria however do not take into account the crystallographic texture and hence fail to predict certain mechanical behaviour such as the number of correct ears in deep drawing. The crystallographic anisotropy of polycrystals are closely related to the orientations of the constituent grains which depend on the deformation texture and recrystallization texture described in Chapter 9. Whitely and Wise[14] were able to relate the dependence of R to the densities of $\{111\}$, $\{110\}$, and $\{100\}$ planes parallel to the sheet rolling plane. High values of R are found with high densities of $\{111\}$ and $\{110\}$ and low densities of $\{100\}$ planes. Methods of developing textures with favourable orientations and high normal anisotropy have been summarized by Dillamore and Roberts[15].

In the early crystallographic models of calculating plastic anisotropy, the sheet was assumed to be made up of texture components having ideal crystallographic orientations. A high value of R is obtained when the active slip systems which are activated during tensile deformation are suitably oriented for deformation in the width direction but not in the thickness direction. The relation betwen the R-value and the slip systems can be estimated by

$$R = [(\mathbf{b} \cdot \mathbf{p})(\mathbf{b} \cdot \mathbf{d})] / [\mathbf{t} \cdot \mathbf{p})(\mathbf{t} \cdot \mathbf{d})] \tag{10.6}$$

where vectors $\mathbf{b}, \mathbf{t}, \mathbf{d}, \mathbf{p}$ are unit vectors along the width direction, the thickness direction, the slip direction and the normal direction to the slip plane respectively. Vieth and Whitely[16] took one or more equally favourably oriented slip systems with the highest Schmid factor from the applied tensile stress. Fukuda[17] assumed that slip occurs on five slip systems in the order of magnitude of the Schmid factors. The resultant R-value will be infinite if the denominator of (10.6) become zero. To avoid this, the following equation has been proposed by Lee[18]

$$R = \sum [|(\mathbf{b} \cdot \mathbf{p})(\mathbf{b} \cdot \mathbf{d})| S] / \sum [|(\mathbf{t} \cdot \mathbf{p})(\mathbf{t} \cdot \mathbf{d})| S] \tag{10.7}$$

where $\sum$ indicates the summation of all operating slip systems and S is the Schmid factor given by

$$S = |(\mathbf{l} \cdot \mathbf{p})(\mathbf{l} \cdot \mathbf{d})| \tag{10.8}$$

and $\mathbf{l}$ is unit vector along the tension direction. In (10.7), all the slip systems are assumed to contribute to the deformation but their contributions are assumed to be proportional to their Schmid factors. For a sheet texture with ideal orientation $(H_1 H_2 H_3)[K_1 K_2 K_3]$, $\mathbf{l}$, $\mathbf{t}$ and $\mathbf{b}$ can be obtained from the following relation

$$\begin{bmatrix} \cos\alpha & 0 & \sin\alpha \\ 0 & 1 & 0 \\ -\sin\alpha & 0 & \cos\alpha \end{bmatrix} \begin{bmatrix} k1 & k2 & k3 \\ h1 & h2 & h3 \\ n1 & n2 & n3 \end{bmatrix} = \begin{bmatrix} l1 & l2 & l3 \\ t1 & t2 & t3 \\ b1 & b2 & b3 \end{bmatrix} = \begin{bmatrix} \mathbf{l} \\ \mathbf{t} \\ \mathbf{b} \end{bmatrix} \tag{10.9}$$

where $\mathbf{h}(h_1, h_2, h_3)$ is the unit vector normal to the sheet plane and $\mathbf{k}(k_1, k_2, k_3)$ is the unit vector along the rolling diection of the sheet, α is the angle between the rolling direction $\mathbf{k}$ and the tensile direction $\mathbf{l}$, and $\mathbf{n} = \mathbf{k} \times \mathbf{h}$. Therefore if a sheet texture has more than one ideal orientations, the R value of the sheet is determined by the volume fraction of each texture component i, i.e.

$$R = \sum R_i V_i. \tag{10.10}$$

This simple method gives good qualitative explanation of the texture effects on the plastic anisotropy of sheet metals. However, there has been criticism on either the correlation between measured and calculated values or the physical backgrounds for choosing the operating slip systems.

With textured metals, the only complete description of anisotropy is achieved by using the crystallite orientation distribution function ODF (see Sec. 3.5.4). Assuming the same strain to each grain and applying the Bishop and Hill's analysis,

Hosford and Backofen[19] considered the deformation of a polycrystalline specimen under tensile test by an amount

$$E = dE \begin{bmatrix} 1 & 0 & 0 \\ 0 & -q & 0 \\ 0 & 0 & -(1-q) \end{bmatrix} \tag{10.11}$$

where q is the contraction ratio and is related to R by

$$R = q/(1-q). \tag{10.12}$$

A range of values of q is used to calculate the average Taylor factor M for a known ODF $f(g)$ by

$$M(q) = M(q,g)f(g)dg. \tag{10.13}$$

The values of q and hence R will be determined for which M is minimized. This approach has been applied by Bunge and Roberts[20], Parniere and Rosch[21], and Davies et.al.[22]. Kumar[23] has refined the approach by introducing the shear strains in the deformation tensor. In this method, hundreds or even thousands of grains are involved and extensive computations have to be carried out to obtain reasonable predictions. An alternate approach is used by Lequeu et. al.[24] based on the Continuum Mechanics of Textured Polycrystals (CMTP)[25] in which the yield function of a textured aggregate is approximated by a continuum yield function, such as equation (9.20). The stress tensor σ^s referred to the axes of the tensile test specimen are represented by the following:

$$\sigma^s = \begin{vmatrix} \sigma_{11} & 0 & 0 \\ 0 & 0 & 0 \\ 0 & 0 & 0 \end{vmatrix} \tag{10.14}$$

With the assumption of conservation of volume in plastic deformation, the associated strain tensor ϵ^s can then be written as

$$\epsilon^s = \begin{vmatrix} 1 & \epsilon_{12} & 0 \\ \epsilon_{21} & -R/(1+R) & 0 \\ 0 & 0 & -1/(1+R) \end{vmatrix} \tag{10.15}$$

where ϵ_{11} has been set to 1 as a test condition. A procedure similar to the one described in Sec.9.3 is used. The R-value is varied until the loading condition ($\sigma_{22} = \sigma_{33} = 0$) is satisfied. The R-values can be calculated at different angles Θ to the rolling direction of a sheet. In order to take into account of the loading direction, the stress and the strain tensors of a tensile specimen at an angle Θ to the rolling direction of a sheet are converted to the specimen coordinate axes which coincide with the rolling, transverse and normal directions. This is done by means of the transformation matrix $\mathbf{P}_a$

$$\mathbf{P}_a = \begin{vmatrix} \cos\Theta & -\sin\Theta & 0 \\ \sin\Theta & \cos\Theta & 0 \\ 0 & 0 & 0 \end{vmatrix}. \tag{10.16}$$

The converted strain tensor is further transformed to the cube axis of a grain. By applying the flow rule to the crystallographic yield loci $F(\sigma_{ij}^c)$ approximated by a continuous function,

$$\epsilon_{ij}^c = \lambda \partial F(\sigma_{ij}^c)/\partial(\sigma_{ij}^c) \tag{10.17}$$

where $F(\sigma_{ij}^c)$ is the yield loci of the CMTP method, and the superscript (c) refers to the crystal axes. The derived strain components in the crystal axes are then transformed back to the specimen axes.

To calculate the R-value for a textured polycrystals, the full ODF is replaced by a selection of ideal orientations, each representing a specific volume fraction of the grains and having a particular Gaussian distribution of orientation scatter width ω. An ideal orientation $\{hkl\}<uvw>$ can consist of up to four distinct orientations $(hkl)[uvw]$, $(\bar{h}\bar{k}\bar{l})[uvw]$, $(hkl)[\bar{u}\bar{v}\bar{w}]$, and $(\bar{h}\bar{k}\bar{l})[\bar{u}\bar{v}\bar{w}]$. The stress direction $[\sigma_{ij} = 0$ for $(i,j) \neq (1,1)]$ and the strain ϵ_{11} are the same in each of the four sets of grain orientations. Furthermore, the summation of the stresses of individual grains equals to the imposed stress σ_{ij}. The R-value is given by the average of the four sets of grains:

$$R(\Theta) = \sum \epsilon_{yy}\{hkl\}<uvw>/\sum \epsilon_{zz}\{hkl\}<uvw> \tag{10.18}$$

In this manner, the R-value can be obtained as a function of the texture components, the coefficients of the yield criterion and the inclination Θ of the tensile axis to the sheet rolling direction. Fig. 10.6 shows the predicted $R(\Theta)$ curves for three single orientations and an annealed commercially pure copper sheet. Good agreements are found between the predictions and the experimental results quoted from various sources[26,27,28,29]. The shape of the $R(\Theta)$ curve of the copper sheet (Fig. 10.6d) follows a similar shape as that of Fig. 10.6a as the sheet possesses a fairly strong cube texture.

The CMTP method gives a rapid way of assessing the $R(\Theta)$ curves and good approximations of experimental strain ratios as much less grain orientations are sufficient to reproduce the required features[30]. The average R-value is generally better reproduced than the planar variation of R-value. It must be noted that the above R-values are determined prior to the tensile deformation and precautions must therefore be taken to compare these values with the observed ones. Variation of plastic strain ratio with strain level has been reported by Hu[31] in steels. The level of anisotropy always decreased as the test proceeded. This phenmenon was attributed by Hu to differential hardening in different directions in the test piece. Arthey and Hutchinson[32] found that steels which yielded homogeneously, in the absence of interstitial atoms, had R-values which were almost independent of the strain level in the tensile test. Interstial-containing steels with sharp textures show R-values that change with increasing strain level such that the anisotropy decreases. In HCP metals, the R-values may change with strain as the relative contributions from slip and twinning mechanism vary. The R-value is not simply a function of the current state of the deformation of the material but a summation of its behaviour from the onset of straining which may involve changes in the distributions of crystal orientations and the occurrence of yield point.

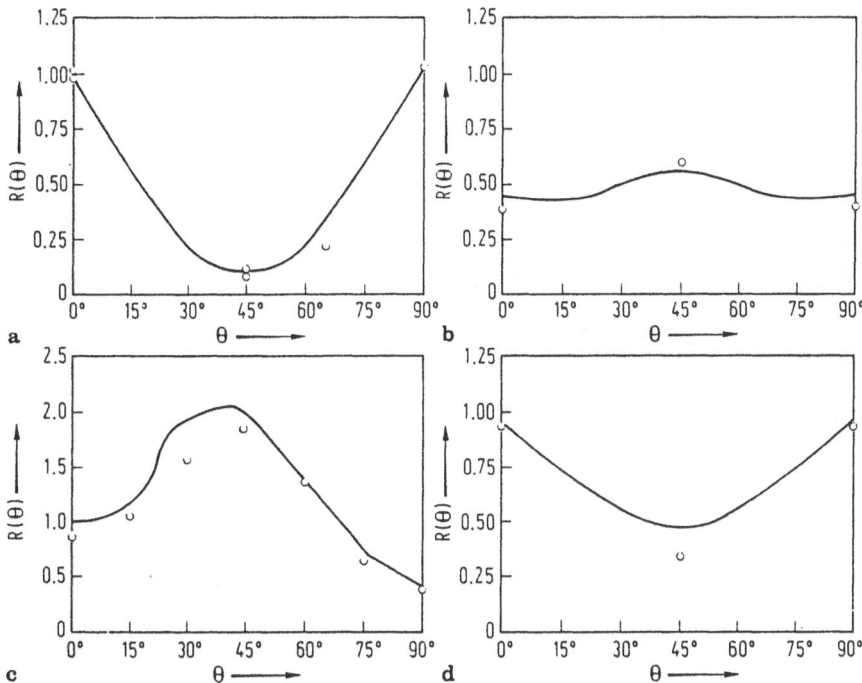

Fig. 10.6. Comparison of the predicted plastic strain ratios and experimental data ( ) for the single orientations. **a** (100)[100]; **b** (100)[012]; **c** (112)[11−1], and for the **d** an annealed copper sheet with {100}<001>(46.96%) + {122}<221>(23.59%) + {437}<184>(8.29%) + {418}<744>(8.29%) + {148}<841>(8.29%) + {123}<634>(4.58%). The R-values are calculated at a tensile strain of 0.025 with $L = 0, q_1 = 0.46, q_2 = 0.51$, and $n = 1.6$

10.3 Formation of Ears in Deep Drawing

When a cylindrical cup is drawn from a circular blank of most metals, undulations often develop on the rim (Fig. 10.7). This phenomenon is known as "earing" and the distinct high points are called ears and the low points called "troughs". The formation of ears leads to wastage of material since additional trimming operation is needed to remove them. The cupping test is the most frequent test for earing because of its direct applicability to deep drawing practice. There are various expressions used for measuring earing. Blade and Pearson[33] suggest that one of the simplest method of representing earing is to express the difference in maximum and minimum height of the cup as a percentage of the mean cup height by the formula:

$$\text{Percentage earing} \quad E = (h_{\max} - h_{\min})/h_a \times 100 \qquad (10.19)$$

$$\text{and} \quad h_a = (h_{\max} + h_{\min})/2 \qquad (10.20)$$

where $h_{\max}$ and $h_{\min}$ are the heights of ears and troughs measured from the bottom of the cup. The extent of earing however is influenced by the deep drawing conditions

Fig. 10.7. A deep-drawn aluminium cup of commercial purity showing 4 ears at 45° to the rolling direction

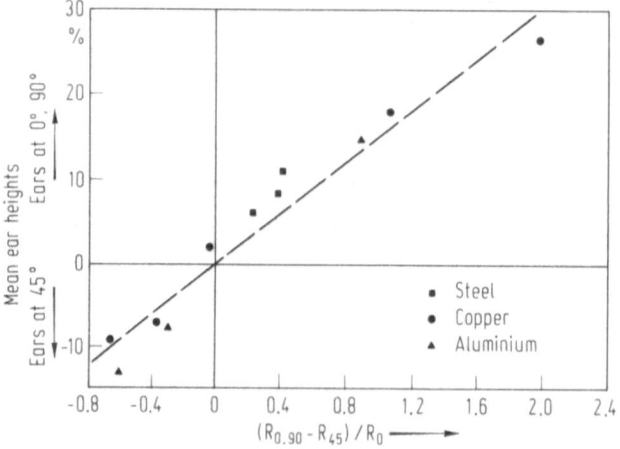

Fig. 10.8. The relation between the percentage earing and the variation in plastic strain ratios R, measured at uniaxial strain of 20%. The punch diameter and nose radius are 50 mm and 9.5 mm respectively. (From Wilson and Bulter [35])

such as the sheet thickness, draw ratio, punch-die clearance, punch diameter, nose radius, blank-holder pressure and lubrication. To get consistent results, the earing test has to be standardized[34].

The formation of ears during deep drawing is a manifestation of the directional properties of anisotropic sheets. There is a linear relationship between the percentage earing and the corresponding variation in the planar anisotropy in various test materials as shown in Fig. 10.8. The earing percentage is considered positive when the ears occurs at 0° and 90° and negative at 45°. Although it was well known in the forties that the earing has its origin in the anisotropic properties of the sheet materials, attempts to correlate between earing and some other mechanical properties which can be measured at various angles to the rolling direction has been unsuccessful as no consistent relationship has been found. The mathematical theory of anisotropic plasticity[36] was applied by Bourne and Hill[37] to predict the formation of ears in directions corresponding to stationary values of yield strength. The theory can explain the 45° and 90° ears but fail to account for the occurrence of 6 ears which are common in brass sheets. Budiansky and Wang[38] and Chiang and Kobayashi[39] have applied Hill's theory to calculate stress-strain distribution in the flange of a deep drawn cup based on planar isotropy. Sowerby and Johnson[40] used the slip-line theory to predict earing pattern of a material with orthotropic anisotropy in the plane of the sheet. None of the above studies were able to account for the plastic flow pattern in sheets possessing different anisotropic properties.

Complex changes in the directional properties of annealed brass and earing behaviour have been extensively reported in the literatures[41,42,43,44]. Cups with four ears and at 45° to the rolling direction, six ears at 0° and 60° to the rolling direction and no ears were produced depending on the combination of penultimate anneal and final rolling reduction in the production of brass sheets. It was observed that the directionality and earing tendency increased with the degree of both penultimate and final reductions, and increased as the penultimate annealing decreased and as the final annealing temperature increased. No earing was observed to occur in 70/30 brass rolled to 40% to 50% reduction in thickness and annealed to small grain size (i.e., 10-15μm). In aluminium, the types of earing commonly encountered are four ears, at 0° and 90° to the rolling direction or four ears at 45° to the rolling direction. Cups with eight ears are also frequently observed in low earing materials. The metallurgical factors in sheet metal fabrication influence earing through changes in the crystallographic textures. It is well known that the cube texture {100}<001> gives rise to 0° and 90° earing whereas 45° earing is attributed to the rolling-type textures.

The relationship between earing and crystallographic textures has been much explored in the literatures[45,46,47]. The early experimental work of Wilson and Brick[48] suggested that ears occur in that azimuthal position on the cupping blank in which the crystals are so oriented that circumferential compressive stresses act in the direction which develops maximum resolved shear stresses on the four sets of octahedral (slip) planes. Ears lie in the direction of low peripheral concentrations of {111} poles and troughs lie in the direction of high peripheral concentrations of {111} poles in a pole figure projected on to the sheet surface. This observation however will not hold for many single crystal orientations when there may be no such preferred orientations.

A calculation of earing in single crystal sheets was proposed by Tucker[49]. The magnitude and type of earing to be expected was assessed from the state of strain along a circumferential direction in the flange of a circular blank. The stress and deformation in a section from a drawn cup are illustrated in Fig. 10.9.

During deep drawing, metal in the outer portion of the blank is drawn radially toward the throat of the die and the outer circumference decreases. There is a compressive stress $(-\sigma)$ circumferentially and tensile stress (σ) radially at any

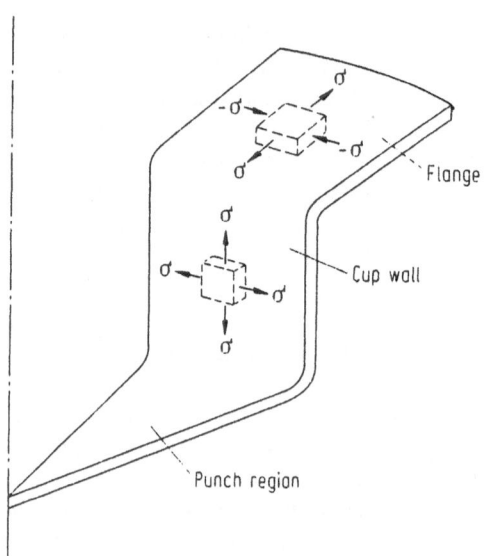

Fig. 10.9. Deformation in a section of a deep drawn cup

point in the flange of a blank. The stress normal to the sheet surface is assumed
to be zero. Ears are found at those places where a circumferential reduction and
radial tension develop first. The directions of the principal stresses at any point
relative to the cubic axes of the crystal may be expressed in terms of Θ, the angle
to the rolling direction of a sheet. If, relative to the direction of principal stress,
the direction cosines of the $\{111\}$ plane are (a_n, b_n, c_n) and the $<111>$ direction
lying in this plane are (d_n, e_n, f_n), then the resolved shear stress on this slip system
τ_n is given by

$$\tau_n = \sigma \mu_n = \sigma (a_n d_n - b_n e_n) \tag{10.20}$$

where μ_n is the Schmid factor, σ is the numerical value of the compressive and
tensile stress. The deformation mode is shear on $\{111\}$ planes in $<111>$ directions
and the change in the crystallographic orientation during drawing is ignored. By
applying the following equation

$$\tau_n = \tau_o + k\tau^n \tag{10.21}$$

where k is a constant, the variation of shear strain in terms of Θ is known i.e,

$$\tau_n = \{[\sigma(a_n d_n - b_n e_n) - \tau_o]/k\}^{1/n} . \tag{10.22}$$

When an element in the flange is subjected to a shear strain τ_n, the strain of the
element ϵ_n in the radial direction will be

$$\epsilon_n = \{1 + 2\tau_n |a_n e_n| + \tau_n^2 b_n^2\} - 1. \tag{10.23}$$

The circumferential strain distribution will be similar to the cup profile. The absolute value of τ_n is not known. τ_n is large in normal drawing operations and $\tau_n > \tau_o$,

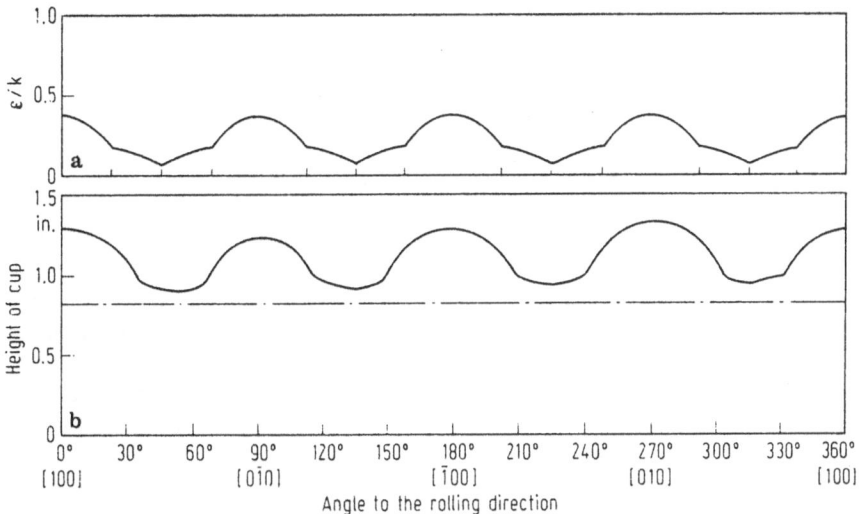

Fig. 10.10. (a) Theoretical and (b) experimental profiles of cup pressed from (100) single crystal (From Tucker, G.E.G., Acta Matall. Vol.9, April, 1961, p.281)

then

$$\epsilon_n \doteq \tau_n |b_n| = \{[\sigma(a_n d_n - b_n e_n) - \tau_o]/k\}^{1/n} |b_n|. \qquad (10.24)$$

When more than one slip system is operating, the radial strain is taken as the average of the strains calculated for each of the twelve possible slip systems.

Examples of calculation for single crystals of various orientations have been shown by Tucker. Fig. 10.10 shows the theoretical and experimental profiles of a cup pressed from a single crystal of (001) orientation. Taking [100] as the reference direction and let Θ be measured positive towards [010], the radial strain is given by

$$\epsilon/k = (2/3\sqrt{3})\cos^2 2\Theta \cos\Theta \qquad\qquad 0° < \Theta < 22.5° \qquad (10.25a)$$
$$\epsilon/k = (1/3\sqrt{3})\cos^2(2\Theta - 45°)\cos\Theta \qquad 22.5° < \Theta < 45°. \qquad (10.25b)$$

Good agreement is found between the experimental and the calculated cup profiles.

The single crystal work of Tucker has been extended by Kanetake et. al.[50] to cover textured polycrystalline sheets which is represented by a crystallographic orientation distribution function $w(\psi, \theta, \phi)$ referred in Sec.3.5.4. The radial strain of an element in the flange for each single crystal orientation ϵ_s is weighted by $w(\psi, \theta, \phi)$ and average over all orientations. The radial strain at a given Θ in the flange of a polycrystalline circular blank ϵ_p is given by

$$\epsilon_p = 1/8\pi \iiint \epsilon_s(\psi, \theta, \phi) w(\psi, \theta, \phi) \sin\theta\, d\psi d\phi. \qquad (10.26)$$

The average height of cup wall h_w has been calculated by the following equation

$$h_w = (R_b^2 - R^2)/(R_p + R_d) \tag{10.27}$$

where R_b, R_p and R_d are the radius of the circular blank, punch and die respectively, and

$$R = 2(r_p + t_o/2)^2 + (r_p + t_o/2)(R_p - r_p)\pi + (R_p - r_p)^2 \tag{10.28}$$

where r_p is the radius of the punch edge and t_o is the initial thickness of the blank. The cup profiles calculated using various values of μ_o and n together with the measured profiles are indicated by the solid line and by the dash lines repectively in Fig. 10.11 for the A1100-H24 aluminium sheet. The principal features of the cup profile are predicted well by the theoretical curves.

A different approach[51] to predict earing pattern is based on the uniaxial mechanical properties. At the edge of a blank during drawing, the state of stress is effectively that of simple compression. The tendency for earing is argued to be present in the form of an orientation dependency of radial velocity at the beginning of drawing of a circular bank into a cup. This earing tendency is assessed from the state of strain rates in compression at various orientations. Assuming the circumferential velocity component is negligible, the hoop strain rate $\dot{\varepsilon}_{\Theta\Theta}$ in the rim of the cup and the shear strain rate $\dot{\varepsilon}_{r\Theta}$ in the flange of the cup are related to the radial velocity v by

$$\dot{\varepsilon}_{\Theta\Theta} = v/r; \quad \dot{\varepsilon}_{r\Theta} = (1/2r)\partial v/\partial\Theta. \tag{10.29}$$

$\dot{\varepsilon}_{\Theta\Theta}$ is also the extensional strain rate in a tensile test at an angle $\Theta \pm \pi/2$ to the rolling direction. The ratio of the shear-to-elongation strain rate ratio Γ is given by

$$\Gamma = (1/2)\partial(\ln v)/\partial\Theta \tag{10.30}$$

and the radial velocity can be determined by integrating (10.30), i.e.

$$v(\Theta) = v(0)\exp\left[2\int_0^\Theta \Gamma(\Theta + \pi/2)d\Theta\right]. \tag{10.31}$$

Let $R(\Theta, t)$ as the outer radius of flange throughout the drawing, the radial velocity is given by

$$v = \dot{R}(\Theta, t). \tag{10.32}$$

For constant radial velocity,

$$v(\Theta)t = R(\Theta) - b; \quad v(0)t = R(0) - b \tag{10.33}$$

where $R(\Theta)$ and $R(0)$ are the flange radii at an angle $\Theta°$ and $0°$ to the rolling direction respectively, and b is the blank radius. The relative change in the flange radii of the drawn cup may be assessed by

$$R(\Theta)/R(0) = 1 + [b/R(0) - 1]\left\{1 - \exp\left[2\int_0^\Theta \Gamma(\Theta + \pi/2)\,d\Theta\right]\right\}. \tag{10.34}$$

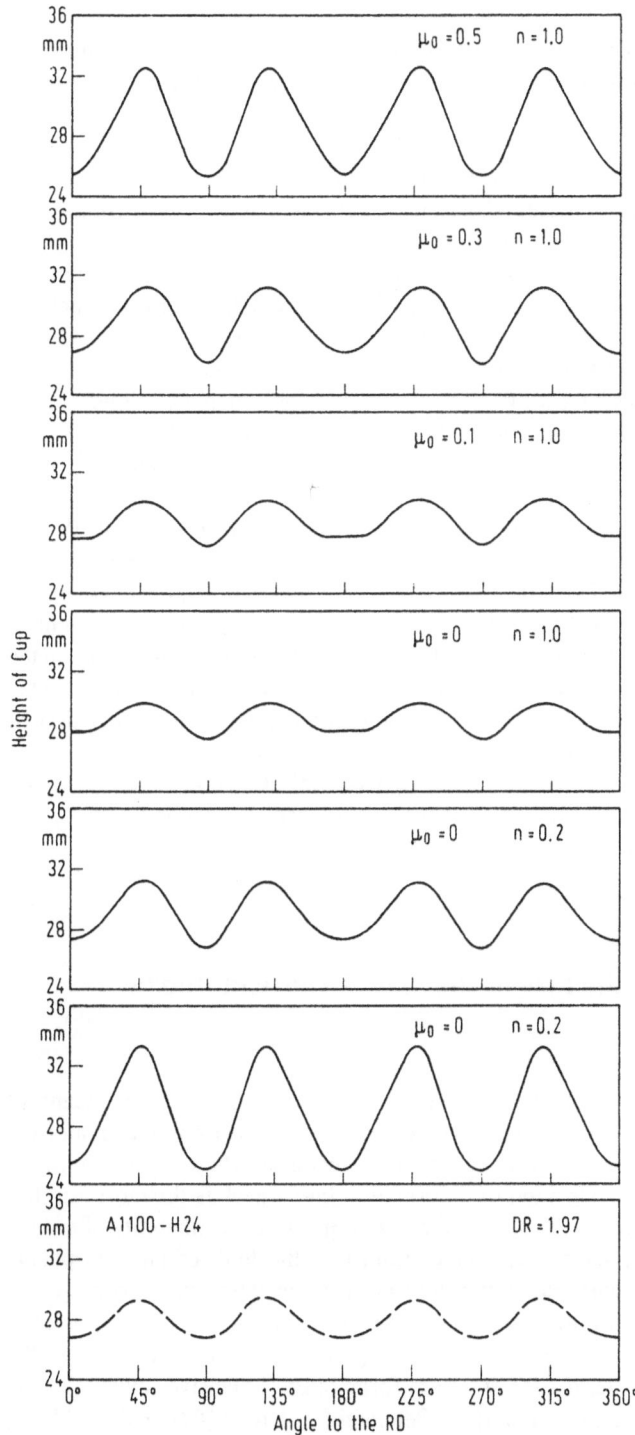

Fig. 10.11. Cup profile calculated using various μ_o and n (solid line) and experimental profile for A1100-H24 with drawing ratio (DR) = 1.97. (From Kanetake et.al.[50])

The associated variation of the shear-to-elongation strain rate ratio Γ with angle to the rolling direction are made from the tricomponent yield surface calculated by the Taylor/Bishop-Hill method[52,53] using experimentally measured orientation distribution function as shown by the work of Barlat and Richmond [54]. This model gives good prediction for sheet metals with sharp rolling textures but poor predictions for partially annealed sheets in which there is a mix of rolling and recrystallization texture components.

The plastic strain ratio or strain rate ratio has also been used to correlate with the earing pattern[55]. Ears are usually formed in the directions in which the plastic strain ratio is a maximum. It follows that the greater the value of ΔR, the more pronounced will be the earing. Absence of earing occurs when ΔR is zero. In aluminium, the types of earing most commonly encountered are four ears at 0° and 90° to the rolling direction or four ears at 45° to the rolling direction. The texture of a commercially pure aluminium sheet is usually approximated by different proportions of $\{100\}<001>$ and $\{112\}<111>$ components. When the sheet has a strong cube texture (commonly found after recrystallization), 0° and 90° ears are observed experimentally. This is in good agreement with the predicted $r(\Theta)$ curve as shown in Fig. 10.6a where the R-value at 0° and 90° are found to be maximum. For a strong $\{112\}<111>$ textured sheet (commonly obtained after cold rolling), 45° ears are observed, which is consistent with the predicted $r(\Theta)$ curve (Fig. 10.6c). In practice, a balance between the cube and $\{112\}<111>$ texture components will be beneficial for a low earing sheet. The production of ear-free aluminium flat rolled products depends on the balancing of these two types of textures. Such earing control is achieved by a control of both the microstructural evolution (see Sec. 3.6) as well as the position of the recrystallization (see Sec. 9.4) in the fabrication schedule.

10.4 Prediction of the Size of the M-K Groove in Biaxial Stretching

The required shapes of sheet metal products are usually obtained by means of uniaxial or biaxial stretching. These processes lead to an increase in the area of the sheet metal surface at the cost of reduction in its thickness. The reaching of a load maximum followed by localization of deformation into a neck is the most familiar form of failure in sheet metal. The general aim to improve the formability of a sheet metal is to improve the largest allowable strain (i.e. the limit of uniform plastic deformation) on every element of the deforming sheet while minimizing strain gradients, thus maximizing total elongation. There has been world-wide research interest in the understanding of sheet metal properties, and their influence on the performance of sheet materials for optimum formability. Extensive reviews of the subjects have been written by Mellor[56] Semiatain and Jonas[57], and Dodd and Bai[58].

Sheet metal formability studies have followed three main directions:

(a) attempts to ascertain fundamental properties of the metal such as the plastic strain ratio (r), the strain-hardening exponent (n) and the strain rate sensitivity (m),

(b) the development of simulative tests, and (c) the assessment of the Forming Limit Diagram (FLD).

Only partial success has been achieved to predict the forming behaviour of sheet metals from fundamental material properties. Thus in operations where drawing is the predominant deformation mode a good correlation has been shown between drawability and the plastic anisotropy ratio[59]. However, the influence of the normal anisotropy on pure stretch-forming is not well established. In the work of Keeler and Backofen[60] they conclude that, for a complex stamping involving both drawing and stretching, a high r and a high n characterize suitable sheets. Schmidt and Lawley[61] have pointed out that by varying the normal anisotropy, while keeping grain size constant in 70/30 brass, the pure drawing or the combined drawing and stretching are enhanced with increasing r-values but the reverse is true for pure stretching. The most important limitation in a tensile test is that it is insufficiently representative of the behaviour under different conditions of straining.

The lack of success of attempts to predict sheet metal formability from fundamental material properties has led to the development of simulative tests such as the Swift Drawing test, the Erichsen Cupping tests and many others[62,63]. The tests indicate whether one sheet is more likely to be successful in an industrial forming operation than another, but they will not predict whether or not a given industrial sheet forming operation can be successfully carried out on a given material, nor what modifications need to be made to the process details. Due to the limitations both of the simulative tests and of the fundamental material properties, Keeler[64] and Goodwin[65] introduced the concept of representing acceptable strain limit during sheet metal forming by a FLD. In the experimental construction of a FLD, a circle grid pattern is first electro-etched or imprinted by a photographical method on the sheet. The circles are then deformed to elliptical shapes. From the measurement of the ellipse axes, the major and minor strains at the point of localized necking can be determined. The details of this technique have been discussed by Keeler[66] and Pearce[67].

A typical FLD is shown in Fig. 10.12. The right portion of the diagram relates to tension-tension deformation and the tension-compression deformation relates to the left region of the diagram. The area above the forming limit curve is a failure region and a safe area is that below the curve. Where the minor strain is zero, deformation is by plain strain which normally corresponds to the lowest limit strain. The major to minor strain induced during the test is varied by using different specimen widths and punch lubricants[68]. There is an uncertainty as to how the limit strains should be defined since none of the test methods provide a clear distinction between the neck and uniform deformation regions. Though the technique of measuring the FLD is still not so satisfactory, the forming limit diagram has been shown to be a valid concept of interpreting the limit strain [69,70].

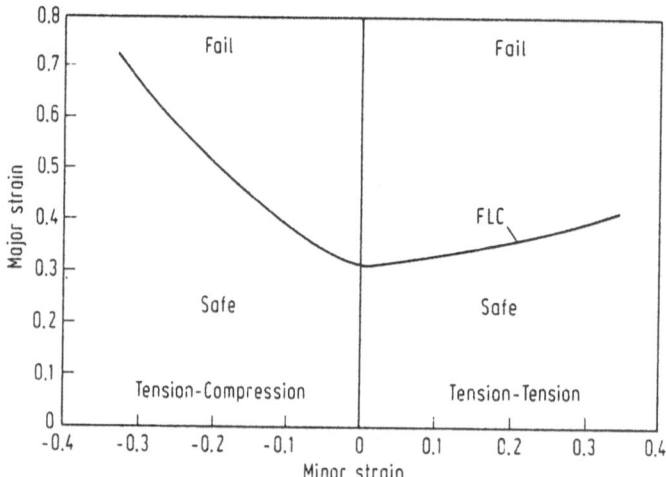

Fig. 10.12. A typical forming limit diagram

Theoretical analysis of localized necking and have been based on different criteria. Hill's theory[71] of deformation instability applies in the tension-compression region only. In biaxial stretching, positive plastic strain increments occur in all directions of the sheet surface. Since no inextensional direction can be found, Hill's theory cannot explain the phenomenon of localized necking in the biaxial strain state. Among the theories of forming limit diagrams, Marciniak and Kuczynski (M-K) have proposed the most influential hypothesis for the limit strains prediction in the stretch forming of sheet metal[72]. They postulated the existence of an initial imperfection in the sheet which will develop into a groove running in a direction perpendicular to the larger principal stress. The model postulates the presence of a material imperfection at which a shift to plane strain state occurs and hence permits a localized necking to form. In the M-K analysis, it is assumed that there is a long groove defect perpendicular to the maximum principal stress, as shown in Fig. 10.13. The initial inhomogeneity factor is given by the thickness ratio

$$f_o = t_B/t_A \qquad\qquad (10.35)$$

where A denotes the area outside the groove and B the groove region. In the groove region force equilibrium gives

$$\sigma_{11B} = \sigma_{11A}/f_o. \qquad\qquad (10.36)$$

It is further assumed a strain compatibility requirement that the groove strain $d\epsilon_{22B}$ is always the same as the corresponding strain outside the groove during straining so that

$$d\epsilon_{22A} = d\epsilon_{22B}. \qquad\qquad (10.37)$$

Due to the force equilibrium and strain compatibility requirement, t_B will decrease more quickly than t_A. Eventually $d\epsilon_{11B} \gg d\epsilon_{11A}$ and hence the deformation in the

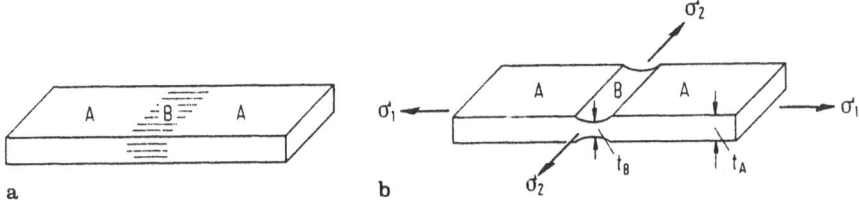

Fig. 10.13. The existence of a groove (thickness inhomogeneity) in a sheet metal

groove approaches plane strain, i.e., $d\epsilon_{22B} = 0$. At this stage, the strain outside the groove is identified as the limiting value for localized necking.

However, Tadros and Mellor[73] have considered the values of the groove depth in the initial sheet which must be chosen to make the experimental limit strain in equibiaxial tension match the prediction of the theory. They concluded that such a large value of defect in the initial sheet is unlikely to exist. Other objections are that (i) computational results are extremely sensitive to the precise size of the postulated groove, and (ii) the theoretical values of limit strain are less than the experimental values when the general straining pattern is near to plane strain and greater than the experimental values when it is near to balanced biaxial tension. Tadros and Mellor[74] suggested that the incipient grooves may appear during stable plastic flow but that a single groove will not exhibit continuing growth until diffuse necking occurs. They therefore proposed that the MK theory should be applied to the material at the time when diffuse necking sets in. Although a better correlation with the shape of the experimental forming limit curve was obtained, the difficulty of explaining and assigning a value to the groove depth remains.

Yamaguchi and Mellor[75] and Parmar et al.[76] have taken explicit account of the influence of grain anisotropy in biaxial stretching. In these models it is assumed that, up to the stage of stretching at which diffuse instability is developed, the amplitude of surface roughening, R, will grow according to the relationship proposed by Fukuda et al.[77]

$$R = R_0 + kd_0\bar{\epsilon} \tag{10.38}$$

where R_0 is the initial surface roughness, k an empirical constant which is dependent on the crystal structure of the material, d_0 the average grain diameter and $\bar{\epsilon}$ the generalized strain. The model also assumed that incipient grooves are formed within the roughened surfaces and that, in stretching beyond diffuse instability, strain localization develops within the deepest groove in accordance with the M-K model. These analyses would predict a thickness and grain size dependence of the limit strains in isotropic sheet metals under biaxial tension. However, Parmar et al.[76] have pointed out that there is no clear physical link between an incipient neck and the surface roughness. Tadros[78] has found from his experimental investigations that a groove does not develop from a single trough in the surface roughness, but that in fact the surface roughness is superimposed on the developing groove.

Various equivalent forms of the "groove" have been suggested which include surface roughness, inclusions or voids. In the damage mechanics model of the forming limit, a damage variable is assigned and associated with the void area fraction in the sheet. These voids are thought to be voids generated by plastic deformation[79,80,81]. The void coalescence model which is capable of describing the behaviour of some steel sheets[82] may not be necessarily true for other metals such as commercially pure aluminium and 70/30 brass in which the failure mode has been related to shear banding[83].

The postulation of a groove defect is the cornerstone of the Marciniak-Kuczynski (M-K) theory in the prediction of the biaxial limit strains of sheet metals. Without the assumption of an initial thickness inhomogeneity, such a groove has been shown by Lee and Chan[84,85] to arise from texture colonies and its size predicted based on a knowledge of the texture segregation in commercial sheets. For a textured material deformed in biaxial tension, the strain path may not be linear. Suppose the local texture in region B of a sheet metal as shown in Fig. 10.13 is different from the mean texture in region A. Upon stretching, region A and B will take up different strain paths. The strain in the transverse direction is assumed to be the same in both the A and B region, and the difference in the thickness strains will show up as a groove lying perpendicular to the rolling direction. The principal strain axes are taken as coincident with the long axis of the groove. In the region A, the imposed plain stress state is denoted by

$$\sigma_A = \sigma_A \begin{bmatrix} 1 & 0 & 0 \\ 0 & 1 & 0 \\ 0 & 0 & 0 \end{bmatrix}. \tag{10.39}$$

The corresponding strain tensor of equation (10.39) is denoted by

$$\epsilon_A = d\epsilon_A \begin{bmatrix} X_A & 0 & 0 \\ 0 & 1 - X_A & 0 \\ 0 & 0 & -1 \end{bmatrix} \tag{10.40}$$

where X_A is a dimensionless parameter. In the groove region B, force equilibrium gives

$$\sigma_{1B} = \sigma_{1A}/f \tag{10.41}$$

where f is the inhomogeneity factor and equals (t_B/t_A). The groove strain $d\epsilon_{2B}$ is assumed to be the same as the corresponding strain outside the groove:

$$d\epsilon_{2B} = d\epsilon_{2A} = d\epsilon_A(1 - X_A) \tag{10.42}$$

The strain tensor in the region B can be written as

$$\epsilon_B = d\epsilon_B \begin{bmatrix} X_B & 0 & 0 \\ 0 & 1 - X_B & 0 \\ 0 & 0 & -1 \end{bmatrix} \tag{10.43}$$

where

$$d\epsilon_B = (1 - X_A)d\epsilon_A/(1 - X_B). \,. \tag{10.44}$$

When the groove is parallel to the rolling direction (RD), equations (10.41), (10.42) and (10.44) will be replaced by

$$\sigma_{2B} = \sigma_{2A}/f \tag{10.45}$$

$$d\epsilon_{1B} = d\epsilon_{1A} = X_A d\epsilon_A \tag{10.46}$$

$$d\epsilon_B = X_A d\epsilon_A / X_B \tag{10.47}$$

respectively. The stress components σ_{ij} and strain components ϵ_{ij} are related by the normality principle,

$$\epsilon_{ij} = \lambda \partial F(\sigma_{ij})/\partial \sigma_{ij} \tag{10.48}$$

where $F(\sigma_{ij})$ is the crystallographic non-integer yield function proposed by Montheillet[25] and λ is the proportionality constant. The biaxial strain path is calculated for the two different regions follows those described in Sec.9.3. For each small increment of deformation, the change in the crystal rotation of each grain is followed and the strain tensor recalculated[85]. The inhomogeneity factor after each small step of deformation is given by

$$f = \exp(\epsilon_{3B})/\exp(\epsilon_{3A}). \tag{10.49}$$

The procedure of determining f is repeated until $d\epsilon_{1B} \gg d\epsilon_{1A}$, ie., the strain state in the groove region approaches plain strain. The major strain outside the groove ϵ_{1A} will give the limit strain of the sheet in equi-biaxial tension.

An numerical example showing the effect of local texture variation on the magnitude of the inhomogeneity factor f is illustrated for a 0.8 mm thick annealed commercial purity aluminium sheet. Samples of the aluminium sheet were stretched in a double action press over an auxiliary steel bank with a central hole diameter of 30 mm. Necking failure were frequently observed in both the rolling and transverse direction of the sheet (Fig. 10.14). The volume fraction of the main texture components is represented by $\{112\}<111>(54\%) + \{100\}<001>(27\%) + \{123\}<634>(19\%)$. In the region B, the texture is assumed to consist of more cube components i.e., $(1+x)V_c + (V_a - x/2) + (V_b - x/2)$, where V_c, V_a and V_b is the volume fraction of the $\{110\}<001>$, $\{112\}<111>$ and $\{123\}<634>$ texture components in the mean textured material respectively, and x is the volume fraction of extra cube components in the B region. Such a segregation of cube components has been reported in aluminium sheets[86].

(a) Groove perpendicular to RD

The simulated strain path of the region outside and within the groove is shown in Fig. 10.15 for $x = 0.05$ and $x = 0.3$. As the cube texture components in the groove region increases the strain paths followed by the region A and B diverge rapidly. The cube texture colony has a tendency to deform towards the plane strain state when ϵ_{2A} reaches about 0.22. The change in the homogeneity factor f with thickness strains for different amounts of cube components in the groove region is shown in Fig. 10.16. Different percentages of cube texture components are assumed in the

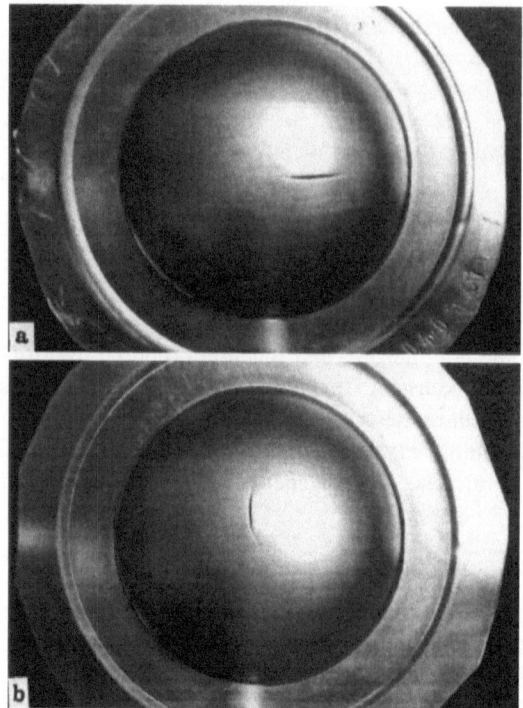

Fig. 10.14. Necking failure in biaxially stretched aluminium sheet. (a) groove parallel and (b) groove perpendicular to the rolling direction. The rolling (horizontal) direction is parallel to the flow lines visible on the surface of the sheet

groove region. The thickness ratio is found to decrease rapidly with straining and the rate of decrease increases with the amount of cube components in the groove.

(b) Groove parallel to RD

The strain path followed by the A and B region is shown in Fig. 10.17. The textures used for the simulations are the same as for the previous case. When the extra cube components in the groove increase from 5% to 30%, the strain along the transverse direction increases more rapidly than that along the rolling direction. A plain strain state is almost reached in the groove as ϵ_{1A} approaches 0.11. The development of the inhomogeneity factor is similar to the previous case but the limit strain is reached earlier when the groove is parallel to the rolling direction.

The above analytical results shows that a M-K groove of sufficient large size can be developed without assuming any initial defects or "equivalent defects" in an initial damage-free sheet. The growth of the groove is sensitive not only to the texture variations but also to the alignment of the texture colony. For a segregation of about 40% cube components in the groove region (i.e., 30% more

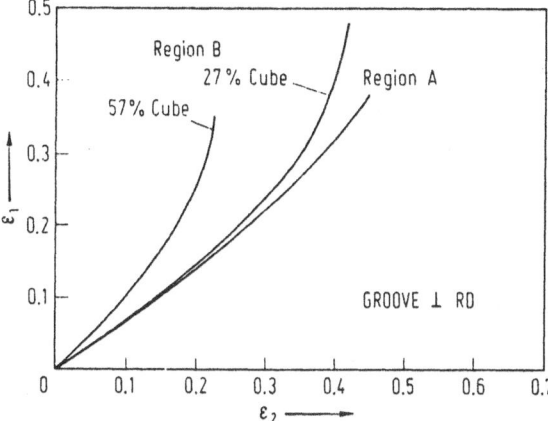

Fig. 10.15. Strain path outside and within the groove with the groove perpendicular to the rolling direction

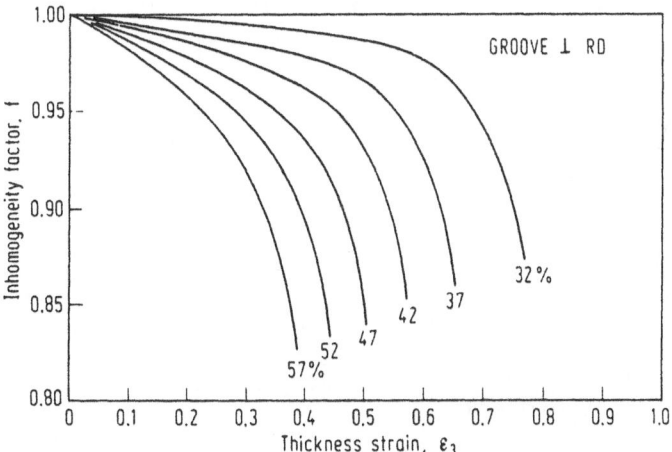

Fig. 10.16. Changes of inhomogeneity factor with different amounts of cube components in region B with the groove perpendicular to the rolling direction

than the average texture) the maximum biaxial limit strain that can be achieved is ($\epsilon_{1A} = 0.109, \epsilon_{2A} = 0.1598$) when the groove is parallel to the RD. When the groove is perpendicular to RD, the same amount of cube segregation will give a biaxial limit strain of ($\epsilon_{1A} = 0.1605, \epsilon_{2A} = 0.2263$). Although the importance of the plastic anisotropy of individual grains is well known in literatures, many theoretical predictions of the forming limits still rely on the hypothetical assumption of an initial defect size. The above mesoplasticity analysis attempts to predict the size of

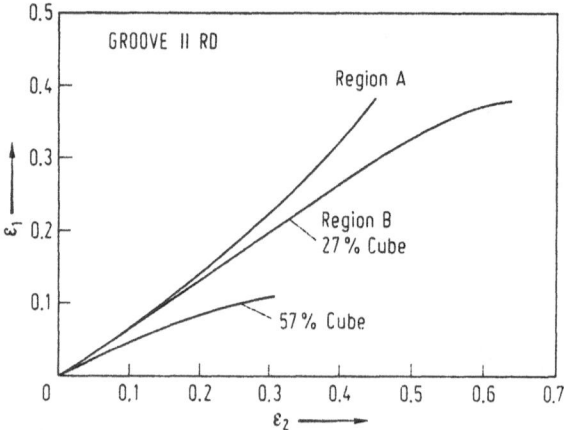

Fig. 10.17. Strain paths outside and within groove with the groove parallel to the rolling direction

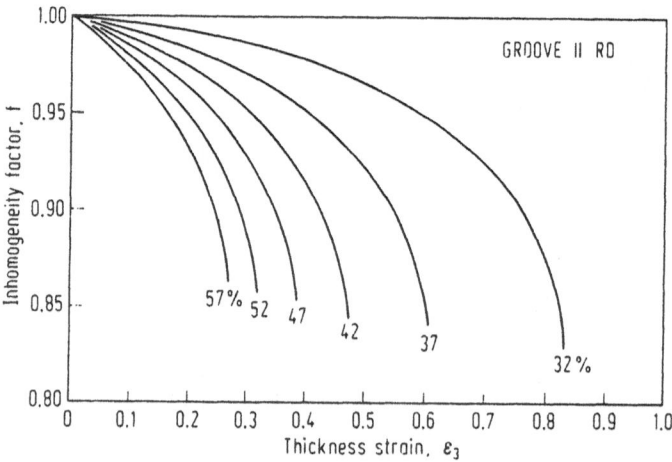

Fig. 10.18. Changes of inhomogeneity factor with different amounts of cube components in region *B* with the groove parallel to the rolling direction

the groove from the amount of segregated texture components without recourse to the original analysis of the M-K model. Such information will be needed for the industrial control of texture variation in sheet metals.

10.5 References

1. Wilson, D.V., in Formability and Metallurgical Structures (edited by Sachdev, A.K and Embury, J.D.), the Metallurgical Society Inc., (1987), p.3.
2. Parker, B.A., ibid., p.71.
3. Thompson, A.W. and Wonsiewicz, B.C., Met.Trans A., 12A(1981), p.531.
4. Bate, P, Int. J. Mech. Sci., 26(1984), p.372.
5. Barlat, F, and Frickle, W.G., 8th Int.Conf. on Texture of Materials, the Metallurgical Society, 1988, p.1043.
6. Lin, D.W., Daniel, D. and Jonas, J.J., Met. Trans., 22A(1991), p.2069.
7. Embury, J.D., in Formability and Metallurgical Structure (edited by Sachdev, A.K. and Embury , J.D.), the Metallurgical Society Inc., (1987), p.101.
8. Atkins, A.G. and Mai, Y.W., Int. J. Mech. Sci., 27(1987), p.1987.
9. McClintock, F.A., J. Appl. Mech., 35 (1968), p.363.
10. Mouild, P.R. and Skeena, C.C., in Formable HSLA and Dual Phase Steels (Edited by Davenport, A.T.), AIME, New York, (1979), p.81.
11. Hill, R., J. Mech. Phys. Solids, 38(1990), p.405.
12. Lankford, W.T., Snyder, S.C., and Bauscher, J.A., Trans. A. Soc. Metals, 42(1050), p.1197.
13. Hosford, W.F. and Backofen, W.A., Fundamental of Deformation Processing, Syracuse University Press, New York, (1964), p.259.
14. Whitely, R.L, and Wise, D.E., in Flat Rolled Products (edited by Earhardt, E.A.), Part 111, Chicago University Press, (1962), p.47.
15. Dillamore, I.L and Roberts, W.T., Metallurgical Reviews, 10(1965), p.271.
16. Vieth, R.W. and Whitely, R.L., IDDRG Colloquim, London, Institute of Sheet Metal Engineering, 1964.
17. Fukuda, M., Trans ISIJ., 8(1964), p.98.
18. Lee. D.N., J. Korean Institute of Metals , 20(1982), p.586.
19. Hosford, W.F. and Backofen, W.A., Proc.9th Sagamore Army Materials Reserach Conf., 1962, New York, p.259.
20. Bunge, H.J. and Roberts, W.T., J. Appl. Cryst., 2(1969), p.116.
21. Parniere, P. and Roesch, L., Deformation and Recrystallization Textures (Third European Colloquium), 1973, p.527.
22. Davies, G.J., Goodwill, D.J. and Kallend, J.S., J. Appl. Cryst. 4(1973), p.67.
23. Kumar, A. and Hutchinson, W.B., Deformation and Recrystallization Textures (Third European Colloquium), 1973 p.527.
24. Lequeu, Ph., Jonas, J.J. and Montheillet, F., Proceedings of International Conference on Computer Modelling of Fabrication Processes and Constitutive Behaviour of Metals, Ottawa, Ontario, May 15-16, 1986, p.237.
25. Montheillet, F., Gilorminin, P., and Jonas, J.J., Acta. Metall., Vol. 33(1985), p.705.
26. Viana, C.S., Kallend, J.S. and Davies, G.J., Int. J. Mech. Sci., 21(1979), p.355.
27. Elias, J.A., Heyer, R.H. and Smith, J.H., Trans. A.I.M.E., 224(1962), p.679.
28. Perovic, B. and Karastojkovic, Z., Metals Technology, 2(1980), p.79.
29, Hirsch, J., Musik, R. and Lucke, K., Proc. Int. Conf. of Textures of Materials, Aachen, 1980, Vol.II, p.437.
30. Lequeu, Ph., Gilormini, P., Montheillet, F., Bacroix, B. and Jonas, J.J, Acta Metall., 35(1987), p.1159.
31. Hu, H., Met. Trans.A, 6A(1975), p.945.
32. Arthey, R.P. and Hutchinson, W.B., Metall. Trans., 1981, 12A(1981), p.31.
33. Blade, J.C and Pearson, W.K.J, J. Inst. of metals, 1962-63, Vol.91, p.10
34. Wright, J.C., J. Inst. Metals, 93(1964-65), p.289.
35. Wilson, D.V. and Bulter, R.D., J. Inst. Metals, 90(1961-62), p.473.
36. Hill, R., The Mathematical Theory of Plasticity, Claredon Press, Oxford, 1950.
37 Bourne, L. and Hill, R., Phil. Mag., 41(1950), p.671.
38. Budiansky, B. and Wang , N.M., J. Mech. Phys. Solids, 14-15(1966), p.357.
39. Chiang, D.C and Kobayashi, S., J. of Eng. for Industry, Nov., 1966, p.443.
40. Sowerby, R. and Johnson, W., J. Strain Analysis, 9(1974), p.102.
41. Cooks, M., J. Inst. Metals, 60(1937), p.159,

42. Burghoff, H.L. and Bohlen, E.C., Trans. A.I.M.E, 147(1947), p.144.
43. Baukloh, U., Heubner, U., and Leogrande, A., Proc. of Int. Conf. on Copper and its Alloys, Amsterdam, September 21-25, 1970, Institute of Metals Monograph No.34, p.149.
44. Schmidt, F.E. and Lawley. A., CDA-ASM Conference on Copper, October 16-19, 1972, Ohio, American Society of Metals.
45. Baldwin, W.M. and Howard, T.S. and Ross, A.W., Metals Technology, Sept. 1945, p.86.
46. Richards, T.L., X-Ray Diffraction of Polycrystalline Materials, Institute of Physics, London. (1955), p.462.
47. Roberts, W.T., Sheet Metals Industries, March, 1966, p.237.
48. Wilson, F.H. and Brick, R.M., Trans AIME, 1945, 161(1945), p.173.
49. Tucker, G.E.G., Acta Metall. 3(1961), p.275.
50. Kanetake, N., Tozawa, Y. and Otani, T., Int.J.Mech.Sci., 25(1983), p.337.
51. Panchanadeeswaran, s., Richmond, O., Frickle, W.G., and Lalli, L.A., 8th International Conference on Textures of Materials, the Metallurgical Society, 1988, p.1103.
52. Taylor, G.I., J.Inst. Metals, 62(1938), p.307
53. Bishop, J.F.W and Hill R., Phil. Mag., 42(1951), p.1298.
54. Barlat, F and Richmond, O., Mat. Sci. Engg., 95(1987), p.15.
55. Hirsch, J., Musick, R. and Lúcke, K., Proceedings of the 5th Int. Conf. Texture of Materials, Japan Iron and Steel Institute, 1978, p.437.
56. Mellor, P.B., International Metals Reviews, 26(1981), p.1.
57. Semiatin, S.L. and Jonas, J.J. (1984), Formability and Workability of Metals, American Socitey of Metals, (1984), p.106.
58. Dodd, B. and Bai, Y, Ductile fracture and Ductility, Academic Press, 1987.
59. Whitely,R.L. Trans.AIME, 52(1960), p.154.
60. Keeler, S.P. and Backofen, A., Met. Trans., Vol. 56A(1963), p.25.
61. Schmidt, F.E. and Lawley, A., Int. Copper Reserach Association, Final Report, Project.N. (1973), p.129.
62. Chung, S.Y. and Swift, H.W., Proc. Inst. Mech. Eng., 68(1951), p.165,
63. Fukui, S., Inst. of Phys. and Chem. Research, Tokyo, Sci. Papers No. 885, (1939), p.373.
64. Keeler, S.P, SAE Paper No.650535, 1965.
65. Goodwin, G.M., SAE Paper N0.680093, 1968.
66. Keeler, S.P., Sheet Metal Industry, 45(1968), p.688.
67. Pearce, R., Int. J. Mech. Sci, Vol.10, (1971) p.995.
68. Hecker, S.S., Metlas Eng. Quart., 13(1975), p.42.
69. Azrin, M. and Backofen, W.A., Met. Trans, 1(1970), p.2587.
70. Ghosh A.K., and Hecker, S.S., Met. Trans., 6A(1975) p.1065.
71. Hill, R., 1952, J.Mech. Phys. Solids, 1(1952), p.19.
72. Marciniak, Z. and Kuczynski, K., Int. J. Mech. Sci., 9(1967), p.609.
73. Tadros, A.K. and Mellor, P.B., Int. J. Mech. Sci. 20(1978), p.121.
74. Tadros, A.K. and Mellor, P.B., Int. J. Mech. Sci. 17(1975), p.203.
75. Yamaguchi, K. and Mellor, P.B., Int. J. Mech. Sci. 18(1976), p.85.
76. Parmar, A., Mellor, P.B. and Chakrabarty, J., Int.J.Mech.Sci. 19(1977), p.389.
77. Fukuda, K., Yamaguchi, K.N., Takakwa, N. and Sakano, Y., J. Japan Soc. Technol. Plasticity, 15(1974), p.994.
78. Tadros, A.K., Doctoral Thesis, Bradford University, (1976).
79. Chu, C.C and Needleman, A., J. Engng. Mater. Technol. 102(1980), p.249.
80. Jalinier, J.M. and Schmitt, J.H., Acta Metall. 30(1982), p.1789.
81. Jalinier, J.M. and Schmitt, J.H., Acta Metall. 30(1989), p.1798.
82. Kim, K.H. and Kim, D.W., Int. J. Mech. Sci. 25(1983), p.293.
83. Wilson, D.V., J. Mech. Working Technol. 16(1988), p.257, (1988).
84. Lee, W.B. and Chan, K.C., Proc. of the 3rd International Conference in Technology of Plasticity, Tokyo, July 1-6, 1990, p.1285.
85. Lee, W.B. and Chan, K.C., Textures and Microstructures, 14-18(1991), p.1221.
86. Wilson, D.V., Roberts, W.T. and Rodrigues, P.M.B, Metall. Trans., 12A(1981), p.1595 and 1604.

11 Grain Boundary Engineering and Related Topics

11.1 Introduction to Grain Boundary Engineering

Grain boundaries play a fundamental role in the plastic deformation of polycrystalline materials. There are two levels of study on the effect of grain boundaries. The first one concerns with the microscopic effect of grain boundaries on the dislocations during plastic deformation, and the second one deals with the macroscopic effect of boundaries on the bulk mechanical properties. In the early stages of deformation, the grain boundary regions are strained more than the grain interior. The grain boundaries are not only effective barriers to dislocation slip, but also act as sinks and sources of dislocations. The localized stress produced by dislocation pile up at grain boundaries will initiate multislip in the vicinity of the grain boundary to accommodate the shape change between differentially deforming grains. These dislocations are called geometrically necessary dislocations as distinct from the statistically stored dislocations[1] which result from homogeneous deformation. The existence of such a zone of misfit of the order of the spacing between the slip bands has been put forward by Hauser and Chalmers[2] to explain the macroscopic influence of grain boundaries on the flow stress in terms of the Hall-Petch relation as described in Sec. 6.4.2.

Many observed dependence of flow stress on grain size have been proposed[4,5,6,7,8]. in a wide range of materials from metals to ceramics. The lack of topological information on the boundary structure in the past has led to the simplified view of the grain boundaries as structureless obstacles to the motion of dislocations. The process of strain transfer from grain-to-grain is complex and involves a detailed knowledge of the interactions of dislocations with grain boundaries. Fig. 11.1 shows a typical example of the perturbation of a grain boundary in an Nimonic alloy when a matrix dislocation enters into it. The capacity to emit dislocations may depend on the character of the grain boundary.

A simple model of slip transfer across grain and phase boundaries in single and dual phase brasses has been proposed by Werner and Prantl[9]. The transfer of slip across a boundary is determined by the misorientation of the slip systems in the adjacent grains (Fig. 11.2). For a given orientation relationship, the angle δ can be determined if the orientation of the boundary plane is known. In cases where the slip planes have a common zone axis close to the boundary plane, slip can be

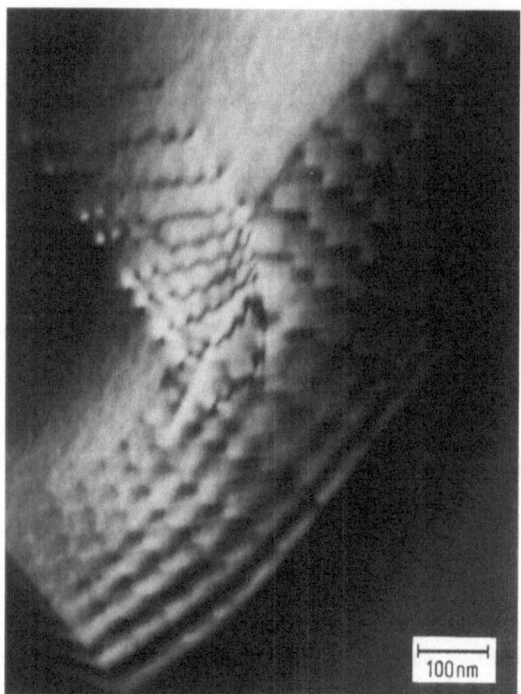

Fig. 11.1. Transmission electron micrographs of a dislocation pile up entering a near $\Sigma = 9$ grain boundary. (From Randle and Ralph[3])

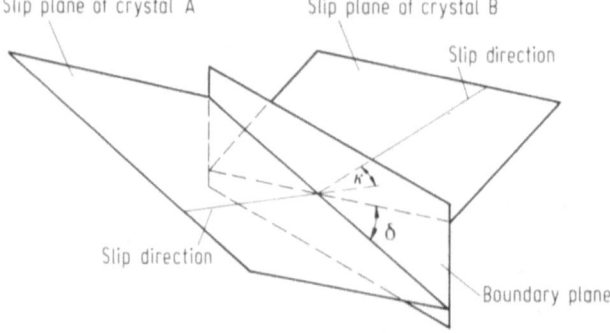

Fig. 11.2. Transfer of slip between two grains. δ is the angle between the intersection lines of the slip planes of the two grains in the boundary plane and κ is the angle between the slip directions lying in the slip planes of the two adjacent grains. (After Werner and Prantl[9])

continuous across the boundary if the grains are in special orientation relationship, i.e. they possess special boundaries (see Sec. 3.3.2). From the bicrystal experiments of Hingwe and Subramanian [10] it is known that the efficiency of grain or phase boundaries as dislocation obstacles is influenced by the character of the boundary between neighbouring grains. The result of calculation from the above geometrical model is the same as that obtained from the Hall-Petch analysis of the yield stress, and phase boundaries are found to be stronger obstacles than grain boundaries.

The influence of grain structure on the ductility and fracture process is complex. If failure occurs below maximum load a brittle form of failure (either cleavage or intergranular failure) is involved. At low temperatures, the material resistance to failure increases according to a (grain size)$^{-1/2}$ relationship. For cubic metals, the slope of this relationship is not large, indicative of the fact that additional shear systems required to maintain grain-to-grain continuity are readily activated. For hexagonal metals, grain refinement is particularly advantageous, both raising strength and increasing ductility. Cleavage fracture follows preferred crystallographic planes, whose position relative to the loading axis is dependent on the boundary orientation. Intergranular fracture arises from intercrystalline stresses, which are also affected by the relative grain misorientation and therefore by the crystallographic texture. The influence of texture on fracture is underexplored and a high benefit can be obtained from the control of fracture anisotropy[11]. The grain size dependence of flow stress disappears as the temperature is raised. Under creep conditions, the creep rate at constant stress increases with decreasing grain sizes due to the availability of grain boundary sliding (see Sec. 5.3.4) and migration as stress-relaxation mechanisms which are prerequisites for the super-plasticity phenomenon. The creep cavitation damage is also found to be dependent on the intercrystalline misorientation[12]. Certain special boundaries are preferentially damaged as opposed to high-angle boundaries which are nearly damage free.

The design of materials brings into consideration different properties of grain boundaries for different applications. At the simplest level, a fine grain structure is preferred for combination of high strength and toughness at low temperatures. For a variety of materials subjected to temperatures above one-third of their absolute melting point, the ability to resist grain boundary sliding and migration is important. This can be achieved by going into specific grain geometries (e.g. in directionally solidified alloys with the grain boundaries parallel to the stress axis) or by using single crystals. To achieve further improvement in material properties, a more subtle control will be on the manipulation of the grain boundary parameter. Designing for strength for instance includes increasing the percentage of special high angle boundaries. To limit grain boundary sliding, an alternative approach is to design the microstructure in the grain boundary region by the incorporation of grain boundary precipitates.

The potential of designing and altering the grain boundary structure offers a challenging area in materials science. This is made possible with the advances in mathematical techniques and instrumentation for the measuring of grain boundary parameters. The crystallite orientation distribution function (see Sec 3.5.4) although represents a successful measure of microstructure in polycrystalline solids, contains no information about morphological measures of the microstructure. The misorientation distribution function on the other hand gives the probability density for the occurrence of specified intercrystalline misorientation between adjacent grains in the polycrystals[13]. In fact, the grain boundary distribution itself may display a texture, which is also known as the *mesotexture* or *grain misorientation texture* (GMT) because it refers to the misorientation between grains. Methods for the acquisition of microtextural (grain specific) data based on electron backscattering diffraction technique in a scanning electron microscope, and convergent beam electron diffraction in a transmission electron microscope have been reviewed by Randle and Ralph[14]. From these type of microtextural data the proportion and distribution of grain boundaries which possess special properties can be obtained.

Grain boundary engineering[15,16,17] is a new area of materials science which requires an understanding of the atomic modelling of grain boundary structures, their characterizations, structure/property relationship and the processing routes to obtain the desired boundary types. The eventual aim is to design materials with the required boundaries which may be beneficial to the overall properties of the materials such as strength, ductility, fracture toughness, machinability and corrosion resistance. In the following sections, examples of engineering materials and products where the role of grain boundary (or its absence) has played a decisive role in their developments are discussed. These include the design of materials with micrograin superplasticity, high temperature turbine blades, and optimum substrates for micromachining. The underlying principles of the technology are explored in the following sections.

11.2 Micrograin Superplasticity

Superplasticity is the capacity of certain polycrystalline materials to deform at low applied stresses to exceptionally large strains without failure by localized necking. Such capacity was demonstrated for the first time by Pearson[18] in a Sn-Bi eutectic alloy with an elongation of 1950%. The research interest in superplasticity did not re-start until Underwood's 1962 review of work carried out in the USSR[19] together with the pioneering work of Backofen and his co-workers[20]. Since then considerable research has been conducted worldwide on the phenomenon. A superplastic elongation of 4850% has been demonstrated in a Pb-62% Sn alloy[21] and 5,500% in a commercial aluminium bronze[22]. Despite these large elongation values, most superplastic alloys have tensile elongation of about 300 to 1,000%. This range of values is sufficient for superplastic forming technology. The advantages of superplastic forming are:

(1) Complex shapes can be made in one piece instead of several pieces which have to be joined. This is expecially true for parts with relatively deep sections and complex curves.
(2) Simple tooling can be used as the forming pressure are very low compared with conventional forming processes.

There are two processing routes for superplastic materials: sheet forming process and isothermal forging. Forming of superplastic sheet is usually achieved by internal pressure so that sheets can be expanded into mould cavity. In forging, the superplastic alloys are slowly shaped in heated dies. The applications for superplastic forming are widespread throughout industry. Examples are cladding panels, various boxes and covers, ejector seats, sport-car bodies, and reinforced wing assemblies in aerospace industry. Compressor wheel integrated with blades has also been made in one operation by isothermal forging in heated dies. Extensive reviews of the subject have been written in the literatures[23,24,25].

There are two types of superplasticity: micrograin superplasticity or fine structure superplasticity (FSS) and internal stress superplasticity (ISS). Micrograin superplasticity depends on the properties of the grain boundary regions while internal stress superplasticity is derived from the relaxation of internal stress by plastic deformation, which contributes to low applied external stress. In materials undergoing polymorphic changes, internal stresses arise from the difference in volume between the two phases during phase transformation. High strain can be achieved by repeatedly cycling the material through its transformation. In this section, the focus will be on fine grain superplastic materials and the role of grain boundaries in achieving the desired properties.

Micrograin superplasticity occurs in a range of metals, alloys and ceramics under favourable conditions of temperature and strain rate. Superplasticity occurs above approximately half the melting point (Tm) and over a specific range of strain rates, usually $\sim 10^{-2}$ to 10^{-4} sec^{-1}. Under isothermal condition, the relationship between true stress σ, strain ϵ and strain rate $\dot{\epsilon}$ can be written as

$$\sigma = \sigma_y + \epsilon^n \dot{\epsilon}^m \tag{11.1}$$

where σ_y is the yield stress and n and m are exponents of strain and strain rate. Neck development is opposed by the high strain rate sensitivity of the flow stress. When the incipient neck starts to form, an increase in the strain rate at the neck causes a large increase in the flow stress and localized necking will be delayed. AT $T > 0.5$ Tm, both σ_y and n is small and

$$\sigma_y = K_1 \dot{\epsilon}^m \tag{11.2}$$

where K_1 is material constant. A true Newtonian-viscous behaviour will result when m approaches 1.

In superplastic flow, there is a sigmodal relationship between the logarithms of stress and strain rate as shown in Fig. 11.3. The slope of the curve $(d \ln \sigma)/(d \ln \epsilon)$ gives the strain rate sensitivity of the flow stress which is largest in the centre of the region II. Superplasticity occurs in region II with m lies in the range from 0.3 to 0.9. When m falls to ~ 0.2, necking becomes the failure mode. The m-values for some common superplastic alloys are shown in Table 11.1.

In superplastic deformation the grain strain is nearly zero and the total strain of the specimen is accounted for by grain boundary sliding. A feature of superplastic flow is that no large scale change in grain shape has occurred. An important structural prerequisite of fine structure superplasticity is that the grain size should

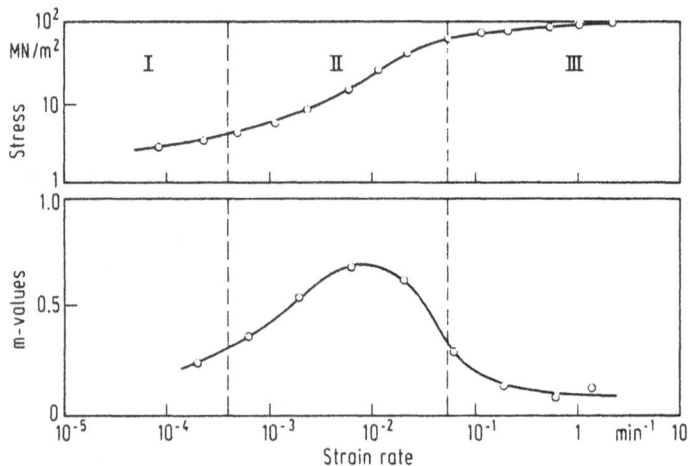

Fig. 11.3. The strain rate dependence of flow stress for the superplastic Mg-Al alloy, grain size 10.5 μm deformed at 350°C (After Lee[26])

Table 11.1. Properties of some common superplastic alloys

Alloy	°C	m	% elongation
Al-6Cu-0.25Zr	350-475	0.5	>1000[27]
Cu-9.8%Al	700	0.7	700[28]
Mg-33.6Al	400	0.8	2100[29]
Pb-19Sn	20	0.5	>500[30]
Ti-6Al-4V	800-1100	0.85	1000[31]
Zn-22Al	200-300	0.45-0.6	500-1500[32]

be small as grain boundary sliding will be inhibited when the grains are coarse. A model of the grain boundary sliding is shown in Fig. 11.4. As the grains at a depth Z deforms, gaps appear between the initially compact grains. Grains from both the adjacent layers (dark grains) fill the gap and the thickness of the material decreases. The grain shape remains equiaxed during the large plastic deformation.

Grain boundary sliding (GBS) if occurring alone will lead to cavitation and overlapping of grain ledges. Superplastic deformation while homogeneous on the macroscopic scale is heterogeneous on the microscopic scale. The sliding of a grain boundary must involve a mechanism that will allow for relaxation of boundary stresses and maintain some structural integrity by atomistic transfer across the interface. To maintain grain contact, extensive material transports are necessary and these take places by the following two mechanisms.

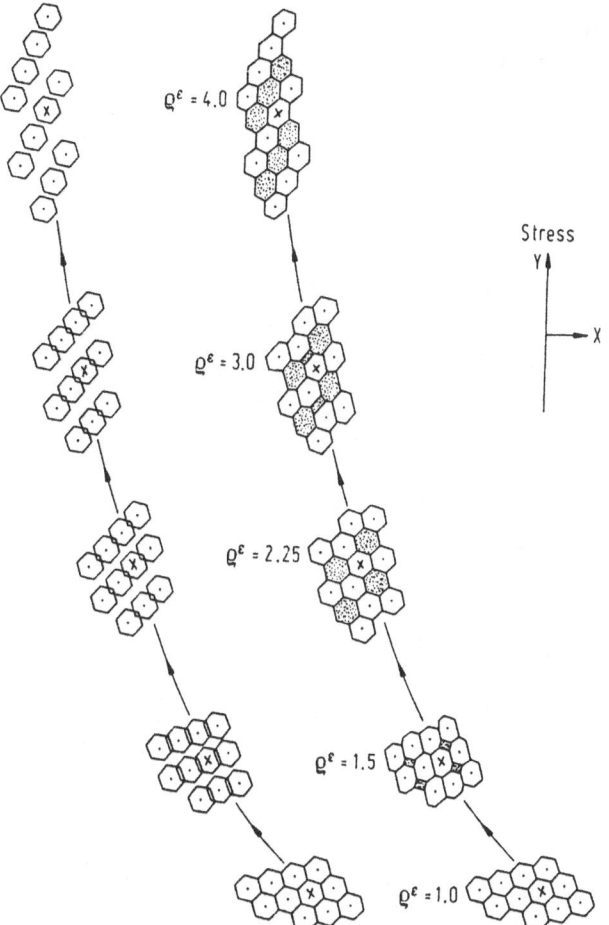

Fig. 11.4. A three dimensional model of grain boundary sliding (From Hazzlendine and Newbury[33]) Displaced left: grain motion without accommodation, the grains initially overlap and finally spread apart. Right: dark grains from adjacent Z planes fill the gap

(1) GBS with diffusional accommodation[34]

The diffusional processes involved are lattice diffusion and grain boundary diffusion. The activation energy for superplastic flow is equal to that for grain boundary diffusion at intermediate temperatures and equal to that for lattice diffusion at higher temperatures. As the rate of diffusion increases at high temperature, superplasticity is basically a high temperature phenomenon. At low strain rate, the diffusion accommodated GBS predominates as the major deformation mechanism. There is very little extension of the grains and the grain shape remains equiaxed. The relative displacement of the grains and the accompanied rotation will destroy any crystallographic texture that may exist before the deformation. No work hardening of the grain occurs due to the dislocation activities. When the strain rate increases, dislocation creep then takes the place of GBS.

The effect of grain size can be incorporated in the following phenomenological equation

$$\dot{\epsilon} = K_2 \sigma^{1/m} D_{\text{eff}} / L^p \tag{11.3}$$

where L is the grain size and p its exponent, D_{eff} is the effective diffusion rate which can be lattice diffusion, grain boundary diffusion or a mix of the two. The grain size exponent p and the type of diffusion for some superplastic alloys are listed in Table 11.2

Table 11.2. The grain size exponent for some superplastic alloys (abstracted from Wadsworth, et al.[35])

Alloy	Observed grain size exponent (p)	Diffusion control type
Al-33Cu	2.1 - 2.4	lattice
Cu-7P	1.98 - 2.04	lattice
Cr-39Ni-26Fe	2	lattice
Pb-5Cd	2.6	grain boundary
Sn-38Pb	2.3	grain boundary
Sn-5Bi	3	grain boundary

(2) GBS accommodated with dislocation slip

Ball and Hutchinson[36] proposed in 1969 that groups of grain could slide until it was obstructed by an unfavourably oriented grain. The stress concentration set up in the blocking grain was released by dislocations as shown in Fig. 11.5. These dislocations pile up against the grain boundary and continuation of sliding requires climb by the leading dislocation toward annihilation sites. The strain rate for the sliding of group of grains is given by

$$\dot{\epsilon} = K_3 (\mathbf{b}/d)^2 D_{gb} (\sigma/E)^2 \tag{11.4}$$

where K_3 is a material constant, $\mathbf{b}$ is the Burgers vector, D_{gb} is the grain boundary diffusion coefficient, (σ/E) is the modulus compensated stress. The above view however has not been substantiated by experimental observation as no pile up of dislocation has been found. To overcome the difficulty, Gifkins[37] proposes that GBS occurs in a mantle-like region adjacent to the grain boundaries (Fig. 11.6). The core of the grain remains free of dislocations which were confined to the mantle, i.e, the near grain boundary region. When slip occurs within the core of each grain, superplasticity ceases and normal ductility is expected. Many modifications of the core-and-mantle model have then been proposed on the details of the strain accommodation mechanism across the interphase or grain boundaries.

There is evidence that superplastic flow is sensitive to the degree of misorientation at grain boundaries (Fig. 11.7) in a manner similar to the sensitivity of grain boundary free energy to misorientation as shown in Fig. 5.14 of Chapter 5. For a high degree of superplasticity, the grain boundaries between adjacent grains should be high angles (i.e. disordered). Grain boundaries corresponding exactly to coincidence-site lattice would contain too few grain boundary dislocations which are needed to accompany the GBS. Sliding of a twin boundary is extremely difficulty as all atoms would have to move at the same time. In a Zn-Al superplastic alloys with a strong texture, grains which are oriented for easy basal slip would require a lower stress for GBS which indicates slip occurs along specific crystallographic planes near the grain boundary[38]

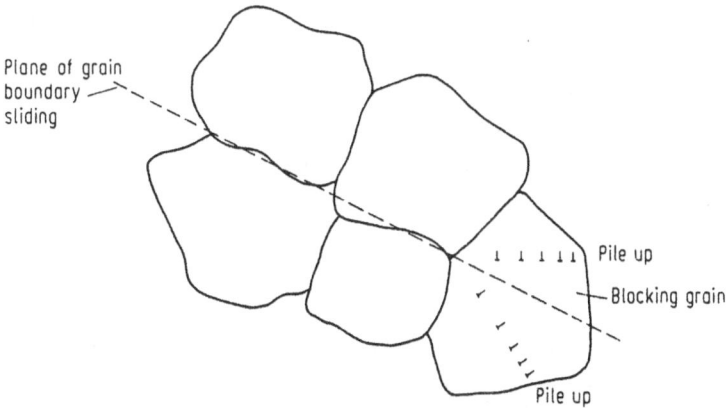

Fig. 11.5. Ball and Hutchinson's model of grain boundary sliding with dislocation accommodation.(From Ball and Hutchinson[36])

Besides diffusional flow and dislocation motion, grain boundary migration and recrystallization also provides means grain strain can be accommodated during plastic deformation. Grain boundaries in superplastic alloys should be made mobile in order to reduce stress concentrations developed at obstructions along the grain boundaries. The ability of the grain boundary to migrate depends on the structure of the boundary, the effect of solute segregation, vacancy absorption and emission, temperature, and the applied stress. For ceramic materials, a high strain rate sensitivity is not a sufficient conditions for superplastic flow. The limited ductility observed in most fine-grained ceramics at high temperatures is that the grain boundaries are not very mobile.

Superplastic flow depends strongly on the grain boundary properties. This offer scope for designing grain boundaries in a fine grained materials that will be conducive to superplasticity. Some examples of modifying the grain boundary

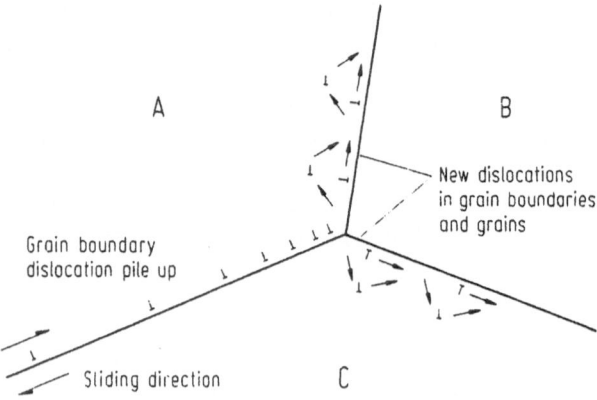

Fig. 11.6. The Core-and-Mantle model of grain boundary sliding with dislocation accommodation (From Gifkins[37])

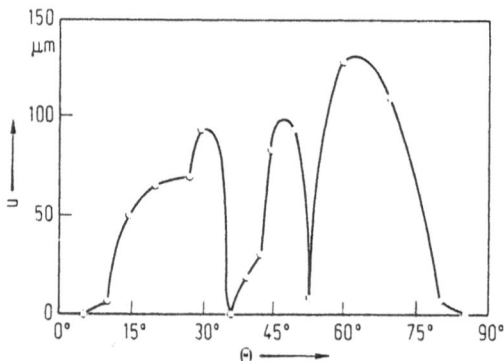

Fig. 11.7. Grain boundary sliding displacement, u, versus misorientation for <001. aluminium tilt boundaries. (After Biscondi and Goux[39])

structure or composition in an alloy system to obtain superplasticity are illustrated below for the case of a tool steel, an aluminium-lithium based alloy and an Al_2O_3 particulate composite.

(a) Tool steel

An oil-hardening commercial tool steel can be made superplastic by appropriate thermomechanical treatment[40]. The required structure is a uniform distribution of spheriodized cementite particles (0.1-0.5 μm) in a fine-grained ferrite matrix of grain size of about 1 μm containing only high angle boundaries. The fine structure is obtained by rolling the commercial steel from the solution treatment temperature of 1050°C to about 600°C followed by isothermal rolling at 650°C. The as-rolled steel consists of both high and low angle boundaries, which can be converted to one containing low angle boundaries by thermally cycling the steel between the lower critical temperature and 650°C[41]. The tool steel in this condition exhibited a strain rate sensitivity exponent of 0.5 and a maximum elongation to failure of 1200% at 650°C.

(b) Aluminium-lithium based alloy

Superplasticity in Al-3Li-1.5Cu-1Mg-0.2Zr alloy made from powder metallurgy route can be developed by thermomechanical processing[42]. The very fine powders were made using Rapid Solidification Processing (RSP), forged at 382°C, reheated and then extruded at the same temperature. The as-extruded structure consists of medium angle subgrains with misorientation in the range 3°-10°. These boundaries are not mobile. For the development of superplastic properties, the as-extruded alloy is solution treated at 538°C followed by aging at 400°C. The aged alloy is then rolled at 300°C and further recrystallized from 400°C to 520°C. After the thermomechanical treatment, the microstructure contains very fine equiaxed grains (2-3 μm) with large angle, disordered boundaries (i.e. $> 10°$ misorientation) which are mobile and able to slide.

(c) Al-Al$_2$O$_3$ particulate composite

A fine grained Al-Al$_2$O$_3$ particulate composite can be made superplastic at 50°C by creating a thin diffusion layer of aluminium-gallium solid solution at grain boundaries[43]. The fine-grained Al-Al$_2$O$_3$ was immersed in liquid gallium at 50°C. The gallium penetrated the grain boundaries and made the material very brittle. Superplasticity was obtained by diffusing the gallium at 50°C for 50 hours to form an Al-Ga solid solution of low melting point in the grain boundaries. In this example, the chemical composition of the boundary is altered whereas in the above two cases, only the boundary structures are altered. Annealing at higher temperatures causes diffusion of gallium to the lattice and the material looses its superplasticity.

11.3 Design of Turbine Blade Materials

The gas turbine engine uses energy generated by the combustion of kerosene to increase the energy of an air stream. The resulting hot high pressure gas is then expanded, first through the turbine and then through the nozzle to provide the thrust. Higher thrust is achieved by a high compression ratio and high turbine entry temperature. Gas turbines work with high temperatures and the turbine blades are exposed to the full temperature, high gas velocities, stress and corrosion. An enormous amount of research has been directed to improve the efficiency of jet engines and fuel economy by increasing the turbine inlet temperature. The choice of the turbine materials is usually based on its creep and fatigue strength. It has been estimated that two third of damage undergone by the material is due to creep.

The mechanism for creep deformation is complex and may involve the following: (a) dislocation gliding (b) dislocation climbing, and (c) diffusion creep. At low temperatures and at high stress edge dislocations glide only in their slip planes. At elevated temperature, dislocations can overcome barriers by thermally assisted mechanisms involving the diffusion of vacancies such as climbing. At low stress and high temperature the deformation is not due to the motion of dislocation but to diffusion of vacancies, which is also known as the Nabarro-Hering creep. The various creep deformation mechanisms have been shown by Ashby[44] in stress-temperature space which indicates the dominant deformation mechanism for that stress-temperature combination.

The creep rate $\dot{\epsilon}$ is controlled by the dislocation density ρ and the average velocity $\bar{v}$(cm/sec) of glide and climb dislocations:

$$\dot{\epsilon} = b\rho\bar{v} \tag{11.5}$$

where b is the Burgers vector. The average velocity $\bar{v}$ is related to the applied stress σ, temperature T and the coefficient of self-diffusion D_{SD} by the following

$$\bar{v} \approx b^2/kT\sigma D_{SD} \tag{11.6}$$

where k is the Boltzmann's constant. The activation energy for self-diffusion has been shown by Sherby et al.[45] to be equal to the activation energy for high-temperature creep Q_c. If solute atoms segregate at dislocations, Q_c will be increased and this explains why does γ-Fe which dissolves more carbon solute atoms will have a higher creep strength than α-Fe. However, the most efficient way to increase the creep strength of an alloy is to impede the motion of dislocations by an evenly dispersion of fine particles. Such dispersions are produced by precipitation, sintering or internal oxidation. For a detailed discussion on the mechanism of dispersion or precipitation hardening, please refer to Sec. 6.3 of this book.

The metallurgical factors which enchance creep resistance are summarized below:

(1) material with covalent bonding and a high melting point (i.e. ceramic materials such as Si_3N_4).

(2) fine particles which are evenly distributed in the matrix (i.e intermetallic compounds such as Ni_3Al, Ni_3Ti, etc. in nickel based superalloys, and carbides such as VC, TiC, Mo_2C in creep resistant steels).

(3) alloys with low stacking fault energy as extended partial dislocations have difficulty to cross slip and climb to avoid obstacles. The stacking fault energy can be lowered by solid-solution alloying additions.

(4) segregation of solute atoms to climbing dislocations.

(5) thermal stability of microstructures, e.g. thermomechanical processing of alloys to avoid coarsening of the precipitates, (Ostwald ripening).

It is well known that metals undergo a transition from transgranular fracture to intergranular fracture as the temperature is increased. At low temperatures, grain boundaries have a strengthening effect (Sec. 6.4), but with respect to creep, grain boundaries are the weakest parts in a microstructure. In the third stage of creep, nucleation and growth of grain boundary voids and cracks occur. Cavities are formed by excessive sliding at the grain boundaries. While grain boundary sliding is a design feature in superplastic alloys, a high rate of grain boundary sliding is undesirable for creep resistant alloy. The compensation of misfits caused by grain boundary sliding can occur by diffusion of atoms, and the grain boundary diffusion coefficients depends on the structure of the grain boundary (see Sec. 3.3.2). Boundaries that deviate from good coincidence have high grain boundary diffusivity that leads to a high rate of sliding and should be avoided.

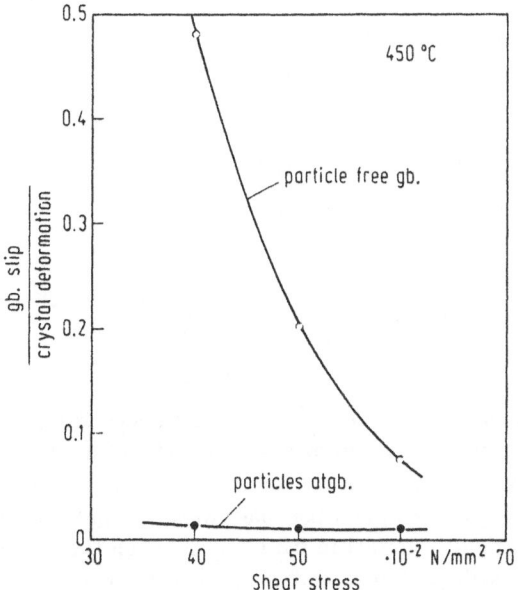

Fig. 11.8. Effect of particles on grain boundary slip in an aluminium alloy. (After Horton[46])

The amount of grain boundary sliding can be controlled by segregation of solute atoms and dispersions of particles to the grain boundaries. The interaction of solute impurities with grain boundaries is strongly dependent on the grain boundary structures. Boundaries preferentially segregated with impurity atoms creates an easy fracture path along those boundaries where the separation energy in the presence of solute atoms is low. The effect of particles on grain boundary sliding is best illustrated in an Al-0.05% Fe alloy (Fig. 11.8). A small addition of 0.08% Fe to aluminium inhibits almost completely the grain boundary sliding. In nickel based superalloys the γ' phase precipitates (which is isomorphic to the Ni_3AL phase) in the γ (Ni) matrix harden it, while the carbides (i.e. $M_{23}C_6$) play a significant part in inhibiting grain boundary sliding when present as discrete particles in grain boundaries. However, the properties will deteriorate when the carbides exist as films in grain boundaries.

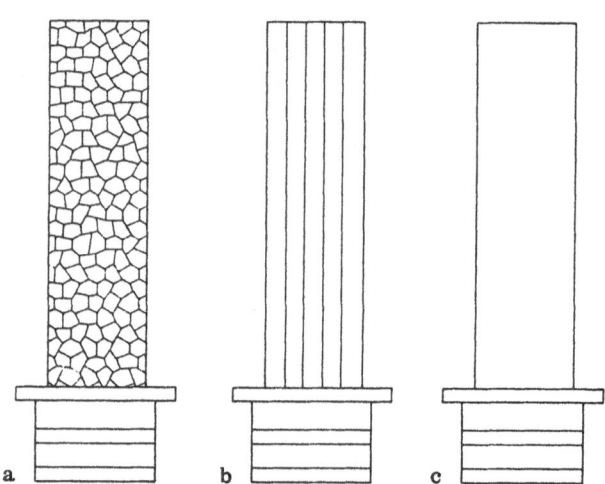

Fig. 11.9. Schematic diagram showing different cast structures of turbine blades for aircraft engines. **a** Equiaxed-grain structure; **b** columnar-grain structure; **c** single crystal in superalloy

For materials with equiaxed structures, creep fractures are associated always with grain boundaries. The requirement of gas turbines operating at higher temperature has been met by recent development in directional solidification and single crystal technology (Fig. 11.9). To solve the grain boundary weakness problem in turbine blades, grain boundaries oriented normal to the stress direction (i.e. along the airfoil direction due to the centrifugal force) were suppressed by using directional solidification. The techniques for producing columnar grain structure were developed during the late 1960s[47]. The basic principle of the method is to remove heat from the investment casting in an undirectional manner and to maintain a positive temperature gradient ahead of the solidifying interface to prevent extraneous nucleation. Grains which possess <001> crystallographic orientations closest to the direction of heat flow grow faster than those grains less favourably oriented, and progressively eliminate less favoured crystals. Another parallel development is the production of unidirectionally solidified eutectic superalloys. Such eutectic system should have a favourable volume of matrix phase to hard second phase or an intermetallic compound which has its length parallel to the direction of solidification. One of the alloy in this family is the Ni-19.7%Nb-6.0%Cr-2.5% Al developed by Lemekey and Thompson [48]. Its microstructure consists of an aligned lamellar structure of an intermetallic compound Ni₃Cb (δ) and a nickel chromium matrix (γ) strengthened by Ni₃Al precipitate (γ'). The rupture stress to produce failure in 300 hours for three nickel based superalloys cast with different structures is shown in Fig. 11.10.

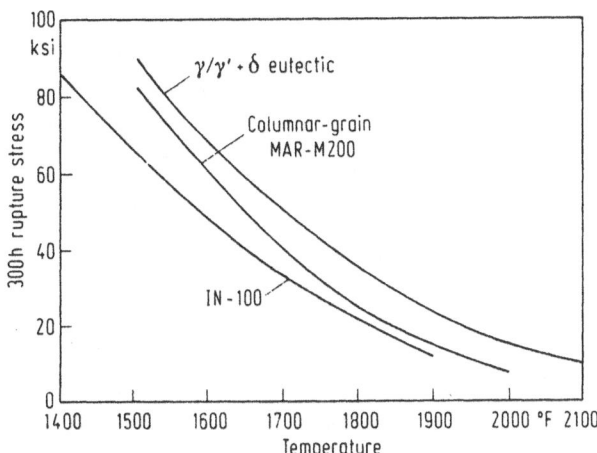

Fig. 11.10.   Rupture stress-temperature curves for three nickel-based superalloys. (After Erickson et al.[49])

   The third generation of turbine blades comes from the introduction of single grain manufacturing of airfoils. Not only grain boundaries are completely eliminated but the alloying additions added to harden the grain boundary can be eliminated as these additions often lowers the melting and solvus temperature of the alloy. Single crystal turbine blades are produced by modification of the directional solidification process. A limited number of grains are allowed to grow into a spiral crystal selector in the casting mould (Fig. 11.11). The cast component has a <100> direction in the growth direction but the orientation is uncontrolled in the plane normal to the growth direction. To control all crystal directions requires the use of a "seed crystal" in the lower part of the mould.

   A considerable increase in the life of the turbine blades can be achieved as a consequence of the crystal anisotropy. The orientation and temperature dependence of the yield stress of nickel-base single crystals is well known. Because of the crystallographic nature of plastic deformation, the creep deformation of nickel-base superalloy single crystal is highly anisotropic. The stress rupture life in uniaxial creep test depends on the direction of the applied stress with regard to the crystallographic directions of the turbine blade. Such an orientation dependence is shown in Fig. 11.12 for a nickel-base single crystal superalloy CM SX2. Longer stress rupture life is found when <100> of the single crystal is parallel to the airfoil direction. The development of air cooled blades while enabling higher turbine inlet

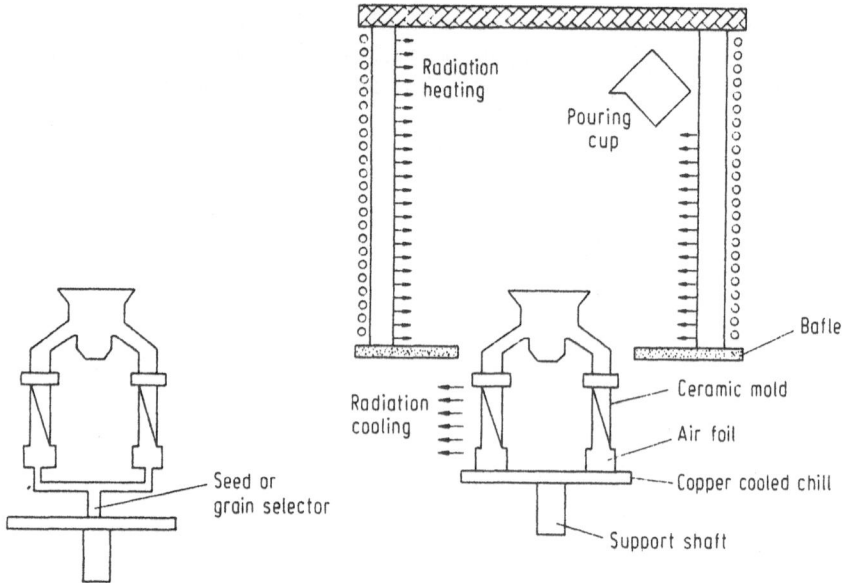

Fig. 11.11. Mould system for a directionally solidification furnace with the mould on a water cooled copper chill. (From Lacaze and Hazottee[50])

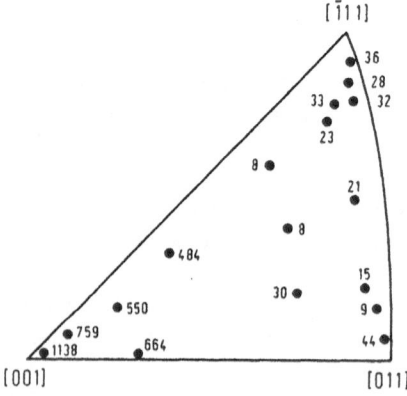

Fig. 11.12. A stereographic triangle showing the orientation dependence of stress rupture life (in hours) of a CM SX2 single crystal nickel-base superalloy at 760°C and 75MPa. (After Caron and Khan[51])

temperature to be used brings about more complex design of the component. As the part goemetry becomes more and more complex, the actual orientation map of the local stress in the blade during service becomes important, and a precise characterization of the material behavior is needed on structures modelization. The stress dependence of the primary creep rate on a nickel-base superalloy has been expressed as[52]

$$\dot{\epsilon} = C \exp \beta \sigma \tag{11.7}$$

where $\dot{\epsilon}$ is the primary creep rate, σ is the applied tensile stress, C is a function of temperature and β is a material constant. Eq.(11.7) is used to develop a set of constitutive equations for every slip systems. The shearing stress components of stress acting on the ith slip plane can be resolved parallel to the jth slip direction are denoted as τ_{ij} (with $i = 1$ to 4 and $j = 1$ to 3). The shearing deformation rate γ_{ij} are then given by

$$\begin{aligned}
\dot{\gamma}_{ij} &= 0, \qquad \text{when} \quad \tau_{ij} = 0 \\
\dot{\gamma}_{ij} &= C'(\tau_{ij}/|\tau_{ij}|)\exp(\beta - 1), \qquad \text{when} \quad \tau_{ij} \neq 0
\end{aligned} \tag{11.8}$$

where c' is a constant in the creep law. All $\dot{\gamma}_{ij}$ referred to the crystallographic axes are superposed linearly to give the total deformation rate. The slip planes in face-centered cubic crystals are the $\{111\}$ planes. At low to moderate temperatures and at high strain rate, the slip directions are $<110>$. At about 700°C, $<112>$ slip directions become effective in addition to $<110>$[53]. While primary octahedral slip was found to be predominate at temperatures below the peak stress temperature, thermally activated cross-slip of $a/2<110>$ dislocations from $\{111\}$ planes into $\{100\}$ planes occurs above the peak temperature[54]. The crystal anisotropy of creep becomes less pronounced at high temperature because of $\{100\}$ slip.

The fast development in the turbine blade technology during the last decades is a successfull example of collaboration between different scientific and engineering disciplines such as aerodynamics, materials science and mechanics, and manufacturing engineering. Metallic materials have now reached the highest level of their capabilities. As the gas inlet temperature of the next generation of gas turbine are required to be several hundred degrees above those of the current ones, a big forward to higher temperature is to adopt alternative materials such as ceramics and composites, an area where a fundamental understanding of the structure and mechanics of advanced materials will be needed.

11.4 Ultra-Precision Machining of Crystalline Materials

Ultra-precision machining has been developing rapidly in the manufacture of optical components and laser mirrors with a surface roughness of a few nanometers. Spherical or aspherical surfaces produced by this method has high quality without requiring any post-machining polishing. It becomes now possible to generate well-defined modulations on workpiece surface in the micron range by means of a fast tool servo (FTS) device[55]. The success of this technology relies on high precision machine tools, advanced control system, laser metrology and single point diamond cutting tool. Materials which have been successfully cut are mainly ductile metals such as copper, aluminium, and non-electrolytic nickel. In micro- or ultra-precision machining, the depth of cut will be less than the average grain size of a polycrys-

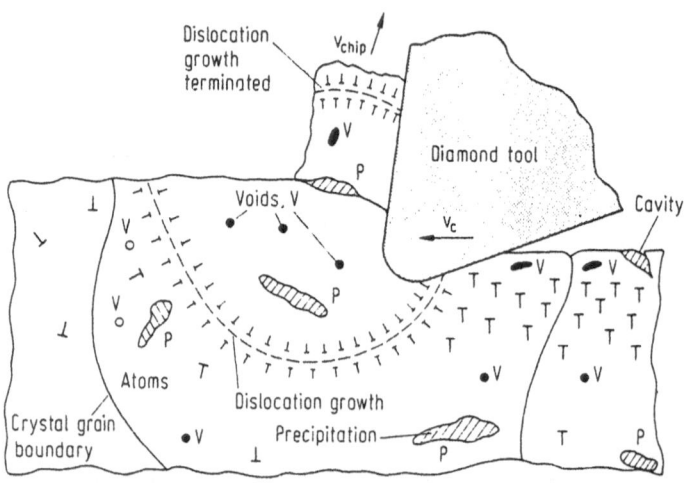

Fig. 11.13. Imperfections in materials that influence micromachining. (From Hashimoto[56])

talline aggregate. The cutting process leads to a multiplication of defects and is affected by the imperfections which exist in the material (Fig. 11.13). An insight in the behaviour of the microcutting can be obtained by considering the process of chip formation from the view point of mesoplasticity.

The chip removal process in diamond machining can be classified into two types. One is due to plastic deformation on the slip planes and the other is due to cleavage fracture on the cleavage plane. In machining a homogeneous material, the shear plane and the cleavage plane coincides with the planes of maximum shear or tensile stress. The criteria of chip removal, however, is more complex for a crystalline material. In the ductile mode of machining, the crystallographic orientation of the substrate material causes systematic variation on the cutting force in micromachining. Crystalline materials are known to be anisotropic in their physical and mechanical properties. The surface quality and cutting force are affected by the crystallographic structure of the substrate material. There is strong experimental evidence that the shear angles and the cutting forces[57,58,59] vary with the crystallographic orientations of the metals being cut. Shear front-lamellar structures at the top of the chip have been reported[60] to correlate with grain orientations. With regard to the quality of the machining process, the differences in the Young's Modulus along different crystallographic directions were considered by some researchers to be responsible for the differences in the machined surface. Although shear direction has been shown to be very sensitive to crystallographic orientation, no analytical relationship between shear direction and crystallography has been established, and past attempts to correlate the shear stress and shear angle based on Hill's macroscopic anisotropic function have been unsuccessful[58].

In machining ductile metals, the magnitude of the shear angle (or the idealized orientation of the shear zone) indicates the machinability of the work materials and

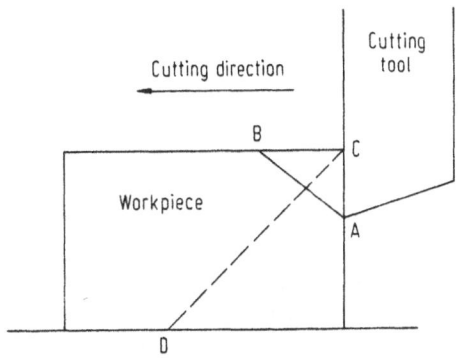

Fig. 11.14. Activation of shear in metal cutting.

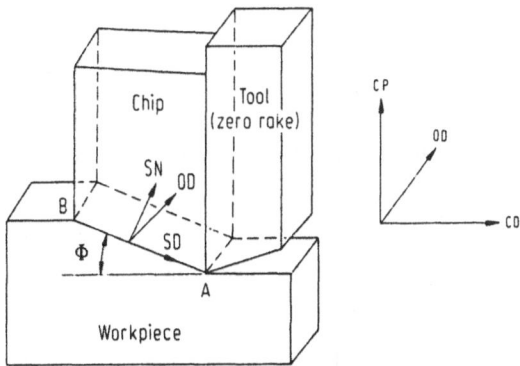

Fig. 11.15. The cutting geometry. ϕ = Shear angle. Workpiece coordinate system: CD = Cutting direction; CP = Cutting plane normal; OD = Observation direction. Shear band coordinate system: SD = Shear direction; SN = Shear plane normal; OD = Observation direction

the efficiency of the process. The shear zone angle has been found to vary with the tool geometry, the cutting conditions and the workpiece materials. A large shear angle is associated with continuous chip formation, good surface finish and low cutting forces. Various theoretical shear angle equations have been derived in the past but met with partial success only[61]. Most studies of the cutting mechanism are performed under the assumption that the material is isotropic and is a homogeneous continuum. The effect of material anisotropy is often not included in the theoretical analysis. One important source of material anisotropy lies in the crystallographic nature of the metallic substrate. A single crystal cutting experiment was carried out by Ueda and Iwata[59] based on a micromachining device inside a scanning electron microscope. The shear angles were measured from SEM micrographs. A 0° rake angle was used in the cutting. No built-up edge was reported in the cutting experiment. Ueda and Iwata calculated the shear angles based on a simple analysis of the Schmid factor of the slip systems in different orientations. The predicted cyclic variation in shear angle for the two single crystals were different from the experimental data and the predicted values were out of place with each other.

A predictive model has been proposed by Lee[62] to analyze the orientation of the shear zone in single crystal cutting. In the model, plane strain orthogonal cutting is assumed and the rake angle is zero. As the tool advances, the material ahead of it is being compressed in the cutting direction and a shear band (a zone between parallel shear planes) joining the top of the tool to the surface of the work material develops. The shear band occurs along the direction AB as shearing along CD is prohibited by the geometry of the tool and workpiece as shown in Fig. 11.14. The effect of friction is not considered here as it can be included in the classical shear angle equation of the Merchant type[63]. The workpiece is assumed to be a good conductor of heat and cutting is performed under isothermal conditions.

Referring to the shear band coordinate system (SD-SN-OD) as shown in Fig. 11.15, the strain in the shear band is described by the displacement gradient E_s

$$E_s = \begin{bmatrix} 0 & 0 & -d\gamma \\ 0 & 0 & 0 \\ 0 & 0 & 0 \end{bmatrix} \tag{11.9}$$

where $d\gamma$ is the shear strain in the band. The same strain in the workpiece coordinate (CD-CP-OD) is given by

$$E_w = AE_sA^{-1} \tag{11.10}$$

where A is the transformation matrix from the shear band coordinate system (SD-SN-OD) to the workpiece coordinate system (CD-CP-OD), and

$$E_w = d\gamma \begin{bmatrix} \cos\phi\sin\phi & 0 & -\sin^2\phi \\ 0 & 0 & 0 \\ \cos^2\phi & 0 & -\sin\phi\sin\phi \end{bmatrix}. \tag{11.11}$$

The symmetric strain tensor in the shear band (ϵ_w) can be derived from (11.11). The increment of plastic work dW is given by

$$dW = \sigma d\epsilon_w \tag{11.12}$$

where σ is the equivalent stress or the plastic work per unit volume and strain, and $d\epsilon_w$ is the macroscopic effective strain.

During machining, the zone of workpiece material in contact with the tool tip acts as strong source of dislocations. Fine cracks are produced near the vicinity of the tool tip and trigger the primary shearing process[64]. It must be noted that the shear band may not be parallel to a particular crystallographic slip plane of the crystal, and are not dislocation glide planes as implied by other workers [58, 60]. However, the shear in the band has to be accomplished by homogeneously distributed slip, i.e. all operative slip systems co-operate in the shear band development. This is an important physical basis for the use of the Taylor theory in the analysis of the shear angle problem. The finding of a direction at which shearing will occur will be similar to the analysis of the shear banding problem discussed in Sec. 5.2.

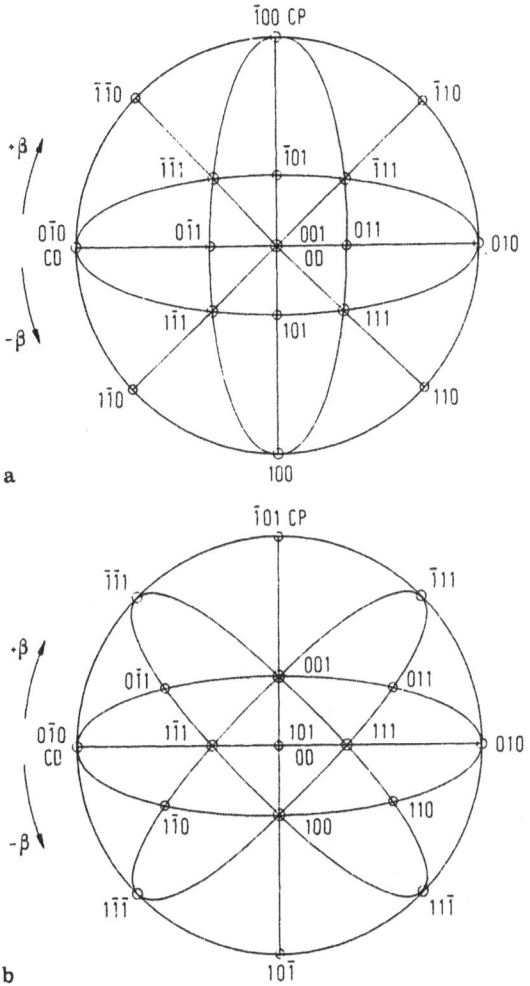

Fig. 11.16. Crystallographic orientation of the β-brass crystal used in the cutting experiment of Ueda et. al.[59]. **a** crystal with [001] axis parallel to OD; **b** crystal with [101] axis parallel to OD

A theoretical analysis of single crystal cutting for different crystallographic cutting directions of two β-brass single crystals (i.e. $A(001)$ and $B(101)$) with either the [001] or [101] axis parallel to the observation direction (OD) is illustrated below. In β-brass, slip is assumed to occur on the {110} planes and along <111> directions[65]. The crystal is rotated successfully by an angle β about the direction OD as shown in Fig. 11.16.

For each crystallographic cutting direction, the variation of the effective Taylor factor with the possible shear angle is calculated. The inputs to the programme are the incremental shear strain $d\gamma$ and the crystallographic orientation. When

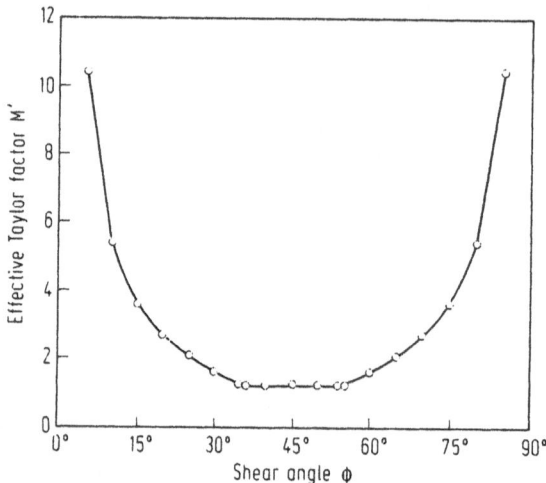

Fig. 11.17. Variation of effective Taylor factor *M*- with shear angle in crystal with [101] parallel to CD and [010] parallel to CP

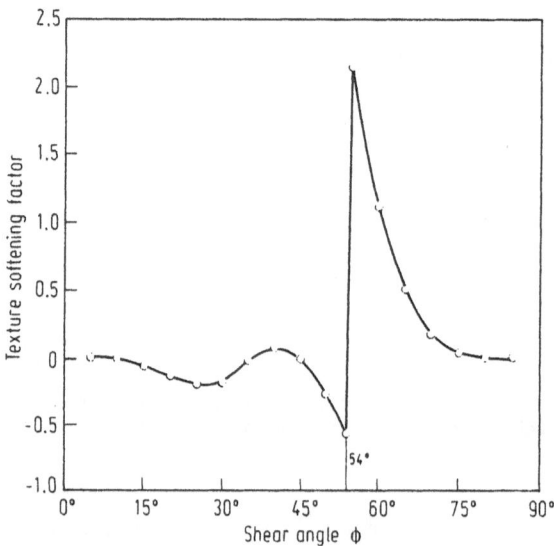

Fig. 11.18. Variation of texture softening factor with shear angle in the crystal with [101] parallel to CD and [010] parallel to CP

no unique minimum M-value is found, the texture softening factor is calculated for each shear angle having the same minimum shear strength. The most likely shear angle is the one which had the most negative texture softening factor. An

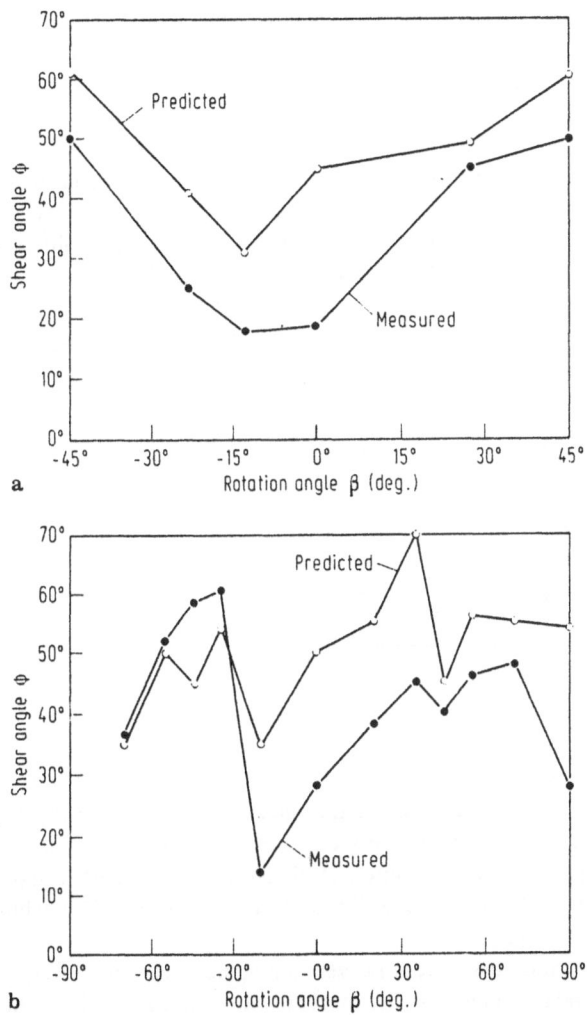

Fig. 11.19. Variation of shear angle with rotation angle in (a) crystal A[001] and (b) crystal B[101].
The measured values are taken from the experimental data of Ueda and Iwata[59]

example of the variation of M-value with the shear angle ϕ is shown for the crystal
orientation with <101> parallel to CD and [010] parallel to CP (Fig. 11.17). The
M-factor shows a minimum value for shear angles ranging from 36° to 54°. The
corresponding texture softening factor is shown in Fig. 18. For this cutting direction,
the angle at which the texture softening factor is most negative is found to be
54°. The predicted shear angles determined for different crystallographic cutting
directions of the two β-brass single crystals are shown in Fig. 19a and Fig. 19b
respectively.

In both crystals, there are good agreements between the theoretical shear angles and the experimental values with respect to the periodic variation, and the maximum and minimum values. The shape of the calculated and experimental curves are strikingly similar. The predicted values in the model are higher than the experimental ones by an average of 15°. This difference could be caused by the presence of shear between the tool and chip interface which can be taken into consideration by the macroscopic type of analysis. The analysis at the meso-level of the workpiece is significant as it demonstrates for the first time the correct qualitative change in the shear angles with different crystallographic cutting directions.

Table 11.3. Variation of cutting force and effective Taylor factor with different crystallographic orientation in copper crystal.

Cutting plane	Cutting direction	Cutting force (MN)	Effective Taylor factor M'
(111)	[$\bar{1}$10]	290	2.04
	[$\bar{2}$11]	220	1.83
(110)	[$\bar{1}$11]	300	2.04
	[$\bar{1}$11]	250	2.45
	[001]	170	1.22

The effect of the crystallographic cutting direction on the static cutting force levels has been recently reported by Konig and Spernath[57] for a copper crystal and their experimental results are tabulated in Table 10.3. In copper, the operative slip system is {111}<110>. The variation in the shear strength of the crystal as indicated by the magnitude of calculated effective Taylor factor is calculated. Except for the crystallographic cutting direction [$\bar{1}$10] on the (110) plane, a high cutting force is reflected by a high effective Taylor factor. A linear relationship exists between these two parameters.

In ultra-precision machining, the effect of crystallographic orientation of crystalline materials exerts a great influence on the mechanism of chip formation. The systematic variation in micro-cutting force is not caused by machine tool chatter alone but has its origin in the varying crystallographic orientations of the crystallites the tool traverses during a revolution of cut. As most engineering materials are polycrystalline, micro-cutting force variation cannot be avoided. However, the crystallographic nature of cutting force variation can be minimized by the proper choice of substrate materials and machining processes. When turning is required, a workpiece with an ultra-fine grain size and a random crystallographic texture would be preferred. In nonaxial-machining processes such as milling, a substrate material with a strong crystallographic texture which behaves like a single crystal would be desirable to reduce the variation in shear angle and hence achieve better surface finish. The pattern of variation in microcutting force can be predicted if the change in the crystallographic orientation of the substrate material with respect to the cutting direction is known.

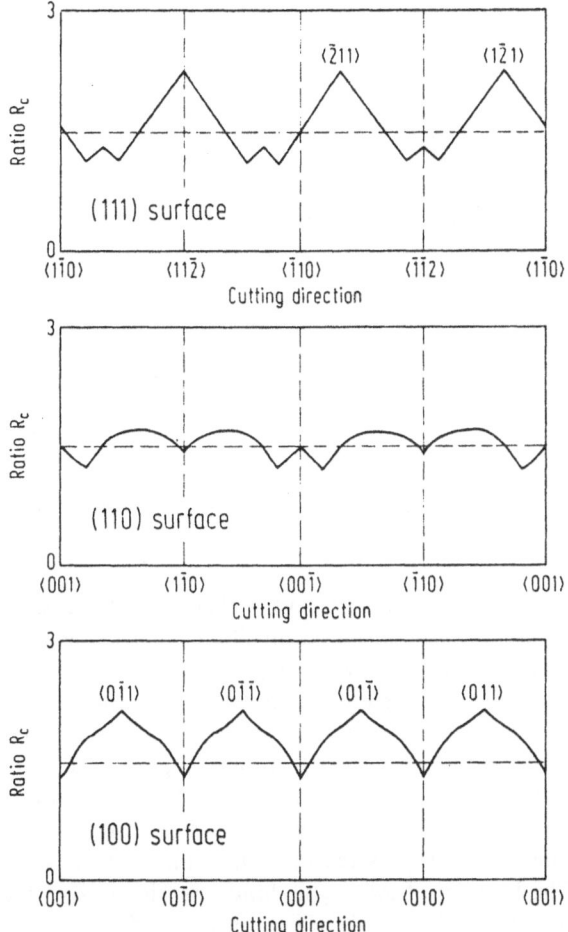

Fig. 11.20. The variation of the ratio of maximum resolved tensile stress to maximum resolved shear stress with cutting directions in different crystallographic planes. (From Nakasuji et al.[67])

The shear mode of chip formation will be conducive to good surface finish in precision machining and this explain that non-ferrous metals, such as aluminium and copper, were the first to be diamond turned. Brittle materials, on the other hand, exhibit discontinuous chip in machining and are susceptible to fragmentation and uncontrolled cracking which causes problems for surface finish and integrity. The critical stress for clevage depends on crystal structure, intrinsic strength and defects distribution. As the resolved shear stress on the clevage plane vary with crystallographic orientation, the critical chip thickness also depends on crystal orientation. Anisotropy in surface finish of a machined germanium were observed when the cutting direction relative to the crystal orientation varies successfully[67]. The ratio

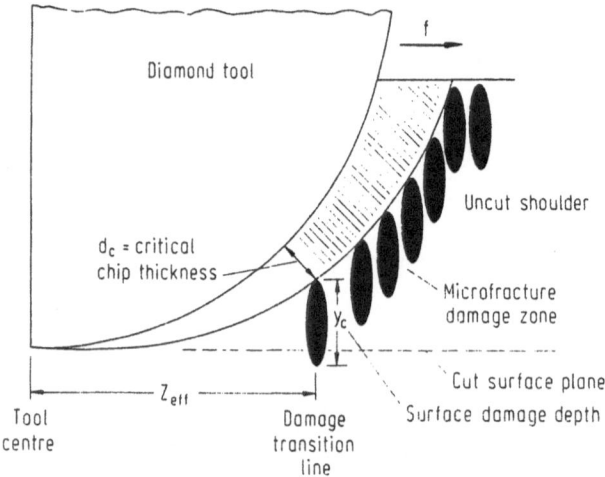

Fig. 11.21. Machining geometry in the micromachining of brittle (From Blackley and Scattergood[68])

R_c of the maximum resolved tensile stress on the cleavage plane to resolved shear stress on the slip plane varies in a systematic manner with different crystallographic cutting directions in a germanium crystal as shown in Fig. 11.20.

For a brittle material, there exists a critical depth of cut dc where the energy of propagating cracks equals the energy of plastic deformation. Fracture would not occur if the cutting depth is less than d_c. A model of ductil-regime machining for diamond turning of brittle materials is shown in Fig. 11.21 from the work of Blackley and Scattergood [68] . The flank of the new groove created by the tool radius is free from damage until a point y_c is reached where the chip thickness exceeds a critical value d_c. The location of the ductile-brittle transition (Z_{eff}) can be obtained by the interrupted cutting test as described by Blake[69]. During a cut, the tool was moved from the surface of the workpiece leaving a shoulder in the shape of the tool radius (R). The ductile-regime response is obtained when the depth of damage y_c, initiated at cutting depth d_c does not replicate below the cut surface plane. y_c represents the average fracture depth after initiation. The location of the end of the fracture area is called the ductile-brittle transition. As the feed rate (f) is increased, this transition will move down the shoulder region. The mechanistic damage parameters (d_c and y_c) are related to the machining geometry by the following

$$\left(Z_{eff}^2 - f^2\right)/R^2 = dc^2/f^2 - 2\left[(d_c - y_c)/R\right]. \tag{11.13}$$

In the limiting case $Z_{eff} = 0$ and the maximum feed rate f_{max} becomes

$$f_{max} = d_c \left[R/2\left(d_c + y_c\right)\right]^{1/2}. \tag{11.14}$$

In machining silicon and germanium, d_c is in the order of a few tenths of a micron. If Z_{eff} can be made large enough, a brilliant machined surface will result. The type of equation given in (11.13) gives only a machinability criterion from the experimental determination of the parameters d_c and y_c. There has been evidence that the critical depth also depends on crystal orientation and the size of the plastic zone at the tool tip. The basic physical understanding underlying ductile-regime machining of brittle materials is still not clearly understood, and further research will be needed to increase the range of materials that can be successfully cut using the single point diamond turning (SPDT) technology.

The above type of analysis is applicable for micromachining a ductile or brittle material with a typical chip cross section less than 20 μm. As a result of these small chip cross section, the chip formation takes place inside individual grains of a polycrysatlline material. Ultra-precision machining has now reached such a level that a chip thickness of 1 nm has become possible[70]. No direct experimental observation of the nanometer cutting process is possible as cutting is performed at the atomic scale. Whether there is a shear zone formation or chip flow direction is not known.

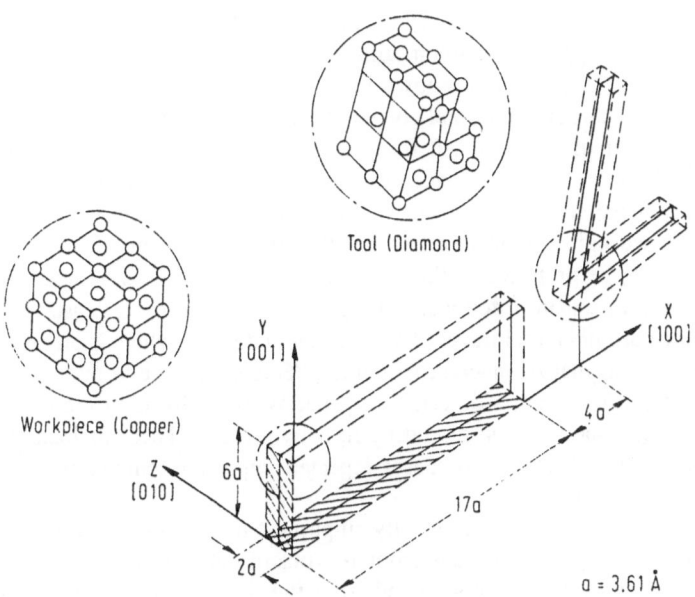

Fig. 11.22. The atomic cutting model. (After Inamura[71])

A computer simulation model of the nanometer cutting process based on non-linear finite-element formulation has been demonstrated by Inamura[71]. Fig. 11.22 shows a crystal model of a copper workpiece and a diamond tool composed of carbon atoms. The process of cutting is described by the atomic motion of a tool and the workpiece, where atoms move to their minimum-energy position. The interatomic potential energy is assumed to be described as the sum of the pairwise

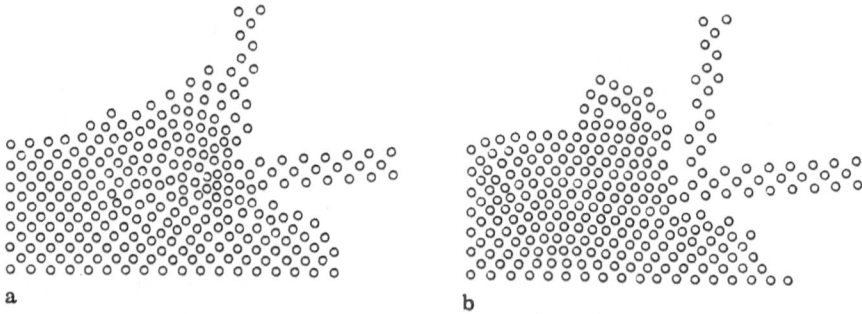

Fig. 11.23. Computer simulation of the chip formation in nanometer cutting. **a** With interaction between copper and carbon atoms; **b** without interaction between the copper and carbon atoms (After Inamura[71])

empirical potentials which are regarded as the elements and the atoms as nodes. The computer simulation results reveal the process of chip formation is susceptible to the affinity between tool and workpiece atoms. When a difference in the potentials between the copper and carbon atom is assigned, chip formation with or without a shear zone may result as shown in Fig. 11.23. In Fig. 11.23a, active interaction between the carbon atoms and copper atoms is assumed, and the copper atoms swell up in front of the tool and form successive lumps. In Fig. 11.23b, when the potential between the copper and carbon atoms is assumed to be the one producing only repulsive forces which decays smoothly as the distance, a shear deformation of the workpiece atoms in front of the tool then occurs. The shear deformation mode is to be preferred as the force distribution on the rake force will be compressive whereas tensile force would be predicted in the former case. The specific cutting force around a tool travel of 4 nm is about 43 GPa which is an order of magnitude larger than that of macroscale cutting. This difference is due to the well known size effect in cutting a defectless crystal. When dislocation motion is generated by the tool, a sudden drop of cutting force will result. These dislocation motion generates atomic vibration and causes repeated impulsive temperature rise on the tool face during cutting.

Our understanding of the technologically important machining process progresses gradually over the past decades from the macroscale, mesoscale to the present nanosacle of analysis. Insights gained from the deeper level of study always throw light on problems which would not have been solved at the macroscale of observation. The contribution of different disciplines such as material science, plasticity theories (applied to appropriate scales), numerical methods, and molecular mechanics to the machining study is indispensable to the advancement of the subject.

11.5 References

1. Cottrell, A.H., The Mechanical properties of Matter, Wiley, New York, (1964), p.277.
2. Hauser, J.J. and Chalmers, B., Acta Metall., 9(1961), p.802.
3. Randle, V. and Ralph, B., Revue Phys. Appl.23 (1989) p.501.
4. Armstrong, R., Codd, I., Douthwaite, R.M. and Petch, N.J., Phil. Mag., 7(1962), p.45.
5. Conard, H., Feuerstein, S. and Rice, L., Mat.Sci.Eng., 2(1967), p.167.
6. Ashby, M.F., Phil. Mag., 21(1970), p.339.
7. Hirth, J.P., Met. Trans., 3 (1972), p.3047.
8. Thompson, A.W., Work Hardening in Tension and Fatique (edited by Thompson, A.W.), AIME, New York,, (1977), p.89.
9. Werner, E. and Prantl, W., Acta Metall., 38(1990), p.533.
10. Hingwe, A.K and Subramanian, K.N., J. Mater.Sci., 10(1975) p.183.
11. Gil Sevillano, J. and Meizoso, A.M., 8th International Conference on Textures of Materials, the Metallurgical Society, 1988, p.897.
12. Fiels, D.P., Zhao, J. and Adams, B.L., Texures and Microstructures, 14-18(1991), p.977.
13. Haessner, F., Pospitsech, J., and Sztwiertnia, Mater. Sci. Engng, 57(1983), p.1.
14. Randle, V. and Ralph, B., Texture and Microstructures, 8&9(1989), p.531.
15. Watanabbe, T. Res.Mech., 11(1984), p.47.
16. Sickafus, K.E. and Sass, S.L., Acta Met., 35(1987), p.69.
17. Randle, V. and Ralph, B., Revue Phys, Appl., 23 (1989), p.501.
18. Pearson, C.E, J.Inst.Metals, 54(1934), p.111.
19. Underwoood, E.E., J.Metals, 14(1962), p.914.
20. Backofen, W.A., Turner, I.R. and Avery, D.H., Trans ASM., 57(1964), p.980.
21. Ahmed, M.M. and Langdon, T.G., Metall.Trans., 8A(1977), p.821.
22. Higashi, K., Ohnishi, T. and Nakatani, Y., Scr.Metall, 19(1985), p.821.
23. Edington, J.W., Mclton, K.N. and Culter, C.P., Prog. Mater. Sci, 21(1976), p.61.
24. Sherby, O.D. and Wadswoth, J., Prog.in Mater.Sci, 33(1989), p.169.
25. Paton, N.E. and Hamilton, C.H. (editors), Superplastic Forming of Structural Alloys, the Metallurgical Society of AIME, Warrendale, PA, (1982).
26. Lee, D., Acta Metall., 17(1967), p.1057.
27. Watts, B.M., Stowell, M.J., Baikie, B.L. and Owen, D.C.E, Met.Sci., 10(1976), p.189.
28. Taplin, D.M.R. and Sagat, S., Metal.Sci.Eng., 9(1972), Vol.53.
29. Lee, D., Acta. Metall., 17(1969), p.1057.
30. Cline, H.E. and Alden, T.H, Trans. A.I.M.E, 239(1967), p.710.
31. Griffiths, P. and Hammond, C., Acta Metall., 20 (1972) p.935.
32. Lee, E.H. and Underwood, E.E., Met.Trans., 1(1970), p.1399.
33. Hazzlendine, P.M. and Newbury, D.E., 3rd I.C.S.M.A., Institute of Metals, London, (1973), p.202.
34. Ashby, M.F. and Verrall, R.A., Acta Metall., 21(1973), p.149.
35. Wadsworth, J., Oyama, T. and Sherby, O.D., Proc. Conf. Advances in Materials Technoloy in America, ASME, New York, 2(1980), p.29.
36. Ball, A. and Hutchinson, M.M., Met. Sci. J., 3(1969), p.1.
37. Gifkins, R.C, Met.Trans, 7A (1976), p.1225.
38. Johnson, R.H., Packer, C.M., Anderson, L.J. and Sherby, O.D., Phil.Mag., 18(1968), p.1309.
39. Biscondi, M. and Goux, C., Mem. Sci. Rev. Met., 65(1968), p.167.
40. Wadsworth, J., Lin, J.H. and Sherby, O.D., Metals Technology, May, 1981, p.190.
41. Sherby, O.D., Caligiurl, R.D., Kayali, E.S. and White, R.A., Proc. 25th Sagamore Army Materials Research Conf., (1981), New York, Plenum.
42. Wadsworth, J. and Pelton, A.R., Scripta Metallurgica, 18(1984), p.387.
43. Marya, S.K. and Wyon, G., Proceedings of the 4th Int. Conf. on Strength of Metals and Alloys, Nancy, 1(1976), p.438.
44. Ashby, M.F., Acta Met., 20(1972), p.887.
45. Sherby, O.D., Orr, R.L. and Dorn, J.E., Trans AIME., 200(1954), p.71.
46. Horton, A.A.P., Acta Met., 20(1972), p.477.
47. Versnyder, F.I. and Shank, H.E., Mat. Sci. and Eng., 6(1970), p.213.
48. Lemekey, F.D. and Thompson, E.R., Proc. Conf. on In Situ Composites, Lakeville, Connecticut, September 5-8, 1972, p.105.

49. Erickson, J.S., Sullian, C.P. and Versnyder, S.U., High-Temperature Materials in Gas Turbines (edited by Sahm, P.R. and Spedel, M.O.), Elsevier Scientific Publishing Co., (1974), p.315.
50. Lacaze, J. and Hazottee, A., Textures and Microstructures, 13(1990), p.1.
51. Caron, P. and Khan, T., Superalloys 88 (edited by Duhl et.al. the Metallurgical Society Inc., (1988), p.215.
52. Webster, G.A. and Piearcey, B.J., Metal Science, 1(1967) p.97.
53. Leverant, G.R. and Kear, B.H., Met.Trans., Vol.1(1970), p.491.
54. Takeuchi, S. and Kuramoto, E., Acta.Met., 21(1973) p.415.
55. Weck, M. and Pyra, M., Proced. of ASPE Annual Meeting, 23-28 September 23-28, 1990, Rorchester, New York, USA.
56. Hashimoto, Proceedings of 4th Int.Precision Engineering Seminar, Monterey, USA, 1989.
57. Konig, W. and Spenrath, N., Progress in Precision Engineering (edited by Seyfried et.al.), Springer-Verlag, (1991), p.141.
58. Sato, M., Kato, Y and Tuchiya, K., Trans. JIM., 9(1978), p.530.
59. Ueda,K., and Iwata, K., Annals of CIRP., 29(1980), p.41.
60. Black, J.T., J.Engng.Ind., 2(1972), p.307.
61. Shaw, M.C., Metal Cutting Principles, p.176, Clarendon, Oxford (1984).
62. Lee, W.B., Precision Engineering, 12(1990), p.25.
63. Merchant, M.E., J.Appl.Phys. 16(1945), p.318.
64. Iwata, K. Osakada, K. and Terasaka, Y., J. Engng Mater. Technol., 106(1984), p.132.
65. Yamagata, T., Yoshida, H. and Fukazawa, Y., Trans. Japan Inst.Metals, 17(1976), p.393.
66. Hill, R., J. Mech. Phys. Solids, 3(1954), p.47.
67. Nakasuji, T., Kodera, S., and Hara, H., Annals of CIRP., 39(1990), p.89.
68. Blackley, Q.S. and Scattergood, R.O., Precision Engineering. 13(1991), p.95.
69. Blake, P.N., J.Amer. Ceram. Soc., 4(1990), p.73.
70. Ikawa, N., Shimada, S., Ohmori, Y., Cry, C.K., Taylor, J.S. and Donaldson, Preprint of the Spring Annual Meeting of JSPE, (1989), p.761.
71. Inamura, T., Progress in Precision Engineering (edited by Seyfried, P. et al.), Springer-Verlag, (1991), p.232.

Subject Index

Author Index

Springer-Verlag
and the Environment

We at Springer-Verlag firmly believe that an international science publisher has a special obligation to the environment, and our corporate policies consistently reflect this conviction.

We also expect our business partners – paper mills, printers, packaging manufacturers, etc. – to commit themselves to using environmentally friendly materials and production processes.

The paper in this book is made from low- or no-chlorine pulp and is acid free, in conformance with international standards for paper permanency.